T5-CVH-973

Ecologies and Politics of Health

Human health exists at the interface of environment and society. Decades of work by researchers, practitioners, and policy-makers has shown that health is shaped by a myriad of factors, including the biophysical environment, climate, political economy, gender, social networks, culture, and infrastructure. Yet while there is emerging interest within the natural and social sciences on the social and ecological dimensions of human disease and health, there have been few studies that address them in an integrated manner.

Ecologies and Politics of Health brings together contributions from the natural and social sciences to examine three key themes: the ecological dimensions of health and vulnerability, the socio-political dimensions of human health, and the intersections between the ecological and social dimensions of health. The thirteen case study chapters collectively present results from Africa, Asia, Latin America, the United States, Australia, and global cities. Part I interrogates the utility of several theoretical frameworks and conventions for understanding health within complex social and ecological systems. Part II concentrates upon empirically grounded and quantitative work that collectively redefines health in a more expansive way which extends beyond the absence of disease. Part III examines the role of the state and management interventions through historically rich approaches centering on both disease- and non-disease-related examples from Latin America, eastern Africa, and the United States. Finally, Part IV highlights how health vulnerabilities are differentially constructed with concomitant impacts for disease management and policy interventions.

This timely volume advances knowledge on health–environment interactions, disease vulnerabilities, global development, and political ecology. It offers theoretical and methodological contributions that will be a valuable resource for researchers and practitioners in geography, public health, biology, anthropology, sociology, and ecology.

Brian King is an Associate Professor at the Department of Geography, Pennsylvania State University, USA.

Kelley A. Crews is an Associate Professor at the Department of Geography and the Environment, University of Texas at Austin, USA and currently is on leave as a Visiting Scientist and Program Director at the National Science Foundation, USA.

Routledge Studies in Human Geography

This series provides a forum for innovative, vibrant, and critical debate within Human Geography. Titles will reflect the wealth of research which is taking place in this diverse and ever-expanding field. Contributions will be drawn from the main sub-disciplines and from innovative areas of work which have no particular sub-disciplinary allegiances.

Published:

1 **A Geography of Islands**
Small island insularity
Stephen A. Royle

2 **Citizenships, Contingency and the Countryside**
Rights, culture, land and the environment
Gavin Parker

3 **The Differentiated Countryside**
Jonathan Murdoch, Philip Lowe, Neil Ward and Terry Marsden

4 **The Human Geography of East Central Europe**
David Turnock

5 **Imagined Regional Communities**
Integration and sovereignty in the global south
James D. Sidaway

6 **Mapping Modernities**
Geographies of Central and Eastern Europe 1920–2000
Alan Dingsdale

7 **Rural Poverty**
Marginalisation and exclusion in Britain and the United States
Paul Milbourne

8 **Poverty and the Third Way**
Colin C. Williams and Jan Windebank

9 **Ageing and Place**
Edited by Gavin J. Andrews and David R. Phillips

10 **Geographies of Commodity Chains**
Edited by Alex Hughes and Suzanne Reimer

11 **Queering Tourism**
Paradoxical performances at Gay Pride parades
Lynda T. Johnston

12 **Cross-Continental Food Chains**
Edited by Niels Fold and Bill Pritchard

13 **Private Cities**
Edited by Georg Glasze, Chris Webster and Klaus Frantz

14 **Global Geographies of Post-Socialist Transition**
Tassilo Herrschel

15 **Urban Development in Post-Reform China**
Fulong Wu, Jiang Xu and Anthony Gar-On Yeh

16 **Rural Governance**
International perspectives
Edited by Lynda Cheshire, Vaughan Higgins and Geoffrey Lawrence

17 **Global Perspectives on Rural Childhood and Youth**
Young rural lives
Edited by Ruth Panelli, Samantha Punch, and Elsbeth Robson

18 **World City Syndrome**
Neoliberalism and inequality in Cape Town
David A. McDonald

19 **Exploring Post-development**
Aram Ziai

20 **Family Farms**
Harold Brookfield and Helen Parsons

21 **China on the Move**
Migration, the state, and the household
C. Cindy Fan

22 **Participatory Action Research Approaches and Methods**
Connecting people, participation and place
Edited by Sara Kindon, Rachel Pain and Mike Kesby

23 **Time-Space Compression**
Historical geographies
Barney Warf

24 **Sensing Cities**
Monica Degen

25 **International Migration and Knowledge**
Allan Williams and Vladimir Baláž

26 **The Spatial Turn**
Interdisciplinary perspectives
Edited by Barney Warf and Santa Arias

27 **Whose Urban Renaissance?**
An international comparison of urban regeneration policies
Edited by Libby Porter and Katie Shaw

28 **Rethinking Maps**
Edited by Martin Dodge, Rob Kitchin and Chris Perkins

29 **Rural–Urban Dynamics**
Livelihoods, mobility and markets in African and Asian frontiers
Edited by Jytte Agergaard, Niels Fold and Katherine V. Gough

30 Spaces of Vernacular Creativity
Rethinking the cultural economy
Edited by Tim Edensor, Deborah Leslie, Steve Millington and Norma Rantisi

31 Critical Reflections on Regional Competitiveness
Gillian Bristow

32 Governance and Planning of Mega-City Regions
An international comparative perspective
Edited by Jiang Xu and Anthony G.O. Yeh

33 Design Economies and the Changing World Economy
Innovation, production and competitiveness
John Bryson and Grete Rustin

34 Globalization of Advertising
Agencies, cities and spaces of creativity
James Faulconbridge, Peter J. Taylor, J.V. Beaverstock and C Nativel

35 Cities and Low Carbon Transitions
Edited by Harriet Bulkeley, Vanesa Castán Broto, Mike Hodson and Simon Marvin

36 Globalization, Modernity and the City
John Rennie Short

37 Climate Change and the Crisis of Capitalism
A chance to reclaim, self, society and nature
Edited by Mark Pelling, David Manual Navarette and Michael Redclift

38 New Economic Spaces in Asian Cities
From industrial restructuring to the cultural turn
Edited by Peter W. Daniels, Kong Chong Ho and Thomas A. Hutton

39 Cities, Regions and Flows
Edited by Peter V. Hall and Markus Hesse

40 The Politics of Urban Cultural Policy
Global perspectives
Edited by Carl Grodach and Daniel Silver

41 Ecologies and Politics of Health
Edited by Brian King and Kelley A. Crews

42 Producer Services in China
Economic and urban development
Edited by Anthony G.O. Yeh and Fiona F. Yang

43 Locating Right to the City in the Global South
Tony Roshan Samara, Shenjing He and Guo Chen

Forthcoming:

44 Gender, Development and Transnational Feminism
Edited by Ann M. Oberhauser and Ibipo Johnston-Anumonwo

45 Fieldwork in the Global South
Ethical challenges and dilemmas
Edited by Jenny Lunn

Ecologies and Politics of Health

Edited by Brian King and Kelley A. Crews

LONDON AND NEW YORK

First published 2013
by Routledge
2 Park Square, Milton Park, Abingdon, Oxon OX14 4RN

Simultaneously published in the USA and Canada
by Routledge
711 Third Avenue, New York, NY 10017

Routledge is an imprint of the Taylor & Francis Group, an informa business

British Library Cataloguing in Publication Data
A catalogue record for this book is available from the British Library

Library of Congress Cataloging in Publication Data
King, Brian (Brian Hastings), 1973-
Ecologies and politics of health / Brian King and Kelley A. Crews.
p. cm.
Includes bibliographical references and index.
ISBN 978-0-415-59066-2 (hbk: alk. paper) 1. Environmental health.
2. Medical policy. I. Crews, Kelley A. II. Title.
RA565.K56 2013 362.1–dc23
2012016552

ISBN: 978–0–415–59066–2 (hbk)
ISBN: 978–0–203–11552–7 (ebk)

Typeset in Times New Roman
by Bookcraft Ltd, Stroud, Gloucestershire

Printed and bound by CPI Group (UK) Ltd, Croydon, CR0 4YY

Contents

Figures

Tables

Contributors

M. Eric Benbow is a community and disease ecologist with research interests in microbe-invertebrate interactions at multiple spatial and temporal scales. As an Assistant Professor of Biology at the University of Dayton, his studies focus on understanding how microbe–invertebrate interactions structure larger-scale communities and landscapes by both incorporating and testing ecological theory in these systems. Ultimately, his research is based in basic science and ecological theory with the goal of applications in ecosystem management, conservation, disease ecology, and forensics.

Lindsay P. Campbell is a Ph.D. student in the Department of Ecology and Evolutionary Biology at the University of Kansas. Her research interests involve modeling the spatial distribution of disease vectors, reservoirs, and pathogens under changing ecological conditions using geospatial technologies. She has been involved with research projects in several regions of the world, with a specific focus on Buruli ulcer disease in west Africa.

Marcia C. Castro is Associate Professor of Demography in the Department of Global Health and Population, Harvard School of Public Health, and Associate Faculty of the Harvard University Center for the Environment. Her research focuses on modeling determinants of malaria transmission through an interdisciplinary approach, with particular emphasis on generating evidence for better control strategies; expansion of the Brazilian Amazon frontier, and the social and environmental impacts of large-scale development projects implemented in the region; urbanization and health; use of spatial analysis in the social sciences; and on population dynamics and mortality models. Castro earned her Ph.D. in demography from Princeton University.

Kelley A. Crews, Ph.D., is Associate Professor at the Department of Geography and the Environment, University of Texas at Austin, USA and currently on leave from there as a Program Officer in the Geography and Spatial Sciences Program at the National Science Foundation in Arlington, Virginia, USA. She has over fifteen years of experience in the global tropics, where she focuses on the vulnerability and resilience of socio-ecological systems in developing states via the integration of satellite-derived and field-acquired datasets. Her current work examines the fluidity of human reactions to environmental disturbance as a means of mapping the interplay of both human as well as environmental health in the Okavango Delta and central Kalahari of Botswana. She is currently appointed as a Visiting Scientist at the National Science Foundation where she is a Program Director in the Geography and Spatial Sciences program.

Alex de Sherbinin is a Senior Researcher at the Center for International Earth Science Information Network (CIESIN), an environmental data and analysis center within the Earth Institute at Columbia University. He has published widely on population–environment interactions and the human aspects of environmental change at local, national, and global scales.

Michael E. Emch, Ph.D., a medical geographer, is Professor of Geography at the University of North Carolina at Chapel Hill. He is also Professor of Epidemiology and a Fellow of the Carolina Population Center and also directs the Spatial Health Research Group at the University of North Carolina. He has published widely in the subfield of disease ecology, mostly of infectious diseases of the tropical world.

Carolyn A. Fahey is an undergraduate student majoring in International Health at Georgetown University with research interests in the political ecology of agriculture, nutrition, and health.

Andrew O. Finley is an Assistant Professor at Michigan State University with a joint appointment in the Departments of Forestry and Geography, and adjunct in the Department of Statistics and Probability. His research interests lie in developing methodologies for monitoring and modeling environmental processes, Bayesian statistics, spatial statistics, and statistical computing.

Gilvan Guedes is an Associate Professor in the Graduate Program on Interdisciplinary Territorial Studies and Director of the Interdisciplinary Center on Territorial Studies at the Vale do Rio Doce University. His research interests concentrate upon livelihood strategies and land use/cover change in small-scale agricultural frontiers, mixed method approaches, and interactions between socio-demography and environmental behavior.

Mary Hayden is a Scientist at the National Center for Atmospheric Research (NCAR) in the Research Applications Laboratory/Integrated Science Program, adjoint faculty at the University of Colorado School of Public Health, and a Guest Researcher with the Centers for Disease Control and Prevention (CDC). Her research interests are societal impacts and climate change, particularly weather-, climate-, and health-related links, community participatory research, and social mobilization.

James R. Hull is a Postdoctoral Research Associate at the Environmental Change Initiative and the Population Studies and Training Center at Brown University. His primary research focuses on sociological and social network aspects of the population–environment–development nexus, and on understanding the role of non-economic factors in economic development. Current projects in Thailand and Brazil explore relationships between social structure and shifting systems of social and economic exchange, mapping historical transformations in the distribution and returns to diverse forms of capital, and developing new methodologies for the measurement of multidimensional poverty.

Brian King is an Associate Professor in the Department of Geography at the Pennsylvania State University. He received his Ph.D. in geography from the University of Colorado in 2004 with a certificate in Development Studies. His research interests concentrate upon the impacts of conservation and development within southern Africa, social and environmental justice, and health–environment interactions. His work has recently been extended into other thematic and geographic areas, with two separate projects examining how environmental variabilities shape livelihood responses in the Okavango Delta of Botswana, and how livelihood systems in South Africa are being transformed by HIV/AIDS.

Paul F. McCord is a Research Scientist at Michigan State University's Center for Global Change and Earth Observations, where he examines outcomes of human–environment interactions as they impact health and resource access across sub-Saharan Africa. He begins a geography Ph.D. program at Indiana University in the fall of 2012.

Kendra McSweeney is an Associate Professor of Geography at the Ohio State University. Working with forest dwellers in Ecuador, Honduras, and Ohio, she is primarily interested in how marginalized rural societies respond to, and shape, environmental change. These dynamics are always power-laden, and can be expressed in surprising ways, including through the health of individuals and communities.

Richard W. Merritt is currently a University Distinguished Professor of Entomology at Michigan State University. His major research interests focus on animal–microbial interactions, population dynamics, and the influence of environmental factors on immature aquatic insects, especially the Diptera. His most recent research has concentrated on the biomonitoring of streams, rivers, and marshes and the effects of pollutants on aquatic ecosystems and the macroinvertebrates that occur in these water bodies. Dr Merritt is also interested in the field of Forensic Entomology and assists police departments with crime scene investigations involving insects.

Joseph P. Messina is a Professor in the Department of Geography, the Center for Global Change and Earth Observations, and Michigan AgBioResearch at Michigan State University. He explores questions on land change science, disease ecology, access to healthcare, and spatial methodologies.

Jacob C. Miller is a Ph.D. candidate in the School of Geography and Development at the University of Arizona, where he also holds an MA. His research focuses on urban experience, affect, identity, and consumption under conditions of economic transition and crisis.

Zoe Pearson is a Ph.D. student in geography at the Ohio State University, with a graduate minor in Women's, Gender, and Sexuality Studies. She has studied the politics of resource use and management in Ecuador and Honduras, and is currently researching the political, ecological, and health implications of international drug control policy for coca leaf users, growers, and sellers in Bolivia.

Cynthia Pope is Professor of Geography at Central Connecticut State University. Her research centers on the interplay of disease, gender, and national security, particularly in Latin America and the Caribbean. She has published in international journals, such as the *Journal of International Women's Studies*, as well as geography journals, such as the *Annals of the Association of American Geographers* and *GeoJournal*. She has co-edited two volumes on the changing landscapes of HIV: *HIV/AIDS: Global Frontiers in Prevention/Intervention* (Routledge, 2008) and *Strong Women, Dangerous Times: Gender and HIV/AIDS in Africa* (Nova Science, 2009).

Jiaguo Qi is Director of the Center for Global Change and Earth Observations and Professor in the Department of Geography at Michigan State University. His research focuses on two areas: integrating biophysical and social processes and methods in understanding land use and land cover change; and transforming data into information and knowledge. Dr Qi strives to use case studies in different parts of the world to understand the nature of the coupled nature–human systems. The geographic extent of his research is global; with projects in North America, southeast Asia, east Asia, central Asia, east and west Africa, South America, and Australia.

Paul Robbins is Director of the Nelson Institute for Environmental Studies at the University of Wisconsin, Madison. He holds a Ph.D. in geography from Clark University (1996) and researches the relationship between environmental actors (mosquitoes, invasive plants, or elk), diverse publics (homeowners, herders, or foresters), and the institutions that connect them (health offices, conservation organizations, or planning boards).

Elisabeth D. Root, Ph.D., is an Assistant Professor of Geography and Faculty Research Associate at the Institute of Behavioral Science at the University of Colorado at Boulder. Her research explores geographical patterns of health and disease using quantitative spatial methodologies to understand the complex interactions between demographic, socio-economic, and environmental factors that influence human health. She is particularly interested in how we can quantify these contextual factors and interactions to better evaluate the impact of public health interventions. Dr Root combines traditional epidemiological study design, complex spatial statistical methods, as well as geographic information systems (GIS) to explore spatio-temporal patterns of chronic and communicable diseases.

Larry Sawers is Professor of Economics at American University in Washington, DC. He has written extensively on developing economies with special attention to Latin America and sub-Saharan Africa. He authored *The Other Argentina:*

The Interior and National Development (Westview Press, 1996) and numerous articles on the Argentine economy. His publications also include articles on Ecuador, Tanzania, and Lithuania, plus works on economic history and urban economics. In recent years, his research has focused on the HIV epidemics in sub-Saharan Africa, analyzing and modeling the role that endemic diseases play in explaining the extraordinary prevalence of HIV in the region.

Burton H. Singer is Adjunct Professor in the Emerging Pathogens Institute and Department of Mathematics at the University of Florida. He has served as Chair of the National Research Council Committee on National Statistics and as Chair of the Steering Committee for Social and Economic Research in the World Health Organization Tropical Disease Research (TDR) program. He has centered his research in three principal areas: identification of social, biological, and environmental risks associated with vector-borne diseases in the tropics; integration of psychosocial and biological evidence to characterize pathways to alternative states of health; and health impact assessments associated with economic development projects. He was elected to the National Academy of Sciences (1994), the Institute of Medicine of the National Academies (2005), and was a Guggenheim Fellow in 1981–1982. Singer earned his Ph.D. in statistics from Stanford University.

Eileen Stillwaggon is Professor of Economics and Harold G. Evans-Eisenhower Professor at Gettysburg College, USA. Author of *AIDS and the Ecology of Poverty* (Oxford University Press, 2005), Stillwaggon has written extensively on HIV/AIDS in developing nations for scholarly journals and popular media. She also wrote *Stunted Lives, Stagnant Economies: Poverty, Disease, and Underdevelopment* (Rutgers University Press, 1998), based on her work in informal settlements and public clinics and hospitals in Argentina. Her current work involves modeling strategies to improve outcomes and lower costs in preventing and treating infectious diseases and congenital infections.

Jenni van Ravensway is a Research Associate at the Center for Global Change and Earth Observations at Michigan State University. Her research interests involve the development and application of geospatial technologies to analyze global change issues that integrate social, terrestrial, and climate systems, with a focus in remote sensing and GIS. She has been involved with research projects around the globe and has performed Global Positioning System (GPS) fieldwork in the US, west Africa, southeast Asia, and Japan.

Leah K. VanWey is an Associate Professor in the Department of Sociology, core faculty in the Environmental Change Initiative, and Associate Director of the Population Studies and Training Center, all at Brown University. She has studied and continues to research processes of migration and household livelihoods, primarily in developing countries. Leah studies population and environment relationships in the Brazilian Amazon, using in-depth interview and social survey data linked with satellite imagery. She also is currently studying social and economic drivers and consequences of large-scale agricultural intensification in Brazil, social and ecological consequences of the

Belo Monte Hydroelectric Project in Brazil, and the impacts of climate variability on migration and interhousehold transfers in Mexico.

Olga Wilhelmi is a geographer whose research interests are geographic information science, societal impacts, vulnerability, and adaptive capacity to extreme weather events and climate change. She works at the National Center for Atmospheric Research in Boulder, Colorado as a Research Scientist and the head of the GIS program.

Kenneth R. Young is a Professor in the Department of Geography and the Environment of the University of Texas at Austin. He is interested in the biogeography and conservation of the Global Tropics. He has done much of his research in the Andes and Amazon, using landscape and regional approaches to evaluate environmental change. He was co-editor, with Thomas Veblen and Antony Orme, of *The Physical Geography of South America* (Oxford University Press, 2007).

Acknowledgments

This book is the product of the efforts of multiple individuals who deserve our sincere thanks. The development of this project began with a series of paper sessions, titled "Ecologies and Politics of Health," that were held during the 2009 Association of American Geographers Annual Meeting. The participants in those sessions generated thoughtful and challenging insights that informed the approach we took with *Ecologies and Politics of Health*. We are deeply appreciative for our contributors who were tireless in meeting deadlines while ensuring the highest quality of work. We would like to thank Routledge for providing the supportive and stimulating environment that allowed this project to take shape from initial proposal to final product. Our managing editor, Faye Leerink, was remarkably helpful as the various stages unfolded. Because the majority of the chapters contained herein were peer-reviewed, we benefited from the generosity of numerous anonymous referees who provided extremely helpful and timely feedback.

Brian King would like to thank particular individuals who assisted with several of the chapters at various stages. Jamie Shinn helped with literature reviews for some of the material on structural vulnerabilities to health. Kayla Yurco was extremely helpful with a literature search on ecologies and politics of health that informed the writing of Chapter 1, and also provided critical feedback on Chapter 14. The nature society working group at the Pennsylvania State University helped review earlier versions of Chapters 1 and 14, and their time and insights are greatly appreciated. Some of the research that informed Chapter 14 was advanced by a National Science Foundation CAREER grant (BCS/GSS 1056683).

Kelley A. Crews would like to gratefully acknowledge the support of the Humanities Institute at the University of Texas at Austin (UT). That Faculty Fellowship not only provided time to migrate into health-based research but, moreover, offered constructive feedback for the development of that endeavor, leading to support as a US Fulbright Faculty Research Fellow in the Africa Regional Research Program for HIV/AIDS. During that period, interactions with Botswana-posted US Peace Corps workers Patrick Gallagher, Richard Rain, and Stacy Wallick were invaluable in disentangling locally relevant human health issues. The UT Population Research Center's NIHCD R24 Center's Grant (5 R24 HD042849) provided critical seed money for field research resulting in generous support from the National Science Foundation (BCS/GSS SGER, BCS/GSS RAPID 0942211, and BCS/GSS 0964596), facilitating fieldwork and concomitant research that over the years have honed her thinking on the intersections of environmental health and human health processes and local outcomes, and the importance of these to socio-ecological systems, function, and theory.

Foreword

Integration has become the hallmark of much research and scholarship in the first part of the twenty-first century, with few topics of study benefiting more than that of human–environment relationships. This integration runs contrary to the dominant analytical and theoretical trends of the twentieth century, which separated the human (social) and environmental (biophysical) domains of these relationships. This separation was a response, in part, to nineteenth-century attempts to reduce understanding of the human subsystem to an environmental one, and, subsequently, the conceptual and theoretical successes emerging from the study of the human subsystem independently. The current return to integrative understanding of human–environment relationships follows from breakthroughs in genetics and evolutionary-led sociobiology, and from a growing concern with the Anthropocene, in which humankind has become a force approaching nature in affecting the Earth system. In either case, the need to put the pieces of the human–environment puzzle back together again is increasingly recognized in both research and practice. For the last concern, advances in complex adaptive systems research have helped to formalize characteristics and outcomes of the system as a whole, not found in an examination of its individual parts, such as legacy effects, non-linear relationships, thresholds, and emergent properties. Real-world applications seeking more sustainable human–environment conditions reveal that system complexity leads to different outcomes in different contexts, indicating that panaceas are unlikely.

Multiple approaches, or topical domains associated with an approach, seek to understand phenomena and processes that are the outcome of human–environment relationships—risk–hazard/vulnerability studies, political ecology, human ecology, sustainability science, disease/health ecology, to name a few. *Ecologies and Politics of Health* adds another: "hybrid sociopolitical–ecological health." As a cluster, these approaches examine the systemic whole through various lenses of integration. This orientation, however, does not render reductionist approaches mute. We are not at some moment in which "normal science" has become passé. Indeed, much of it underpins complex systems analysis and forms the foundation for understanding the processes at play in either of the human–environment subsystems, as it does in the medical sciences informing human health research. Caution is warranted, however, for those cases in which system principles or lessons may be gleaned from one subsystem and applied uncritically to the other or the system as a whole—a tendency, if unintended, emanating from some venues in the natural and complex system sciences. Likewise, caution is warranted regarding "confirmational bias," which Andrew Vayda (2009) finds recurrent in some social–environmental systems approaches anchored in the social sciences, especially those employing the label "political." This bias involves uncritical acceptance of a metatheory that diverts attention away from alternative factors and processes at play in the outcome (event or phenomenon) of the system. As I understand them, the hybrid approaches in this volume heed these cautions.

The label hybrid typically carries several meanings: (i) the use of quantitative and qualitative data and analytics, or mixed methods; (ii) an appreciation for multiple explanatory perspectives used to frame and inform the human–environment problem; and (iii) the development of an explanatory framework that fuses or mixes post-positivist, structural, and constructivist perspectives. The last meaning is largely implied in calls for integrated theory and is sufficiently ambitious, if not logically impossible, to deter attention from it. Rather, hybrid human–environment approaches, more often than not, employ alternative understandings to inform the base research approach for the practitioner. Hybrid practice is intended to enlarge the phenomena and processes considered in the research problem by exploring multiple ways of identifying them and understanding their possible roles.

What does all this mean for research on human health? Conceptually, a wide range of factors, processes, principles, and theories are open to consideration and actually explored. For example, the suspicious P (population), at least from the lens of some political ecologists, is considered along with unequal entitlements and empowerment, factors that may be omitted in some conventional assessments. The attributes of P, including its sheer size and density, matter for many human health problems, but these attributes also include age, gender, and social status, among others that affect entitlements to healthcare. Analytically, alternative ways of knowing are employed as checks and balances on one another. For example, quantitative tests or models are informed by qualitative understanding, altering their configuration and, presumably, their performance. Likewise, metatheory-led narratives may be put to test, when applicable, potentially revealing the role of variables and processes otherwise muted by the metatheory. In short, hybrid human–environment approaches—or at least those that capture my attention and are represented in *Ecologies and Politics of Health*—appear to be less concerned with championing an explanatory perspective or metatheory and more concerned with understanding the multiple dimensions of complex relationships through mixed methods.

This understanding resides firmly in Pasteur's quadrant of a pure-to-applied research matrix (Stokes 1997), in which basic research is undertaken on societally framed problems with an eye towards informing real-world solutions. This link of research to practice reinforces the need to address the multiple dimensions of human health systems because their complexity leads to variance in place-based outcomes of otherwise similar systems. Such variance can prove problematic for setting policy or programs, an impediment that can be amplified where decision-makers are circumscribed by mission, political boundaries, or other limits that mute their interest or capacity to deal with the health problem holistically. How might such bounded decisions be made more congruent with the complexity of sociopolitical–environmental system associated with the health problem? One current popular response is through the co-production of the problem and its analysis by the scientist–researcher and the practitioner–decision-maker.

Human health science has always involved co-production, at least in the sense of research undertaken on societally defined problems. Recall the research quadrant named for Pasteur (above) and the research undertaken by the World Health Organization or the US Centers for Disease Control and Prevention, all of which respond to societally defined health problems. Recall as well the number of successes from pasteurization to the global eradication of smallpox and rinderpest. These successes, as I understand them, are founded largely on techno-methodological breakthroughs or panaceas. In contrast, this volume points to co-production that, more often than not, moves well beyond simple panaceas. The problem framing and solutions account for the concerns of different socio-economic units (e.g., stakeholders), especially those actually affected by the health problem, the roles of different levels of decision-makers, institutions and governance, and the participation of a range of research specialists (medical to social). Just as the hybrid approaches in this volume inform us that understanding human health issues systemically requires an expansive perspective, they also point to complex co-production linkages for successful health applications. This, I believe, is a major take-away lesson of *Ecologies and Politics of Health*.

Billie Lee Turner, II
Gilbert F. White Professor of Environment and Society
School of Geographical Sciences and Urban Planning
School of Sustainability
Arizona State University

References

Stokes, D. E. (1997) *Pasteur's Quadrant: Basic Science and Technological Innovation.* Washington, DC: The Brookings Institution.

Vayda, A. P. (2009) *Explaining Human Actions and Environmental Change.* Lanham, MD: AltaMira Press.

1 Human health at the nexus of ecologies and politics

Kelley A. Crews and Brian King

Human health exists at the interface of environment and society. Decades of work by researchers, practitioners and policy-makers has shown human health to be shaped by a myriad of factors, including the biophysical environment, climate, political economy, gender, resource access, immune systems, social networks, culture, and infrastructure. Research has concentrated upon infectious and non-infectious disease patterns to examine exposure and vulnerabilities within diverse contexts. Other studies have analyzed the ways that environmental and ecological factors contribute to the spread of disease, reduce quality of life and well-being, and shape the possibilities for healthy decision-making. The growth of the environmental justice movement in the United States, and its subsequent expansion around the world, is a reminder of inequitable exposure to carcinogens and other pollutants that are a product, at least in part, of political and economic systems. The environmental justice concept has broadened since its origins by addressing how disproportionate access to social and environmental amenities, such as green space and recreational opportunities, contribute to human health. A major theme of these research studies and social movements is that human health is increasingly understood as being shaped by social and ecological systems that intersect across multiple spatial and temporal scales. Yet while there is continued, and emerging, interest within the natural and social sciences on the social and ecological dimensions of human disease and health, there have been few studies that address them in an integrated manner. The central objective of this volume is to bring together contributions from the natural and social sciences to examine the social and ecological dimensions of human health. *Ecologies and Politics of Health* is intended to make substantive contributions by addressing three key themes: the ecological dimensions of health and vulnerability, the socio-political dimensions of human health, and the intersections between the ecological and social dimensions of health. This volume combines theoretical, methodological, and heuristic contributions from various disciplines to provide a fuller understanding of the multiple and varied dimensions of human health. Investigating the nexus of the ecologies and politics of health necessarily involves moving beyond critique in order to establish a new and integrated realm of theory and practice that is greater than the sum of its parts. Specifically, *Ecologies and Politics of Health* works to demonstrate what this hybrid approach offers to research and policy beyond simply *socializing the ecological* or *ecologizing the social*. That is, to call for integrating the social and natural sciences into a multidisciplinary (or even interdisciplinary) approach is neither new nor helpful. Rather, what is needed is an approach to health that leverages these fields not merely to supplement each other but to fuse them together in ways that are unanticipated, synergistic, and transformative.

An immediate concern at the outset is defining what is meant by human health. A common feature of many studies is that health is understood primarily as the absence of disease. Part of this conceptualization derives from the biomedical model, which focuses upon the pathology, biochemistry, and physiology of a disease. Disease pathology studies the physical locations where disease exists, centering upon organs, body tissues, fluids, and whole bodies. Biochemistry examines the human body as an organism, focusing upon the structure and function of cells and biomolecules. Finally, physiology studies the mechanical, physical, and biochemical functions of living organisms. The central point is that these three components of the biomedical model—pathology, biochemistry and physiology—locate the human body as the site of illness. The result is that a disease such as acquired immune deficiency syndrome (AIDS) is seen as deriving from the human immunodeficiency virus (HIV) that is transmitted through sexual activity, intravenous drug use, blood transfusions or other activities that facilitate the transfer of bodily fluids. West Nile virus (WNV) is spread through mosquitoes that follow particular disease vectors, vectors that are changing due to climatic variabilities and ecological gradient shifts. Absent these infectious diseases, or exposure to carcinogens, extreme heat, or other factors that produce ill-health, the biomedical model suggests an individual body is free from disease and is therefore presumed healthy.

This narrow definition of health has not been as widely employed within the policy and governmental realms. The 1946 Constitution of the World Health Organization defines health as a "state of complete physical, mental and social well-being and not merely the absence of disease or infirmity" (in Grad 2002: 984). The Constitution continues in identifying health as a right of every human being irrespective of social, economic, religious, and ethnic differences, and that the "health of all peoples is fundamental to the attainment of peace and security" (Grad 2002: 984). The Ottawa Charter for Health Promotion (1986) discusses the promotion of health as including the realization of aspirations, satisfaction of needs, and the ability to adjust to the environment. The Ottawa Charter specifically identifies the following as the fundamental conditions for human health: peace, shelter, education, food, income, ecosystem stability, sustainable resources, social justice, and equity. This expansive understanding of human health challenges definitions and approaches that emphasize health strictly as the absence of disease. Additionally, these criteria require a framework that does not reduce its spatial and temporal analysis to center only upon the individual body that is afflicted but rather situates that person within her particular socio-ecological context. Human health exists at the interface of environment and society; therefore, the linkages across spatial and temporal scales, the networks that produce vulnerability and healthy decision-making, and the stability of social and ecological systems, are all critical elements shaping health. The broad conditions for human health identified by the Ottawa Charter are present in the varied contributions of *Ecologies and Politics of Health*. Whether it concerns the ways that increased access to resources improve quality of life and reduce potential vulnerabilities to illness, the importance of ecosystem functioning in providing services and protection from exogenous threats, or the likely future disruptions due to climate change, we understand health more broadly than the absence of disease. Central to this volume, therefore, is a conceptualization of human health that is centered upon social and ecological security, well-being, equity, and sustainability.

In examining human health at the nexus of ecologies and politics, it is important to first consider the reciprocal though asymmetrical feedbacks of social and ecological processes. "Ecologies of global health" is a phrase that simultaneously captures several important dimensions of the complexities of global health research. First, while the discipline of ecology assesses the abiotic and biotic interactions in a given system, there are other "compound ecologies" that supplement this perspective in understanding global health. Landscape ecology (Forman and Gordon 1986; Forman 1995; Turner *et al.* 2001) and spatial ecology (Tilman and Kareiva 1997) recognize the importance of the spatiality of these interactions. The field of political ecology has been influential in interrogating the underlying power structures at work in these systems. Yet even with these contributions, few studies effectively combine both a critical (e.g., political ecology) and scientific (e.g., ecology) approach, and fewer still add a spatial nuance to those assessments. The ecology-driven approach has been critiqued for ignoring socially imperative conditions and potentially deliberate ignorance of constructs that should address such concerns (Peet and Watts 2004), while concomitantly socio-political approaches have been critiqued for failing to address the importance of ecological processes (Walker, P. 2005). While we acknowledge that a thoroughly integrated ecologies and politics approach, or alternatively a coupled natural human framework, is essential for moving health and health research forward (NRC 2010), it is first necessary to disentangle what the ecological and socio-political components offer in supporting their integration.

An ecological grounding facilitates our understanding of human health in two ways. First, factors impacting health, whether positively or negatively, still predominantly move through time and space (Cutter and Solecki 1996; Rushton 2003; Arcury *et al.* 2005; Koch 2005). This spatio-temporal explicitness can be seen, for example, with precipitation trends under climate change scenarios that will impact the carrying capacity of food insecure areas based upon their geomorphological, hydrological, climatological, and topographic position. To be sure, different people, households, and communities can be differentially situated in their abilities to adapt or migrate to processes of change; however, it is likely that the ecological component provides a baseline or potential for basic health maintenance. That is, while there is a biophysical reality that for non-ecologists can be a contested terrain (cf. Latour 1993), it separates "the [health] environment" from individual or community perceptions and from interpretations of both agency and structuration (Callinicos 2004). To illustrate, consider the case of Lois Gibbs and Love Canal (Gibbs 1982). Lois Gibbs's family's and neighbors' disproportionate exposure to underground chemical waste toxins can be assessed from a number of critical theoretical approaches as to causation, as to intent, and as to meaning; but the occurrence, the actual measurable toxins residing in the bodies of Love Canal residents from that exposure, was an observable and quantifiable reality. Rather than supplanting the socio-political factors in any way, an ecological perspective serves to ground our understanding of human health. The Love Canal case reveals that a thorough engagement with the ecologies of health can serve to emphasize the disproportionate spatial and temporal exposures due to race, class, or other socially constructed categories, in addition to highlighting the future challenges for producing healthy societies and environments.

Second, an ecological perspective positions the point of inquiry *in situ*; namely, that the human health condition is inherently tied to that set of systems in which it is placed, be they urban or rural, individual or community, indoors or outdoors. Artificially separating humans from the environment surrounding them ignores the reciprocal relationships between human health and ecosystem health, and between human vulnerability and ecosystem vulnerability. The danger of this separation is to potentially overlook causative and/or preventative measures that come from the insight of placing health in its environmental context (see Snow 1855 for a classic early example on cholera outbreaks in London). While it might be tempting to see the ecological as being more important in infectious disease patterns than it is in non-infectious disease or even in well-being, the reality remains that human health is a complex interplay both manifesting and further impacting ecological, climatic, and socio-political drivers. Ignoring an urban dweller's physical residence in areas of high traffic-related pollution, little green space, or with buildings outfitted with outdated synthetic paints and carpets is what misled diagnoses of chronic health problems including asthma (Whittemore and Korn 1980), obesity (Ebbeling *et al.* 2002), nature deficit disorder (Louv 2005), and sick building syndrome (Redlich *et al.* 1997).

While research on the ecologies of global health has expanded in recent decades, so too has scholarship within the social sciences on the socio-political dimensions of human health. In a related manner to the ecological perspective, work within the social sciences has concentrated upon examining external factors shaping disease vulnerabilities and health decision-making, in addition to revealing how they operate across multiple spatial and temporal scales. As one example, the field of political ecology has been extended to examine how large-scale social, economic and political processes contribute to human disease (Mayer 1996). This approach has expanded in recent years, whether the studies centered upon the effects from violent conflict (Oppong and Kalipeni 2005), exposure to industrial pollution (Cutchin 2007), or the effects of aquaculture development upon indigenous perceptions of economic and environmental health (Richmond *et al.* 2005). Within the discipline of geography, much of the work has concentrated upon disease exposure models and healthcare accessibility, often through the employ of quantitative techniques of data collection and analysis. Gesler (2003) suggested that medical geography expanded into several new areas since the 1980s, including examinations of the distribution of health services (Mohan 1991; Cromley and Albertson 1993), production and persistence of health inequalities (Hayes 1999; Smyth 2005), and the intersections between disease and gender (MacIntyre *et al.* 1996). More recently, research has directed attention towards subjects that are "indicative of a distancing from concerns with disease and the interests of the medical world in favor of an increased interest in well-being and broader social models of health and health care" (Kearns and Moon 2002: 606). In the investigation of socio-political processes shaping disease transmission, healthcare, and the places of health, a number of these studies have drawn from social and cultural theory to engage with new epistemologies and methodologies, thereby productively expanding the frame to address disability (Dear *et al.* 1997), mental health (Parr 1998), and sexuality (Brown 2006).

Outside of geography, other research fields have contributed to understandings of human health with a focus upon how various socio-political processes unfold over time and space. Medical anthropology has attended to the political and economic dimensions of human disease (Baer 1996; Harper 2004) which has been extended with interdisciplinary scholarship on the emergence of health threats due to climate change (Costello *et al.* 2009). The field of public health has been invaluable in demonstrating the importance of social context, while critiques of the biomedical model have asserted that social networks and relationships are vital in understanding health and well-being. Work within social epidemiology has evaluated nested interactive systems that are seemingly produced by exogenous factors, in addition to local structures and relationships (Krieger 2001). While concentrating upon social, political, and economic processes across multiple scales, some of these fields have also incorporated ecological and environmental perspectives in their analyses. In an early promotion of an ecosocial epidemiology framework, Krieger (1994: 896) suggested that: "Different epidemiologic profiles at the population-level would accordingly be seen as reflecting the interlinked and diverse patterns of exposure and susceptibility that are brought into play by the dynamic intertwining of these changing forms." Environmental historians have productively shown how understandings of public health have changed over time thereby reworking perceptions of disease causation and environment–body interactions (Nash 2006). Additionally, an historical framework has been helpful in demonstrating how the geography of an infectious disease, such as malaria, can shift due to aggressive interventions and ecosystem transformation (Packard 2007).

These varied contributions have simultaneously deepened and broadened understandings of the socio-political dimensions of human health. Regardless, less attention has been directed to the ways in which socio-political and biophysical processes interact in producing disease, reworking vulnerabilities to illness, or shaping health decision-making (King 2010). Building upon previous work on the political ecology of disease, emerging studies leverage broader contributions from the field to understand social and ecological interactions and discursive constructions of disease and health (Hanchette 2008; Mulligan *et al.* 2012). Disease vulnerability is often directly connected to environmental factors, and has been shown by the environmental justice movement, can be interlinked with race, ethnicity, class, and other social categories that experience differential exposure to unhealthy conditions (Bullard 2005). The expansion of the concept of environmental justice on the global scale alludes to the unfortunate parallels in these patterns in various settings, even if they vary due to socio-ecological context (Walker, G. 2012). Recognizing the socio-political dimensions of human health, therefore, reveals how age, class, gender, culture, and other social categories intersect with political economies and ecological conditions that produce inequities in human health and variations in exposure to disease.

Embedded in these research fields and conceptual frameworks are constructs and tools for understanding the ecological and socio-political dimensions of human health. We argue, however, that a framework for evaluating the integration of the ecologies and politics of health has yet to emerge as a viable, critical, and necessary means of addressing some of the world's most pressing health issues. As such, we believe *Ecologies and Politics of Health* is unique for several reasons. First, while several research subfields, such as disease ecology, medical anthropology, and social epidemiology, offer insights into the social and ecological dimensions of human health, the intersections between them are often underdeveloped. The consequence is that the ecological processes serving as drivers for human health might not be integrated with analyses attesting to the importance of social and political factors. While the ecological study of disease, and to a lesser extent health more generally, is advancing from primarily biological vector transmission models to explicitly incorporating socio-economic factors, the multiscale heterogeneity of disease vulnerability is rarely acknowledged. Additionally, understanding how human health is differentially vulnerable to environmental changes, including those due to climate change, increasing urbanization, or any number of ecological and environmental factors, is not always integrated into the object of study. As one example of this, in their review of the linkages between HIV/AIDS and the natural environment, Bolton and Talman (2010) conclude that detailed assessments remain largely anecdotal with few studies addressing the environmental and ecosystem impacts of the disease. The consequence is that the growth of studies on the impacts for agricultural production and food security has not been similarly matched by analyses of the long-term ecological impacts of the epidemic.

Similarly, health studies that concentrate upon environmental conditions that produce poor health, or analyze proximity to sites of infection, often say less about how these locations are produced by political and economic processes over time and space. Returning to the example of HIV/AIDS, differential exposure to the disease is shaped by a variety of social *and* ecological factors that interact in producing vulnerabilities to the disease while constraining effective policy responses for those afflicted. One of the distinguishing features of *Ecologies and Politics of Health* is that, rather than approaching socio-political and ecological factors separately, each of the chapters addresses the intersections between the social and ecological dimensions of human health. The contributors were deliberately selected to ensure that their chapters would attend, at least in some measure, to both the ecological and socio-political factors related to their case studies. Second, a primary objective of this volume is to draw together research from both the natural and social sciences addressing the social and ecological dimensions of human health. As such, this collection includes a notable diversity of disciplinary perspectives, normative intentions, theoretical perspectives, and methodological approaches. We believe that in order to understand the realities of human health, research studies within these domains must engage with each other. This volume is structured to do exactly this in order to generate key insights for future research and practice on human health.

Ecologies and Politics of Health draws upon case studies from local to global scales and across the rural–urban gradient. Many of the chapters address components of infectious disease historically and in the contemporary era, ranging from the human-transmitted HIV/AIDS to insect-transmitted diseases, and, further, to a disease of unknown vector ecology. In addition to the focus upon infectious disease, a number of contributors provide analyses of non-infectious disease such as heat-related illness, injury, and a critical overview of the interactions of infectious disease, immune status, nutritional status, and parasite loading. The thirteen chapters that follow collectively present results from Africa, Asia, Latin America, the United States, Australia, and global cities. A theme running throughout all of the chapters is the need for interdisciplinary rather than multidisciplinary research that is spatially explicit, temporally precise, and locally relevant. Time lags, spatial mismatch, and failure to recognize local resident priorities have challenged attempts at appropriate health management, and therefore environmental management, at times leading to unintended negative consequences. This volume serves as a reminder that the natural environment, the human condition, and the world's socio-political structure are not static; they are dynamic phenomena that are best understood as endogenous to a broader health system, rather than exogenous variables or shocks disrupting health periodically. The contributors also collectively demonstrate how socio-political and ecological processes intersect, and are often integrative, in shaping human health across spatial and temporal scales.

Regardless of the disciplinary, theoretical, and methodological diversity, there are some clear themes that emerge from these chapters that were used to structure the four parts of the volume. To address the challenge of a health paradigm with interacting ecologies and politics, *Ecologies and Politics of Health* engages with explorations of theoretical frameworks, empirical assessments, and discursive analyses. Part I, "Health within social and ecological systems," presents three chapters that interrogate the utility of several theoretical frameworks and conventions for understanding health within complex social and ecological systems. Though these chapters are conceptual rather than empirical in their collective approach, each chapter is advanced from the perspective of considerable investment in fieldwork, primarily in tropical developing countries. Crews (Chapter 2) and Young (Chapter 4) examine in turn socio-ecological systems (SESs)[1] and coupled natural–human (CNH) systems as productive lenses working to blend the natural and social sciences. Crews offers a framework adapting SES to health research by explicitly focusing on four domains: environmental endowment, social infrastructure, cognitive resonance, and immune function. Young focuses on the interplay of livelihoods and health in a changing climatic context, and his discussion of livelihoods, as with VanWey *et al.* (Chapter 3), draws examples from decades of work in the tropics. VanWey *et al.* also leverage previous and ongoing research in smallholder frontiers to posit the benefit of a broader definition of health that encompasses the ability to invest in capital(s), whether these are natural, physical, financial, human, or social.

Part II, "Empirical approaches to injury and infectious disease," moves to empirically grounded and quantitative work that collectively redefines health in a more expansive way to move beyond mere presence or absence of disease. These chapters, intent upon considering geographic and even bacterial frontiers, present findings from four continents that exhibit the policy relevance of this line of health research. Qi *et al.* (Chapter 5) compare findings from two ongoing studies in west Africa and Australia that examine ecological, geographic, climatic, and social factors related to Buruli ulcer disease, a devastating illness with a currently unknown disease ecology. Given that climatic and socio-economic conditions differ substantially between Australia and most west African countries, the resulting assumption is that Buruli ulcer disease is likely related to dynamic interactions between humans and the natural environment. Castro and Singer (Chapter 7) present work on malaria, an illness with a well-known disease ecology, and examine the roles of changing land use (see also Young, Chapter 4) and settlement in frontier areas (see also VanWey *et al.*, Chapter 3). Root and Emch (Chapter 6) complement this work with examples from research in Bangladesh but on injury rather than disease. Much of the health literature focuses on infectious and non-infectious disease, yet in the developing world in particular, injury is implicated as one of the greatest and arguably preventable causes of economic loss through missed labor.

Part III, "Disease histories, the state, and (mis)management," examines the role of the state through historically rich approaches centering on both disease- and non-disease-related examples from Latin America, eastern Africa, and the United States. Insect-borne disease (e.g., trypanosomiasis and dengue), human-transmitted disease (e.g., HIV/AIDS), and human reproductive health (e.g, indigenous fertility) each showcase the complexities, consequences, and externalities deriving from varying types and levels of state intervention. McCord *et al.* (Chapter 9), along with Robbins and Miller (Chapter 11), examine the history of insect management (tsetse flies and mosquitoes) and disease management by various state and non-state agencies. The importance of the role of government managers and related path dependencies on technical fixes is a thread running through both of these chapters. The shifting construction of state-centered narratives around these histories forefronts the importance of state control whether in colonial Africa or in the contemporary United States. Interestingly, moving to human-transmitted HIV/AIDS, Pope (Chapter 10) finds the state treatment by Belize of Mayan communities with respect to HIV/AIDS to strongly echo colonial stereotypes and legacies in the contemporary era. McSweeney and Pearson (Chapter 8) extend the analysis of state activities surrounding health by examining how population serves as a mechanism for active interventions on peoples and territory. In their chapter, fertility is considered in terms of political strategies around defending and securing access to land and resources, specifically the survival and well-being of indigenous populations against external threats. Within this framework, external health interventions, whether they be vaccination or family planning, can be seen as part of larger state attempts to secure knowledge and control over indigenous populations and territories.

Part IV, "Health vulnerabilities," begins with two current and salient challenges to health researchers: climate and HIV/AIDS. In each case, a central theme of vulnerability is highlighted in how health matters are constructed by, in, and for different communities. Whether the urban poor, the elderly, or HIV-positive communities, disempowerment remains one of the broadest health challenges faced today. Chapter 12 by Wilhelmi *et al.* presents a non-infectious disease (heat stress) that is tightly linked to climate change. Their global analysis of urban areas indicates that urban heat island (UHI) effects not only are correlated with many aspects of urban heat intensification, but moreover do not impact all populations equally: the elderly, for example, are at greater risk of non-fatal and fatal illness from increased core temperatures. Furthermore, the urban populations typically most vulnerable to heat-related illness often tend to be the poorer demographics, raising concerns of environmental justice. Equity issues are also at the heart of the investigation by Stillwaggon and Sawers (Chapter 13) into the historical legacy of racial stereotypes. The epidemics in sub-Saharan Africa present three-quarters of the world's HIV/AIDS cases, and the behavioral paradigm largely has indicated that concurrency (multiple overlapping sexual partnerships) is the reason for rates so incredibly higher than in the rest of the world. While early epidemiological research included ecological factors, Stillwaggon and Sawers assert that the dominant approach somehow lost the ecological way and focused on (hetero)sexual behavior to the exclusion of most other potential factors. Documented relationships between HIV/AIDS and co-factor diseases prevalent in the region (e.g., malaria, schistosomiasis, and tuberculosis), nutritional status, and parasite loading argue for a return to a more fully ecologized perspective. King (Chapter 14) also centers on HIV/AIDS in sub-Saharan Africa, but does so with the intention of challenging conventional narratives surrounding the disease. Specifically, he argues that the conceptualization of HIV/AIDS as a shock to social and ecological systems is belied by emerging research attesting to the long-term and disparate impacts of the epidemic. Emphasizing alternative methodological strategies for studying the disease, the chapter argues that HIV/AIDS might be better conceptualized as a coupled socio-ecological experience that extends beyond the individual to families, health workers, livelihoods, and natural resource systems.

Chapter 15 concludes by drawing upon shared themes and lessons that are informative for future research and policy interventions. Future research and practice on the ecologies and politics of health needs to deepen our understandings of the ecological dimensions of health and vulnerability, the socio-political dimensions of human health, and the intersections between the ecological and social dimensions of health. We posit that the ecologies and politics of health framework proposed here is a viable, timely, and needed way forward for that challenge. Whether the case studies attend to HIV/AIDS, heat stress, West Nile virus, or Buruli ulcer disease, we believe evaluating them in concert raises again the challenge for researchers, practitioners and policy-makers to move the health agenda forward in a way that is geographically, ecologically, and demographically equitable.

Note

1 Note that the abbreviation SES is used in this book to denote "socio-ecological system," and we have therefore avoided using the same abbreviation for "socio-economic status," in order to avoid confusion. The latter term is therefore given in full throughout the book.

References

Arcury, T.A., Gesler, W.M., Preisser, J.S., Sherman, J., Spencer, J., and Perin, J. (2005) The effects of geography and spatial behavior on health care utilization among the residents of a rural region. *Health Services Research* 40(1): 135–156.

Baer, H.A. (1996) Toward a political ecology of health in medical anthropology. *Medical Anthropology Quarterly* 10(4): 451–454.

Bolton, S. and Talman, A. (2010) *Interactions Between HIV/AIDS and the Environment: A Review of the Evidence and Recommendations for Next Steps*. Nairobi, Kenya: IUCN ESARO Office.

Brown, M. (2006) Sexual citizenship, political obligation and disease ecology in gay Seattle. *Political Geography* 25(8): 874–898.

Bullard, R.D. (ed.) (2005) *The Quest for Environmental Justice: Human Rights and the Politics of Pollution*. San Francisco, CA: Sierra Club Books.

Callinicos, A. (2004) *Making History: Agency, Structure, and Change in Social Theory*, 2nd edition. Leiden, The Netherlands: Koninklijke Brill NV.

Costello, A., Abbas, M., Allen, A., Ball, S., Bell, S., Bellamy, R., Friel, S., Groce, N., Johnson, A., Kett, M., Lee, M., Levy, C., Maslin, M., McCoy, D., McGuire, B., Montomery, H., Napier, D., Pagel, C., Patel, J., Antonio, J., de Oliveira, P., Redclift, N., Rees, H., Rogger, D., Scott, J., Stephenson, J., Twigg, J., Wolff, J., and Patterson, C. (2009) Managing the health effects of climate change. *The Lancet* 373: 1693–1733.

Cromley, E.K. and Albertsen, P.C. (1993) Multiple-site physician practices and their effect on service distribution. *Health Services Research* 28(4): 503–522.

Cutchin, M. (2007) The need for the "new health geography" in epidemiologic studies of environment and health. *Health and Place* 13(3): 725–742.

Cutter, S.L. and Solecki, W.D. (1996) Setting environmental justice in space and place: acute and chronic airborne toxic releases in the southeastern United States. *Urban Geography* 17(5): 380–399.

Dear, M., Wilton, R., Gaber, S.L., and Takahashi, L. (1997) Seeing people differently: the sociospatial construction of disability. *Environment and Planning D* 15(4): 455–480.

Ebbeling, C.B., Pawlak, D.B., and Ludwig, D.S. (2002) Childhood obesity: public-health crisis, common sense cure. *The Lancet* 360(9331): 473–482.

Forman, R.T.T. (1995) *Land Mosaics: The Ecology of Landscapes and Regions*. New York, NY: Cambridge University Press.

Forman, R.T.T. and Godron, M. (1986) *Landscape Ecology*. New York, NY: Wiley Press.

Gesler, W.M. (2003) Medical geography. In G.L. Gaile, and C.J. Willmott, C.J. (eds). *Geography in America at the Dawn of the 21st Century*. Oxford: Oxford University Press, 492–502.

Gibbs, LM. (1982) *Love Canal: My Story*. Albany, NY: State University of New York Press.

Grad, F.P. (2002) The preamble of the constitution of the World Health Organization. *Bulletin of the World Health Organization* 80(12): 981–984.

Hanchette, C.L. (2008) The political ecology of lead poisoning in eastern North Carolina. *Health and Place* 14(2): 209–216.

Harper, J. (2004) Breathless in Houston: a political ecology of health approach to understanding environmental health concerns. *Medical Anthropology* 23(4): 295–326.

Hayes, M. (1999) "Man, disease and environmental associations": from medical geography to health inequalities. *Progress in Human Geography* 23(2): 289–296.

Kearns, R. and Moon, G. (2002) From medical to health geography: novelty, place and theory after a decade of change. *Progress in Human Geography* 26(5): 605–625.

King, B. (2010) Political ecologies of health. *Progress in Human Geography* 34(1): 38–55.

Koch, T. (2005) *Cartographies of Disease: Maps, Mapping, and Medicine*. Redlands, CA: ESRI Press.

Krieger, N. (1994) Epidemiology and the web of causation: has anyone seen the spider? *Social Science and Medicine* 39(7): 887–903.

Krieger, N. (2001) Theories for social epidemiology in the 21st century: an ecosocial perspective. *International Journal of Epidemiology* 30(4): 668–677.

Latour, B. (1993) *We Have Never Been Modern*. Translation by C. Porter. Cambridge, MA: Harvester Wheatsheaf and the President and Fellows of Harvard College.

Louv, R. (2005) *Last Child in the Woods: Saving Our Children from Nature-Deficit Disorder*. Chapel Hill, NC: Algonquin Books

MacIntyre, S., Hunt, K., and Sweeting, H. (1996) Gender differences in health: are things really as simple as they seem? *Social Science and Medicine* 42(4): 617–624.

Mayer, J.D. (1996) The political ecology of disease as one new focus for medical geography. *Progress in Human Geography* 20(4): 441–456.

Mohan, J. (1991) The internationalisation and commercialisation of health care in Britain. *Environment and Planning A* 23(6): 853–867.

Mulligan, K., Elliott, S.J., and Schuster-Wallace, C. (2012) The place of health and the health of place: Dengue fever and urban governance in Putrajaya, Malaysia. *Health and Place* 18(3): 613–620.

Nash, L. (2006) *Inescapable Ecologies: A History of Environment, Disease, and Knowledge*. Berkeley, CA: University of California Press.

National Research Council (NRC). (2010) *Understanding the Changing Planet: Strategic Directions for the Geographical Sciences*. Washington DC: National Academies Press.

Oppong, J.R. and Kalipeni, E. (2005) The geography of landmines and implications for health and disease in Africa: A political ecology approach *Africa Today* 52(1): 3–25.

Ottawa Charter for Health Promotion. (1986) Ottawa Charter for Health Promotion, First International Conference on Health Promotion, 21 November. Document WHO/HPR/HEP/95.1. See www.who.int/hpr/NPH/docs/ottawa_charter_hp.pdf (accessed March 1, 2012).

Packard, R.M. (2007) *The Making of a Tropical Disease: A Short History of Malaria*. Baltimore, MD: Johns Hopkins University Press.

Parr, H. (1998) Mental health, ethnography and the body *Area* 30(1): 28–37.

Peet, R. and Watts, M. (eds). (2004) *Liberation Ecologies: Environment, Development, Social Movements*, 2nd edition. London, UK: Routledge.

Redlich, C.A., Sparer, J., and Cullen, M.R. (1997) Sick-building syndrome. *The Lancet* 349(9057): 1013–1016.

Richmond, C., Elliott, S.J., Matthews, R., and Elliott, B. (2005) The political ecology of health: perceptions of environment, economy, health and well-being among 'Namgis First Nation. *Health and Place* 11(4): 349–365.

Rushton, G. (2003) Public health, GIS, and spatial analytic tools. *Annual Review of Public Health* 24: 43–56.

Smyth, F. (2005) Medical geography: therapeutic places, spaces and networks. *Progress in Human Geography* 29(4): 488–495.

Snow, J. (1855) Dr Snow's report. In *Report of the Cholera Outbreak in the Parish of St. James, Westminster, During the Autumn of 1854*. London, UK: Churchill, 97–120.
Tilman, D. and Kareiva, P. (eds). (1997) *Spatial Ecology: The Role of Space in Population Dynamics and Interspecific Interactions*. Monographs in Population Biology 30. Princeton, NJ: Princeton University Press.
Turner, M.G., Gardner, R.H., and O'Neill, R.V. (2001) *Landscape Ecology in Theory and Practice: Pattern and Process*. New York, NY: Springer Press
Walker, G. (2012) *Environmental Justice: Concepts, Evidence and Politics*. London, UK: Routledge.
Walker, P. (2005) Political ecology: where is the ecology? *Progress in Human Geography* 29(1): 73–82.
Whittemore, A.S. and Korn, E.L. (1980) Asthma and air pollution in the Los Angeles area. *American Journal of Public Health* 70(7): 687–696.

Part I

Health within social and ecological systems

2 Positioning health in a socio-ecological systems framework

Kelley A. Crews

Overview

The socio-ecological system (SES) framework has been articulated as robust, interdisciplinary, integrative, and adaptable (Young *et al.* 2006; Janssen *et al.* 2007; Ostrom 2007, 2008, 2009; Collins *et al.* 2011; Niedertscheider *et al.* 2012). Its strength draws not only from understanding the human and the environment, but by epistemologically placing them as interacting systems that may both be comprised of and comprising other systems as well. This chapter explores the utility of a health-centered SES framework for interrogating and improving health and health studies. Four primary domains are offered for consideration: environmental endowment (EE), social infrastructure (SI), cognitive resonance (CR), and immune function (IF). The pair-wise areas of overlap, or secondary domains, offer synergies for improved understanding as illustrated with examples from fieldwork in the global tropics. Examination of tertiary domains or intersections yields insight into previous systematic omission of key factors in health-related research. The health-centered SES framework supports the utility and adaptability of the general SES framework and provides promise for improved understanding and analysis of health and health studies.

Introduction

The charge of this volume is to address the socio-political dimensions of human health, the ecological dimensions of health and vulnerability, and particularly the intersections, tensions, and synergies between the two (see Chapter 1). Human health is a pressing topic and, as underscored with vulnerability studies (Cutter 2003; Turner *et al.* 2003), one arguably most in need of not multidisciplinary but interdisciplinary approaches, interweaving lessons learned from both the "natural" and the "social" sciences (see also the Foreword to this volume by Billie Lee Turner, II). To this end, this chapter explores the utility of positioning health in the socio-ecological system (SES) framework to improve health-centered research as well as to leverage, test, and extend the this framework, SES posits the multitier interplay of social systems and natural systems through the operationalization as a resource system (RS) comprised of resource units (RU) that, coupled with a governance system (GS) composed of users (U), will interact in the presence of related ecosystems (ECO) and social, economic, and political settings (S) to interact and produce (observable) outcomes (Ostrom 2008). This evolving paradigmatic scheme was neither intended to have each factor fully implemented in every possible study nor, by extension, to be uniformly operationalized across multiple investigations (Ostrom 2008). Part of the power of the SES framework is thus its adaptability to varying systems, places, times, and thematic foci. Ostrom (2008, 2007) articulated the SES framework as a means of studying a system by examining the reciprocal impacts of both social, economic and political settings as well as related ecosystems on a combination of resource system, resource units, governance system, and users where multiple interactions produce the system outcomes. The system, she suggested, was similar to Koestler's 1973 description of a Holon, where it was both a whole unit unto itself but also simultaneously part of a larger entity; this nesting of tiers within tiers is what she described as one of the greatest challenges for research (Ostrom 2008, p. 249). Conceptually this nesting can be understood through the epistemological lens of scaling known as hierarchy theory (Allen and Starr 1982; Allen and Hoekstra 1992; Ahl and Allen 1996), where perhaps the SES "system" maps to hierarchy theory's "scale of interest" and is part of a larger entity (the scale "above," which provides context) and is further comprised of "subsystems" (hierarchy theory's scale "below," which provides mechanics). While Ostrom explicitly mentions nested tiers, hierarchy theory reminds those investigating multiscale phenomena that hierarchies need not necessarily be nested and, in fact, often are not (Peterson and Parker 1998). For the purposes of testing centering the SES framework on health to explicitly ecologize and politicize health, this chapter leaves for now the consideration of spatial scaling, temporal character, and system/subsystem construction that have received recent attention (Ostrom 2009; Gunderson and Folke 2011) to focus on the fundamentals of how the domains of the SES framework might be construed to better position health studies.

In so considering its application, it is useful to briefly visit key departures and outgrowths of previous efforts in the broadly construed area of human–environment interactions with an eye, ultimately, to leveraging it as means of incorporating both an ecologically and politically rich exploration of health. While it is worth noting that early epidemiological work explicitly considered the environment as critical for understanding human health (for a related discussion, see Chapter 13), the social sciences for the most part did not. Early medical geography (e.g., Meade 1977; Meade *et al.* 1988), with a focus on topological relationships, and political science both examined the proximity of people to environmental "goods" (e.g., healthcare services—see e.g. Gesler and Meade 1998; Gesler *et al.* 2000) and environmental "bads" (e.g., toxic waste disposal sites—see e.g. Bowman and Crews-Meyer 1997). Hazards studies added the examination of (disproportionate) exposure to catastrophic events (e.g., Cutter 1996), while the environmental justice literature and interest groups drew attention to the socio-economically (in addition to spatially) uneven nature of (harmful) environmental exposure (Bullard 1990; Lester *et al.* 2001). Meanwhile, some medical geography and spatial epidemiology researchers moved to examine the vectors of infectious disease as a way of tracking, forecasting, and mitigating many (notably tropical) human infectious diseases (for a review, see Albert *et al.* 2000). A critique of much of the above work is that the focus on catastrophic exposure, environmental bads, and disease clusters seems to have in part precluded an examination of positive aspects of human health and its relation the environment, with recent attention by physicians, landscape architects, and city/regional planning to green space, urban pollution, physical activity and cardiovascular/ pulmonary health notwithstanding (e.g., Hancock 1993; Northridge *et al.* 2003). Further, "health" at this time was often presumed to be the absence of disease, rather than a system unto itself meriting attention to genetic, behavioral and environmental components (Ware 1987). Certainly health studies broke into this realm, but social scientists engaged in understanding the interplay between humans and their environment for the most part did not for several years to come.

Primary domains of health

As shown in Figure 2.1, this chapter posits four distinct thematic areas as the primary domains of a health-centered SES: environmental endowment (EE), social infrastructure (SI), cognitive resonance (CI), and immune function (Figures 2.2a–d). In some ways these domains can be seen to reflect what have been described as types of capital (Bebbington 1999; Ellis 2000), though they do not necessarily map directly, exclusively, exhaustively, or completely onto them (for an extended review of the five types of capital, see Chapter 3). EE perhaps most simply maps onto what is termed natural capital as does immune function as a component of human capital. SI maps most directly onto social capital, but

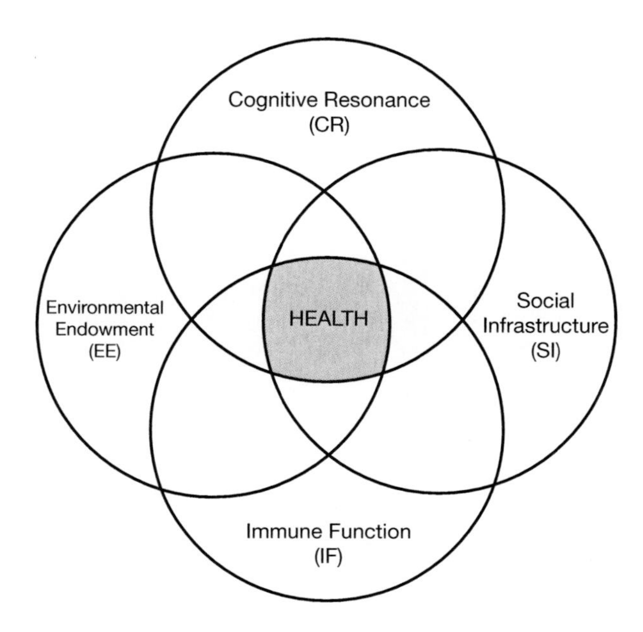

Figure 2.1 Mapping a health-centered socio-ecological system.

as will be discussed here has aspects that could be linked to physical capital (e.g., sidewalks), financial capital (e.g., government financial support), and human capital (e.g., capacity building experience) as well. CR as conceptualized here is not explicitly addressed in the breakdown of capital types and in some ways falls outside that framework, though presumably could partially be considered as human capital. EE refers to the natural resources and ecosystem services in an area at a given time. In a multitemporal perspective these would necessarily have been impacted by a variety of factors including but not limited to climatic, edaphic and hydrological processes, biotic and abiotic interactions, and human use/management. Here, for the sake of clarity, consider the EE at a snapshot in time—perhaps at the moment of human colonization. The EE will have bedrock and often soils of a certain quality: these soils, if present, will have physical and chemical characteristics that make them more or less fertile, able to hold water, and—depending upon the topography—spatially stable. There may or may not be present both surface (e.g., lake or ice cap) or subsurface (e.g., aquifer) waters also of a quality more or less suitable for a variety of consumptive or non-consumptive uses. The area, existing at a particular latitude in a particular topographic (continental position and elevation) with given temperature and precipitation regimes, will (likely) have some combination of flora and fauna corresponding to that biome. Taken together, these amenities and processes further provide not only habitat (for shelter, feeding, and reproduction) but amenities such as shade and ecosystems services such as water filtration. This EE may be used effectively or not, and its amenities may or may not even be accurately perceived by humans in the area; but the characteristics of it exist apart from such perception or use and can be mapped into systems space (e.g., see Walker *et al.* 2004).

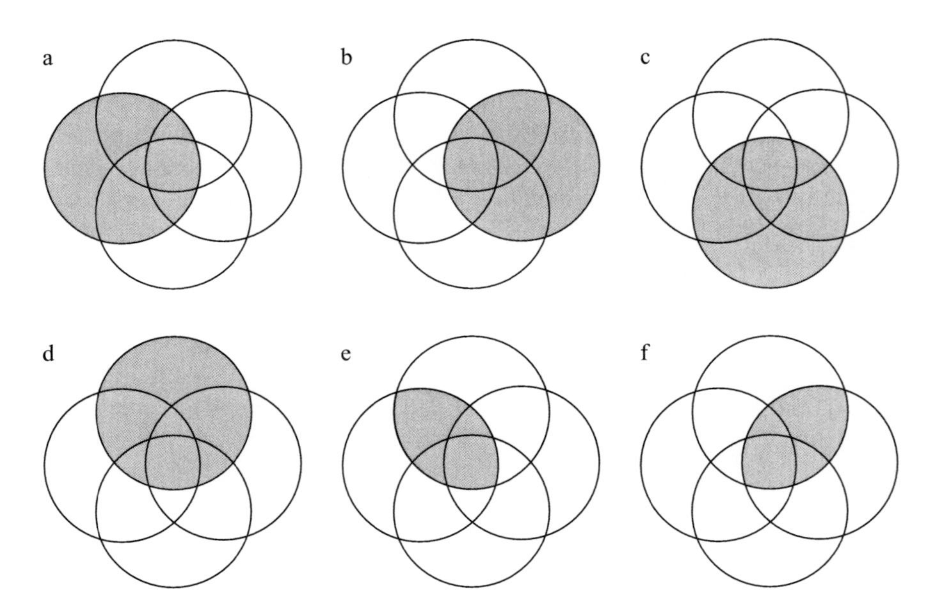

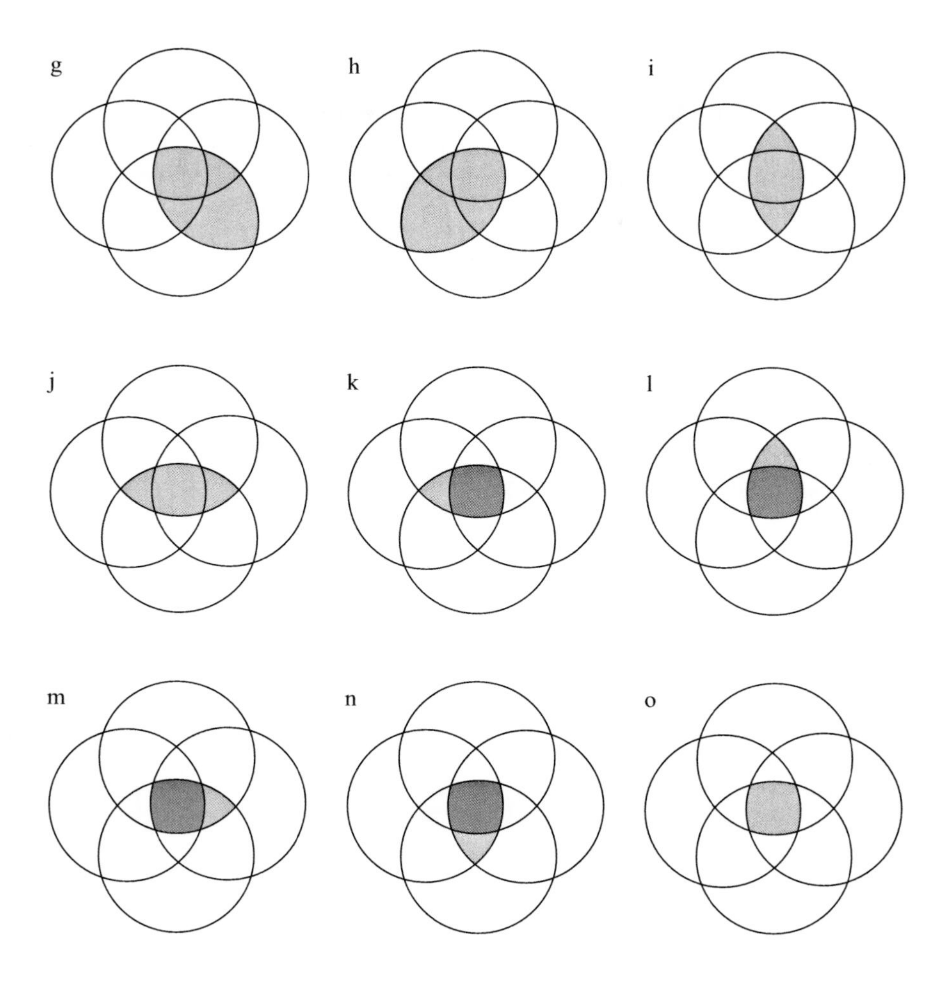

Figure 2.2 Primary, secondary, and tertiary domains of a health-centered socio-ecological system.

Note: (a–d) The primary domains of the health-centered SES, each highlighted in grey: (a) environmental endowment (EE), (b) social infrastructure (SI), (c) cognitive resonance (CR), and (d) immune function (IF). (e–j) Secondary domains, or pair-wise overlaps of the primary domains, each highlighted in grey: (e) EE and CR, (f) CR and SI, (g) SI and IF, (h) IF and EE, (i) CR and IF, and (j) EE and SI. (k–n) The tertiary combinations, with non-white areas representing overlap of three of the four primary domains without respect to the fourth, and lighter grey areas representing that overlap that falls exclusively outside of the fourth area (the portion differing represented in darker grey): (k) all but SI, (l) all but IF, (m) all but EE, and (n) all but CR. (o) Health as the intersection of all four primary domains, highlighted in grey.

The SI of a system (shown in Figures 2.1 and 2.2b) is here defined as including components both tangible (e.g., sidewalks, medical centers, and roads) and intangible (or not directly tangible, though they may produce tangible benefits—e.g., social networks, family support, and village cohesion). Disentangling the mechanics of and interactions within Social Infrastructure certainly presents a ripe field for exploration, much of which has been conducted, for example, under the auspices of human–social dynamics (e.g., Moran 2010), but is beyond the scope of this chapter. SI is meant to represent the social side of the system, as EE represents the ecological side of the system in the SES framework. Taken together, the two represent the incarnation of brackets of a SES as amenable to a focus on health with the two other primary domains meant to further tailor the SES for this purpose. IF (displayed in Figures 2.1 and 2.2c), as the name suggests, represents the ability of an individual (or a group of individuals) to both respond effectively to charges against the immune system and allow for proper physical and mental development. Nutrition status is an important component of IF, both in terms of macro-nutrition (adequate calories consumed to maintain body function and energy levels) and micro-nutrition (vitamins and minerals needed, often in trace amounts; Chandra 1997; Friis and Michaelsen 1998). Macro-nutrition standards promulgated by entities such as the World Health Organization indicate at what levels humans are capable of maintaining vital body function, staying healthy for most of the year, and allowing individuals to perform work. Micro-nutrients are indicated as being linked to targeted immune system and developmental failures, such as selenium and tuberculosis or zinc and childhood development (Friis and Michaelsen 1998; Köhrle *et al.* 2000). IF clearly also is impacted by disease history, genetics/predispositions, general condition, and parasite loading (Chandra 1997; Hayward and Gorman 2004; Stillwaggon 2006).

Lastly, CR (illustrated in Figures 2.1 and 2.2d) is used to represent the complexities of environmental perception and residence time (personal, familial and culture group). CR can be thought of as the set of factors that influence a person's or a group's familiarity with the system. Typically, longer residence time (again, of a person or a group) generates greater system familiarity, presuming the system has remained relatively similar through time. A person who has lived in the same region of a tropical rainforest for several decades is more likely, all else being equal, to be familiar with the inner workings of that system. Similarly, a younger generation raised by an older generation with residence time in the same system will also inherit greater familiarity. Historical (cultural) tradition is another way in which familiarity is increased, as lessons are passed on from generation to generation (and not necessarily by those within the household or family unit) within a cultural group. A person's environmental perception (or the collective perception of a group of people) is necessarily impacted by the availability of information, the accuracy of that information, and—in terms of both environment and health—by residence time. Taken together, these factors comprise CR and are critical for understanding the way in which the system is perceived, decisions made (or not), and actions taken (or not).

Secondary domains

The term "secondary domains" is used here to represent the overlap of pair-wise combinations of the primary domains; these six areas of overlap are represented in Figure 2.2e–j. Understanding these intersections allows for assessing the theoretical and practical synergies of this adaptation of the SES framework. These intersections are critical for the proper assessment of system process in much the same way that interactions are important in statistical representations of relationships of inter-related variables: that intersection, or interaction, presents a situation where the whole is greater than the sum of the parts. That is, when those two elements co-occur (empirically or theoretically), they each act differently to produce an outcome beyond which would normally be expected from either component without their interaction.

The intersection of EE and CR is shown in Figure 2.2e. The environment in any given SES has a certain set of spatio-temporal behaviors that include the first- and second-order variation of that system across time (and space). Following basic denotations of first- and second-order derivatives, here first-order variability is simply the variability in the system (high to low) while second-order variability indicates the variability of that variability. That is, a system could exhibit high first-order variability but low second-order variability, indicating high but predictable variability. The EE may exhibit low (first-order) variability (e.g., rainfall and temperature in the Brazilian Amazon tropical rainforest) or high variability (e.g., rainfall and temperature in temperate forests of Europe). That variability may be consistent over time (consistently variable, that is—e.g., the seasons in the northeastern US) or may remain highly variable but change the nature of its variability (high second-order variability such as flooding cycles in Botswana's Okavango Delta or glacial melt in the Andes; Neuenschwander and Crews 2008; Bury *et al.* 2011). The temporal environmental variability of a system has a signal that can be decomposed into trends (non-cyclic), cycles (e.g., annual and seasonal), and structured residuals that can be further decomposed into transitory (temporary) signals and stochastic (chance) events or disturbances (Jassby and Powell 1990; Rodriguez-Arias and Rodo 2004). Previous work in the Okavango Delta region over a 14-year period reported both seasonal and near-decadal cycles in precipitation and vegetation vigor as well as a slight longer-term trend (Neuenschwander and Crews 2008) punctuated by many discrete disturbances. As multiple cycles go in and out of phase atop longer term trends and disturbances, it is easy to understand how predictability becomes more difficult. Particularly as second-order variability increases, the predictability of the system tends to decline. This unpredictability is separate from a person's or group's perception of it, but certainly impacts the ability of humans to adequately gauge how the system will behave in the future.

Recall, however, that CR also includes personal and group residence time, including cultural familiarity. That is, someone who has experienced a certain set of environmental system behaviors for the previous decades is more likely to understand and be able to forecast how the system may behave in the future, and this may in turn impact the actions or inactions taken (Morrison 1967; Toney 1976; Bell 1992; Lober 1993; Mena *et al.* 2006; Silvano and Begossi 2010). This ability may come from personal residence time, from familial residence time, or from the knowledge of one's culture group even if not experienced first-hand. In 2010 (29 May–1 June), twenty interviews were conducted (in SeTswana) along the Boteti River outflow system of Botswana's Okavango Delta to assess people's perceptions of the river's return after being dry for nearly three decades. Those who had lived in the area for four decades or more not only remembered back to when the river had flown but more interestingly reported many more positive than negative reactions to the river's return (and "life will be better for us"). Similarly, even younger adults who had never seen the river but who were from a fishing (river) culture (the Bayei) found the return overwhelmingly positive. One middle-aged male said "I thought we were going to have to leave to go to Shakawe [part of the Delta's permanently flooded panhandle] so we could live, but now we can stay … the river makes me happy." When specifically prompted for any negatives regarding the river's return, these groups with greater familiarity of "river culture" responded that "there might be crocodiles and alligators, but that is *natural*" (emphasis added). In stark contrast, those interviewed who were not from a river culture and had no household members with residence time enough to have seen the river's earlier days volunteered primarily negative reactions: "the children will die" and "many dangerous animals will come and then they will kill our livestock and eat our cattle". A high percentage of these people also reported fear that the river is what brings lumpskin disease to their cattle, which in effect would knock out the only source of savings most people have. When specifically questioned as to positive reactions to the river, a few reported not having to go so far for water and several were unable to mention benefits of the returning river—an interesting reaction from someone living in a semi-desert. Thus, examining familiarity (CR) of a system along with that system's environmental endowment (EE) and characteristic variability offers insight into people's (perceived) environmental vulnerability (Yanes-Estévez *et al.* 2006) and provides an important backdrop for understanding health in both socio-political and ecological contexts.

CR also plays an interesting role in its overlap with SI, shown in Figure 2.2f. The relationship between these two primary domains is often self-reinforcing and at times reciprocal. That is, for a given population or person in a given system, the person's expectation of SI (e.g., government services) will over time be impacted by the availability or lack thereof of those services. That infrastructure may in turn be changed by [here] the government in response to local demands for such services. In 2007 (8–21 March), ninety interviews were conducted (in Spanish) along the Ucayali River in the Peruvian Amazon, just upstream (south) of its confluence with the Marañon where the Amazon River is formed, as shown in Figure 2.3. The goal of the interviews was to understand both human health and environmental health given the recent change in precipitation cycles bringing drought to many areas of the Amazon just a few years earlier (Asner *et al.* 2004).

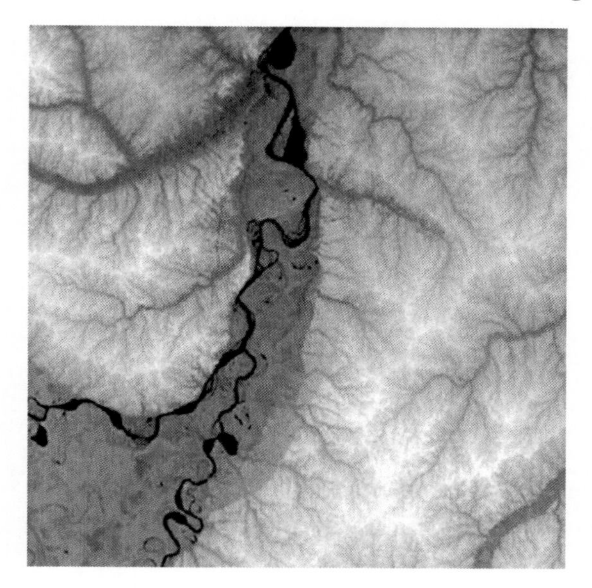

Figure 2.3 Digital elevation model of the Peruvian Amazon.
Source: NASA SRTM (Shuttle Radar Topography Mission)

Of the communities visited, two in particular had evidence of significant external investment in infrastructure. Jenaro Herrera, settled a little over 50 years ago on the eastern banks of the Ucayali River, underwent significant infrastructure improvements throughout the central portion of the 5000-person town with raised concrete sidewalks, paved roads, improved storm drainage and sewer system, and a doctor-staffed medical clinic. In the 39 interviews conducted there, many people reported minor to moderate medical ailments (e.g., asthma) as well as terminal illness (cancer), seemingly despite the area having the best medical care in the region outside Iquitos. In contrast, the smaller village Cedro Isla (over 90 years old, with population of roughly 150) had a non-staffed medical clinic, cracking sidewalks, and non-electrified electrical infrastructure. Yet in this village concerns focused more on malaria and diarrhea, complaints more common to other areas interviewed. One of the primary non-governmental organizations (NGOs) in the area had pulled out because of the local population's refusal to pen their pigs, with resulting high levels of porcine-related communicable disease. The NGO had been out of the area for several years, and people were used to living without a doctor, with free-running pigs, and without access to sewage or electricity. Their level of demand for medical care, and their attention to their own health, had deteriorated since that departure. The residents of Jenaro Herrera, by contrast, had become accustomed to a higher level of overall daily health, reported greater preventative measures (particularly for prenatal patients), and a greater emphasis on "luxury" medical care (e.g., attention to minor ailments or referral to Iquitos and even Lima for addressing grave issues such as cancer—thought by area medical staff to go undiagnosed in most such remote areas). Thus the level of SI can impact an individual's or group's CR, both in general and in particular with health-related matters and services.

Not surprisingly, there is also a compelling and rather well-documented secondary domain or overlap between SI and IF, illustrated in Figure 2.2g. Part of the SI contribution to improving IF comes in the form of preventative care, access to pharmaceuticals, and critical care. But much also comes in the form of education by government and NGO workers, such as HIV/AIDS awareness, notably including the ABC (*a*bstinence first, *b*e faithful second, and use *c*ondoms third) and PMTCT (prevention of mother to child [vertical] transmission). But SI extends well beyond educational campaigns as well. For example, SI as defined here would include access to enriched or fortified foods, which have a clear and documented impact on IF (Chandra 1997; Köhrle *et al.* 2000; Foster 2008). However, it is important to note that the availability or lack thereof of resources may not and, typically, will not be homogeneous among the system's human population (Ribot 1998). Access, both in terms of physical accessibility (how close or far are the goods or services, by what means of transportation) and in terms of institutional accessibility (the standing, ability and power to negotiate through rules of a system) is rarely equivalent across genders, ages, ethnicities, social classes, occupations, citizenship status, and even health status (Ribot and Peluso 2003). Similarly, not all environments are created equal, and the spatial heterogeneity of environmental amenities may help some and harm others. The overlap between IF and EE (illustrated in Figure 2.2h) often means that people are unknowingly exposed to environmental bads or kept from environmental goods without their knowledge (Bullard 1990; Lester *et al.* 2001). Returning to fieldwork along the Ucayali River in the Peruvian Amazon, three settlements offer an interesting juxtaposition: Yanallpa, Jenaro Herrera, and Cedro Isla, all established within roughly 40 years and 50 km of each other, with Cedro Isla on the western banks and the rest on the eastern banks (see Figure 2.3). But interviews and site inspection reveal they are worlds apart: Yanallpa has *terra preta* (black earth) soils so rich that they allow permanent (non-fallowing) agriculture and the cropping of fruit trees (fruits sold downstream to the Iquitos market). Just downstream, Jenaro Herrera has poor soils (planted for 1–2 years and then need fallowing for 4–15 years) and was positioned along the river where flooding caused the complete destruction of their port. Cedro Isla, on the western banks, is so filled with scroll bars that under 10 percent of the area's land is arable (explaining the free-range pigs, since the land is otherwise useless). When settlers came from the Andes Mountains into the Amazon, they lacked any local knowledge to inform them of what the best lands would be. The natural vegetation in each village looks extremely similar, and only digging into the soils reveals their heterogeneous EE. And yet the legacy of those soils (e.g., as observed through the abundance, diversity, and rigor of useful plant species), the impact on the communities' micro- and macronutrition (enhanced by non-toxic and edible plant species), and thus immune response is clear today, as is the legacy of their lack of familiarity with the area when choosing where to establish their new lives. This exemplifies one way in which cognitive resonance (CR) and IF overlap (see Figure 2.2i). Another way is clearly through the traditions and practices of observing what foods to eat or which traditional and non-traditional medicines work (e.g., Rappaport 1968). That is not to say that CI always works in the favor of IF. According to the staff doctor in the Jenaro Herrera medical clinic, the misuse of traditional medicines was one of the leading causes of illness and fatality, after malaria and drowning. In Chapter 14 of this volume, Brian King examines the ways in which how health is theorized matter, especially with respect to envisioning disease generally and HIV/AIDS specifically as a shock to social and ecological systems versus as a socio-ecological *experience* (emphasis added). The literature on perceptions/education and health practices is vast, and will not be recounted here; rather, the point is to acknowledge the important overlap between CR and IF, most especially with regards to human health.

The last secondary domain or area of overlap is that between the EE and SI (see Figure 2.2j). This interplay is really at the heart of the SES framework, and can be traced to several lines of thought deriving in some part from the field of human–environment interactions (e.g., see Moran 2010). Human–environment interactions, whether as a topic of study or an ontological perspective, are interesting in that in essence it posits the separation nature of humans from the environment. Were they not conceptualized as separate, there would be no need to examine their interactions (e.g., see Glacken 1967 for a recounting of this conceptualization for the last few millennia, and see especially the Foreword to the present volume). Cultural ecology's emphasis on the immersion of people in their environment and the relationships among culture groups or ethnicities, tradition, and natural resource use clearly provides a lens for understanding the role of environment in the everyday lives of people, particularly at a localized level (Geertz 1973; Blaikie and Brookfield 1987). Though some early studies focusing on health or nutrition were perhaps misguided in their causal connection of humans and their environment (Rappaport 1968), they did provide needed attention in the social sciences to the explicit consideration of human health and environment beyond simple exposure studies. Political ecology incorporated a more spatially and institutionally expansive means of understanding humans and the environment (Bassett 1988; Batterbury 2001; Zimmerer and Bassett 2003; Peet and Watts 2004; Robbins 2004) and how they changed the landscape (Turner and Robbins 2008), though the increasing focus on institutions and access perhaps came at the expense of an explicit focus on the ecological or environmental sphere (Forsyth 2002; Walker 2005). As with cultural ecology, most studies seemed to focus on the developing world (for a notable contrast, see Chapter 11). It was during this period that two thematically similar but methodologically distinct research traditions emerged: nature–society and population–environment interactions. Again, the legacy of first separating humans as encapsulated and apart from the environment and then examining their interplay wove its way through many studies. Some have referred to the primary difference between the nature–society and population–environment interactions traditions as simply either qualitative–quantitative or theoretical–empirical. Both of these characterizations have elements of truth but are misleading and underestimate the breadth of approach in each. Suffice it to say that these bodies of work have certainly produced both publications and scholars whose work has been influential in the development and application of the SES framework (Lambin *et al.* 2001; Geist and Lambin 2002; Entwisle and Stern 2005; Rindfuss *et al.* 2004; Moran 2010; and for the population–environment interaction/land change science forerunner to SES, see notably Liverman *et al.* 1998; Gutman *et al.* 2004; Turner *et al.* 2004). Today, many SES-like approaches often are viewed under the rubric of coupled natural–human systems or sustainability science (for a comprehensive take on both, see especially Turner *et al.* 2003), depending upon their origin, focus, or even funding. Splitting the hairs of these complex and important bodies of work is beyond the scope of this chapter; rather, their similarities and differences provide entrée into how a SES can be tailored to focus on human health in ecologically and politically compelling ways.

Tertiary domains

Beyond the primary domains (EE, SI, CR, and IF) and their pair-wise areas of overlap ("secondary domains"), there exist a set of trio overlaps ("tertiary domains"), as illustrated in Figure 2.2k–n. With the secondary domains, interrogating their overlap allowed for theorizing potential synergies useful for adapting the SES framework to a health-centered focus. With the tertiary domains, instead it is useful to consider briefly instead what leaving out the fourth primary domain in each case would omit or do differently. Figure 2.2k–n demonstrates these intellectual territories in two ways, showing in two grey tones the complete area of overlap without considering the fourth primary domain and showing in lighter grey only the area of overlap explicitly acknowledging but excluding that fourth domain.

For example, Figure 2.2k indicates the overlap among EE, CR, and IF: that is, excluding SI. Arguably, this is what one era of epidemiological work did when it examined disease outbreaks without regard to social networks, provision of fortified foods, or health education campaigns (Stillwaggon 2006; Foster 2008). Current medical anthropology has reinvigorated the call for the inclusion of these non-medical, non-environmental factors as critical for understanding why some communities are disproportionately affected by infectious disease, such as the tendency of lower-class/inner-city residents to suffer from higher rates of mosquito-borne disease (such as malaria, where micro-breeding habitats are associated with poorer neighborhoods: Fischer and Schweigmann 2008; Grove 2009; Tuiten *et al.* 2009). Moving to Figure 2.2l, the implications of excluding the immune system from health studies seem preposterous: yet as Sawers and Stillwaggon (2010) document (see also Chapter 13 of the present volume), an entire subfield of HIV/AIDS research in effect did just that by failing to account for immuno-suppression stemming from an overtaxed immune system responding to ongoing parasite loading, micro-nutrient deficiencies, and co-factor diseases (Abu-Raddad *et al.* 2006; Foster 2008).

Figure 2.2m illustrates the exclusion of the environment (EE), and harkens to early "cancer cluster" research of the 1970s and 1980s. The tragic exposure, especially of children, to the unmitigated burial of toxic waste in Love Canal underscored to advocates, residents, and scholars that when disease outbreaks exhibit such spatio-temporal clustering there often is an underlying environmental cause (Tobler 1970), made clear by Snow's ground-breaking work on cholera in London (Snow 1855; note also Koch 2005). It is not only infectious disease which implicates environmental factors but chronic disease as well (Hayward and Gorman 2004; see also Chapter 12 of the present volume). Asthma, for example, has relatively recently been linked to air pollution (Whittemore and Korn 1980; Sunyer *et al.* 1997). Figure 2.2n illustrates the last of the tertiary domains whereby the social, the ecological, and the immune-related are considered but without regard to individual and group familiarity with the other system components (CR). Many literatures, academic and medical, have documented the need to include local understanding, culture, and tradition as important components in understanding health and facilitating healthy populations. What is clear, then, is that the SES framework, when morphed to center on health, must make room for people's perceptions and traditions in all other domains. To be clear, it is not that the SES framework excludes that possibility and in fact posits two areas where CR is indeed important: knowledge of SES (factor U7 of the users factors) and information sharing among users (factor I2 of the interactions factors; Ostrom 2008: 250). Further, as stated earlier, the SES framework was neither intended to perfectly represent all systems in all spatio-temporal contexts nor meant to be homogeneously applied with the same factors in every case. What instead it does is offer a powerful basis and jumping off point for theorizing its potential for better envisioning health.

Convergence

At the heart of all the overlapping primary domains is of course health (see Figure 2.2o). It has been argued here that health can and should be placed at the center of an SES framework, and that by doing so previous omissions in health studies are filled and new perspectives emerge from leveraging topics, theories, and applications from both natural and social sciences. The environmental, the social, the cognitive, and the immune: together these provide fertile ground for complementing studies focused on exposure (e.g., with the means to understand the motivation behind people's perceptions, decisions, and actions that may exacerbate or ameliorate the initial environmental exposure), vulnerability (e.g., recognizing intra-, inter-, and extra-personal threats to or adaptations to vulnerability), and well-being (notably the linkages between healthy ecosystems and healthy human populations). Taken together, these provide a tent large enough to consider factors as seemingly diverse as soil quality and sexual networks because in fact there are theoretical and material means for hypothesizing and testing their interaction(s). In concert, these illustrate the powerful adaptability, and thus utility, of the SES framework, and underscore its potential for properly positioning health and health research.

Acknowledgments

Personal observations from fieldwork cited in this chapter were supported in part by grants from the National Science Foundation's (NSF) Geography and Spatial Science program (BCS/GSS RAPID 0942211 and BCS/GRS SGER), NASA's Earth Systems Science Graduate Fellowship (NNG04GR09H), and the University of Texas's Population Research Center's NICHD R24 Center Grant (5 R24 HD042849, NICHD).

References

Abu-Raddad, L., Patnaik, P., and Kublin, J.G. (2006) Dual infection with HIV and malaria fuels the spread of both diseases in sub-Saharan Africa. *Science* 314: 1603–1606.

Ahl, V. and Allen, T.F.H. (1996) *Hierarchy Theory: A Vision, Vocabulary, and Epistemology.* New York, NY: Columbia University Press.

Albert, D., Gesler W., and Levergood, B. (eds). (2000) *Spatial Analysis, GIS, and Remote Sensing Uses in the Health Sciences.* Chelsea, MI: Ann Arbor Press.

Allen, T.F.H. and Hoekstra, T.W. (1992) *Toward a Unified Ecology.* New York, NY: Columbia University Press.

Allen, T.F.H. and Starr, T.B. (1982) *Hierarchy: Perspectives for Ecological Complexity.* Chicago, IL: University of Chicago Press.

Asner, G.P., Nepstad, D., Cardinot, G., and Ray, D. (2004) Drought stress and carbon uptake in an Amazon forest measured with spaceborne imaging spectroscopy. *Proceedings of the National Academy of Science* 101: 6039–6044 (doi:10.1073/pnas.0400168101).

Bassett, T.J. (1988) The political ecology of peasant-herder conflicts in the northern Ivory Coast. *Annals of the Association of American Geographers* 78(3): 453–472.

Batterbury, S. (2001) Landscapes of diversity: a local political ecology of livelihood diversification in southwestern Niger. *Ecumene* 8(4): 437–464.

Bebbington, A. (1999) Capitals and capabilities: a framework for analyzing peasant viability, rural livelihoods and poverty. *World Development* 27(12): 2021–2044.

Bell, M.M. (1992) The fruit of difference: The rural–urban continuum as a system of identity. *Rural Sociology* 57(1): 65–82.

Blaikie, P. and Brookfield, H. (eds). (1987) *Land Degradation and Society.* London: Methuen.

Bowman, A.O. and Crews-Meyer, K.A. (1997) Locating southern LULUs [locally unwanted land uses]: race, class, and environmental justice. *State and Local Government Review* 29: 110–119.

Bullard, R.D. (1990) *Dumping in Dixie.* Boulder, CO: Westview Press.

Bury, J., Mark, B., McKenzie, J., French, A., Baraer, M., Huh, K., Luyo, M., and Gómez, J. (2011) Glacier recession and livelihood vulnerability in the Cordillera Blanca, Peru. *Climatic Change* 105: 179–206.

Chandra, R.K. (1997) Nutrition and the immune system: an introduction. *American Journal of Clinical Nutrition* 66(2): 460S–463S.

Collins, S.L., Carpenter, S.R., Swinton, S.M., Orenstein, D.E., Childers, D.L., Gragson, T.L., Grimm, N.B., Grove, J.M., Harlan, S.L., Kaye, J.P., Knapp, A.K., Kofinas, G.P., Magnuson, J.J., McDowell, W.H., Melack, J.M., Ogden, L.A., Robertson, G.P., Smith, M.D., and Whitmer, A.C. (2011) An integrated conceptual framework for long-term socio-ecological research. *Frontiers in Ecology and the Environment* 9(6): 351–357 (doi 10.1890/100068).

Cutter, S.L. (1996) Vulnerability to environmental hazards. *Progress in Human Geography* 20(4): 529–539.

Cutter, S.L. (2003) The vulnerability of science and the science of vulnerability. *Annals of the Association of American Geographers* 93(1): 1–12.

Ellis, F. (2000) *Rural Livelihoods and Diversity in Developing Countries*. Oxford, UK: Oxford University Press.

Entwisle, B. and Stern, P.C. (eds). (2005) *Population, Land Use, and Environment: Research Directions*. Washington, DC: National Academies Press.

Fischer, S. and Schweigmann, S. (2008) Association of immature mosquitoes and predatory insects in urban rain pools. *Journal of Vector Ecology* 33: 46–55.

Forsyth, T. (2002) *Critical Political Ecology: The Politics of Environmental Science*. New York, NY: Routledge.

Foster, H.D. (2008) Host-pathogen evolution: implications for the prevention and treatment of malaria, myocardial infarction and AIDS. *Medical Hypotheses* 70(1): 21–25.

Friis, H. and Michaelsen, K.F. (1998) Micronutrients and HIV infection: a review. *European Journal of Clinical Nutrition* 52(3): 157–163.

Geertz, C. (1973) *Deep Play: Notes on the Balinese Cockfight: The Interpretation of Cultures*. New York, NY: Basic Books.

Geist, H.J. and Lambin, E.F. (2002) Proximate causes and underlying driving forces of tropical deforestation. *Bioscience* 52(2): 143–150.

Gesler, W., Arcury, T.A., Preisser, J., Trevor, J., Sherman, J.E., and Spencer, J. (2000) Access to care issues for health professionals in the mountain region of North Carolina. *International Quarterly of Community Health Education* 20(1): 83–102.

Gesler, W. and Meade, M.S. (1988) Locational and population factors in health care seeking behavior in Savannah, Georgia. *Health Services Research* 23(3): 443–462.

Glacken, C.J. (1967) *Traces on the Rhodian Shore: Nature and Culture in Western Thought from Ancient Times to the End of the Eighteenth Century*. Berkeley, CA: University of California Press.

Grove, J.M. (2009) Cities: managing densely settled socio-ecological systems. In *Principles of Ecosystem Stewardship: Resilience-Based Natural Resource Management in a Changing World*, F.S. Chapin III, G.P. Kofinas, and C. Folke (eds). New York, NY: Springer-Verlag, 281–294.

Gunderson, L. and Folke, C. (2011) Resilience 2011: leading transformational change. *Ecology and Society* 16(2): 30.

Gutman, G., Janetos, A., Justice, C., Moran, E., Mustard, J., Rindfuss, R.R., Skole, D., and Turner II, B.L. (2004) *Land Change Science: Observing, Monitoring, and Understanding Trajectories of Change on the Earth's Surface*. New York, NY: Kluwer Academic Publishers.

Hancock, T. (1993) The evolution, impact and significance of the healthy cities/healthy communities movement. *Journal of Urban Health* 14(1): 5–18.

Hayward, M.D. and Gorman, B.K. (2004) The long arm of childhood: the influence of early-life social conditions on men's mortality. *Demography* 41(1): 87–107.

Janssen, M.A., Anderies, J.M., and Ostrom, E. (2007) Robustness of social-ecological systems to spatial and temporal variability. *Society and Natural Resources* 20: 1–16.

Jassby, A.D. and Powell, T.M. (1990) Detecting changes in ecological time series. *Ecology* 71(6): 2044–2052.

Koch, T. (2005) *Cartographies of Disease: Maps, Mapping, and Medicine*. Redlands, CA: ESRI Press, 388pp.

Köhrle J., Brigelius-Flohé, R., Böck, A., Gärtner, R., Meyer, O., and Flohé, L. (2000) Selenium in biology: facts and medical perspectives. *Biological Chemistry* 381: 849–864.

Lambin, E.F., Turner, B.L., Geist, H.J., Agbola, S.B., Angelsen, A., Bruce, J.W., Coomes, O.T., Dirzo, R., Fischer, G., Folke, C., George, P.S., Homewood, K., Imbernon, J., Leemans, R., Li, X.B., Moran, E.F., Mortimore, M., Ramakrishnan, P.S., Richards, J.F., Skanes, H., Steffen, W., Stone, G.D., Svedin, U., Veldkamp, T.A., Vogel, C., and Xu, J.C. (2001) The causes of land-use and land-cover change: moving beyond the myths. *Global Environmental Change–Human and Policy Dimensions* 11(4): 261–269.

Lester, J.P., Allen, D.W., and Hill, K.M. (2001) *Environmental Injustice in the United States: Myths and Realities*. Boulder, CO: Worldview Press.

Liverman, D., Moran, E.F., Rindfuss, R.R., and Stern, P.C. (eds). (1998) *People and Pixels*. Washington, DC: National Academy Press.

Lober, D.J. (1993) Beyond self-interest: a model of public attitudes towards waste facility siting. *Journal of Environmental Planning and Management* 36(3): 345–363.

Meade, M.S. (1977) Medical geography as human ecology: the dimension of population movement. *Geographical Review* 67(4): 379–383.

Meade, M.S., Florin, J., and Gesler, W. (1988) *Medical Geography*, 1st edition. New York, NY: Guilford Press.

Mena, C.F., Barbieri, A.F., Walsh, S.J., Erlien, C.M., Holt, F.L., and Bilsborrow, R.E. (2006) Pressure on the Cuyabeno Wildlife Reserve: development and land use/cover change in the northern Ecuadorian Amazon. *World Development* 34(10): 1831–1849.

Moran, E.F. (2010) *Environmental Social Science: Human Environment Interactions and Sustainability*. Malden, MA: Wiley-Blackwell.

Morrison, P.A. (1967) Duration of residence and prospective migration: the evaluation of a stochastic model. *Demography* 4(2): 553–561.

Neuenschwander, A.L. and Crews, K.A. (2008) Disturbance, management, and landscape dynamics: wavelet analysis of vegetation indices in the Lower Okavango Delta, Botswana. *Photogrammetric Engineering and Remote Sensing* 74(6): 753–764.

Niedertscheider, M., Gingrich, S., and Erb, K.-H. (2012) Changes in land use in South Africa between 1961 and 2006: an integrated socio-ecological analysis based on the human appropriation of net primary production framework. *Regional Environmental Change* 12: www.springerlink.com/content/q7uu5377x05513x6/fulltext.pdf (doi:10.1007/s10113-012-0285-6).

Northridge, M.E., Sclar, E.D., and Biswas, P. (2003) Sorting out the connections between the built environment and health: a conceptual framework for navigating pathways and planning healthy cities. *Journal of Urban Health: Bulletin of the New York Academy of Medicine* 80(4): 556–568.

Ostrom, E. (2007) A diagnostic approach for going beyond panaceas. *Proceedings of the National Academy of Science* 104(39): 15181–15187.

Ostrom, E. (2008) Frameworks and theories of environmental change. *Global Environmental Change* 18: 249–252.

Ostrom, E. (2009a) A general framework for analyzing sustainability of social-ecological systems. *Science* 325(5939): 419–422 (doi:10.1126/science.1172133).

Ostrom, E. (2009b) *A Polycentric Approach for Coping with Climate Change*. Background Paper to the 2010 World Development Report. Washington, DC: World Bank.

Peet, R. and Watts, M. (eds). (2004) *Liberation Ecologies: Environment, Development, Social Movements*, 2nd edition. London: Routledge.

Peterson, D.L. and Parker, V.T. (eds). (1998) *Ecological Scale: Theory and Applications*. New York, NY: Columbia University Press.

Rappaport, R.A. (1968) *Pigs for the Ancestors*. New Haven, CT: Yale University Press.

Ribot, J. (1998) Theorizing access: forest profits along Senegal's charcoal commodity chain. *Development and Change* 29(2): 307–342.

Ribot, J.C. and Peluso, N.L. (2003) A theory of access. *Rural Sociology* 68: 153–181.

Rindfuss, R.R., Walsh, S.J., Turner II, B.L., Fox, J., and Mishra, V. (2004) Developing a science of land change: challenges and methodological issues. *Proceedings of the National Academy of Science* 101(39): 13976–13981.

Robbins, P. (2004) *Political Ecology: A Critical Introduction*. Malden, MA: Blackwell

Rodriguez-Arias, M.A. and Rodo, X. (2004) A primer on the study of transitory dynamics in ecological series using the scale-dependent correlation analysis. *Oecologica* 138: 485–504.

Sawers, L. and Stillwaggon, E. (2010) Concurrent sexual partnerships do not explain the HIV epidemics in Africa: a systematic review of the evidence. *Journal of the International AIDS Society* 13(34): 13–34.

Silvano, R.A.M. and Begossi, A. (2010) What can be learned from fishers? An integrated survey of fishers' local ecological knowledge and bluefish (*Pomatomus saltatrix*) biology on the Brazilian coast. *Hydrobiologia* 637(3): 3–18.

Snow, J. 1855. Dr Snow's Report. In *Report of the Cholera Outbreak in the Parish of St. James, Westminster, during the Autumn of 1854*. London: Churchill, 97–120.

Stillwaggon, E. (2006) *AIDS and the Ecology of Poverty*. New York, NY: Oxford University Press.

Sunyer, J., Spix, C., Quénel, P., Ponce-de-León, A., Pönka, A., Barumandzadeh, T., Touloumi, G., Bacharova, L., Wojtyniak, B., Vonk, J., Bisanti, L., Schwartz, J., and Katsouyanni, K. (1997) Urban air pollution and emergency admissions for asthma in four European cities: the APHEA Project. *Thorax* 52(9): 760–765.

Tobler W. (1970) A computer movie simulating urban growth in the Detroit region. *Economic Geography* 46(2): 234–240.

Toney, M.B. (1976) Length of residence, social ties, and economic opportunities. *Demography* 13(3): 297–309.

Tuiten, W., Koenraadt, C.J.M., McComas, K., and Harrington, L.C. (2009) The effect of West Nile virus perceptions and knowledge on protective behavior and mosquito breeding in residential yards in upstate New York. *Ecohealth* 6: 42–51.

Turner II, B.L. and Robbins, P. (2008 Land-change science and political ecology: similarities, differences, and implications for sustainability science. *Annual Reviews of Environment and Resources* 33: 295–316.

Turner II, B.L., Matson, P.A., McCarthy, J.J., Corell, R.W., Christensen, L., Eckley, N., Hovelsrud-Broda, G.K., Kasperson, J.X., Kasperson, R.E., Luers, A., Martello, M.L., Mathiesen, S., Naylor, R., Polsky, C., Pulsipher, A., Schiler, A., Selin, H., and Tyler, N. (2003) Illustrating the coupled human–environment system for vulnerability analysis: three case studies. *Proceedings of the National Academy of Science* 100: 8080–8085.

Turner II, B.L., Geoghegan, J., and Foster, D.R. (2004) *Integrated Land-Change Science and Tropical Deforestation in the Southern Yucatán: Final Frontiers*. Oxford: Clarendon Press of Oxford University Press.

Walker, B., Holling, C.S., Carpenter, S.R., and Kinzig, A. (2004) Resilience, adaptability and transformability in social–ecological systems. *Ecology and Society* 9(2): 5 www.ecologyandsociety.org/vol9/iss2/art5 (accessed 27 September 2012).

Walker, P. (2005) Political ecology: where is the ecology? *Progress in Human Geography* 29(1): 73–82.

Ware, Jr, J.E. (1987) Standards for validating health measures: definition and content. *Journal of Chronic Diseases* 40(6): 473–480.

Whittemore, A.S. and Korn, E.L. (1980) Asthma and air pollution in the Los Angeles area. *American Journal of Public Health* 70(7): 687–696.

Yanes-Estévez, V., Oreja-Rodríguez, J.R., and Alvarez, P. (2006) Mapping perceived environmental uncertainty. *Rasch Measurement Transactions* 19(3): 1033–1034.

Young, O., Berkhout, F., Gallopin, G., Janssen, M., Ostrom, E., and Vanderleeuw, S. (2006) The globalization of socio-ecological systems: An agenda for scientific research. *Global Environmental Change* 16(3): 304–316.

Zimmerer, K.S. and Bassett, T.J. (eds). (2003) *Political Ecology: An Integrative Approach to Geography and Environment-Development Studies*. New York, NY: Guilford.

3 Capitals and context

Bridging health and livelihoods in smallholder frontiers

Leah K. VanWey, James R. Hull, and Gilvan Guedes

Introduction

This chapter approaches the complex, dynamic, and multiscale relationships entailed in the politics and ecologies of health holistically, treating health as one of multiple competing investment priorities for households occupying smallholder frontiers. We draw upon insights from theory viewing health as a form of human capital (Grossman 1972) and poverty as a lack of ability to invest (Reardon and Vosti 1995), and from demographic research pointing to the long-term payoffs to early life health (Hayward and Gorman 2004). Health for us is thus not the lack of an event (e.g. disease, malnutrition) but a dynamic process of household investment in nutrition, preventative medicine, and appropriate treatment of disease (Berman *et al.* 1994). Our approach merges these insights with the livelihoods research tradition (Ellis 1998; de Sherbinin *et al.* 2008) and considers the multiple capitals influencing the capabilities of smallholder families to avoid illness and improve health (Bebbington 1999).

Household investments change the context in which future decisions are made, a process we call endogenous evolution of context. We hypothesize that early in frontier development, smallholders pursue livelihood diversification to diversify their capital portfolios and increase resilience, while in mature frontiers it becomes possible through conversion of financial capital into other capitals to pursue specialized livelihoods while maintaining diverse capital portfolios. Thus, livelihood diversification is predictive of success early in frontier development, but a signal of hardship in later frontier stages. These predictions are typical of a complex system: context-specific, sensitive to initial conditions, and displaying emergent properties.

Smallholder frontiers provide empirical and theoretical leverage to examine complex systems by allowing us to enter at a moment when settlers have not previously been influenced by the environment that they are entering. We can therefore isolate the effects of the humans on the biophysical environment without these effects being contaminated by reciprocal relationships. In addition, the co-location of people and environment in smallholder frontiers closely ties decision-makers to environmental impacts (Axinn *et al.* 2010; Moran 2010).

We present a flexible model of livelihoods and investments based on dynamic changes in (i) the investment returns to different type of capital, (ii) the stocks of each capital class, and (iii) the diversity of household portfolios. To ground our theory empirically, we focus on land use, off-farm employment and migration as key smallholder livelihood strategies in frontiers, tracing their implications for household health and well-being. Land use by farmers in frontier regions is an essential factor mediating the relationships between humans, ecological well-being, and health (Saxena *et al.* 2005). In the cases detailed here, food and financial income from farm production, off-farm employment, and migrant remittances represent strategies that allow households to invest in health (e.g. through nutrition, medications, or immunizations).

We marshal empirical evidence for our theory from published and unpublished results from long-term collaborative research projects involving five frontier areas in Brazil and Thailand. This chapter begins by describing smallholder frontiers and these five specific frontiers, then presents our framework and discusses the interaction between elements as they affect health at the household level. We conclude with a discussion of the significance of our framework for the study of many different phenomena in coupled human and natural systems, focusing on implications for health policies.

Health and household dynamics on frontiers

Smallholder frontiers

Our definition of frontier is purposefully minimal, to emphasize the broad applicability of our theory. The essential characteristics of frontiers are dramatic increases in population density via in-migration, improved accessibility, an absence of established social institutions, and an abundance of natural resources relative to other areas. Frontiers provide opportunities for achieving general well-being and amassing financial wealth with small stocks of human, social, and physical capital. Only natural capital and the human labor to exploit it are required. This acts as a strong incentive for the poor to migrate there (Bilsborrow 2002). Young adults, some with children, are typically over-represented in these migrant streams, further fueling rapid population increase at frontier opening (McCracken *et al.* 2002).

Frontiers represent opportunity to young settlers, but the breadth of migrant origins creates challenges to social organization, requiring creation of new social institutions. Settlers are separated from kin, tradition-based hierarchical social relations are leveled through mixing of ethnicities and backgrounds, and affinity-based social ties are usually reformed on the frontier. Typically remote from government centers, settlers on frontiers must also create their own political institutions. Markets for goods, labor, and land must be created, usually through cooperation among settlers. As new families enter and begin converting natural capital to other forms of capital, ecosystem functioning, human institutions, and population structure respond. In turn, families respond to these shifts in context, altering the decisions they make about livelihood strategies and forcing readjustments to their capital portfolios. Through these endogenous feedbacks, the ecosystem and human environment are continuously modified.

We use findings from four sites in the Brazilian Legal Amazon and one in Northeastern Thailand to illustrate and motivate our theory. Details of the data collection and analyses appear elsewhere (D'Antona and VanWey 2007; Moran *et al.* 2005; Rindfuss *et al.* 2009); here we emphasize site characteristics relevant to theory building and comparison. Chief among these is the process of frontier settlement that began in all five regions during the mid–20th Century. Figure 3.1 shows the timing of in-migration to each state and the subsequent increase in population density.

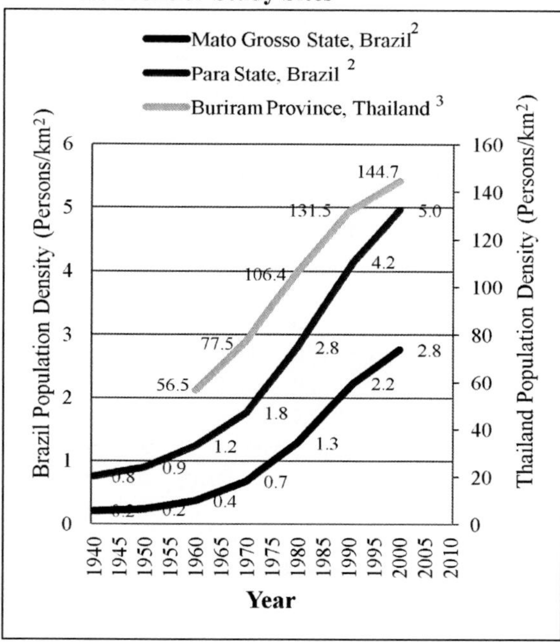

Figure 3.1 Population density changes across frontier study sites.

Notes:

1 Two of the Brazilian sites are located in the state of Pará and two in the state of Mato Grosso. We present population densisties at the state or *changwat* scale rather than the district (*municipio* or *amphoe*) level due to instability in the political boundaries of these frontier districts. Thus, we have three time series, not five.

2 Source: Brazilian Demographic Censuses of 1940, 1950, 1960, 1970, 1980, 1991, and 2000. Instituto Brasileiro de Geografia e Estatistica. Available at http://www.sidra.ibge.gov.br/bda/

3 Source: Thai Population and Housing Censes of 1960, 1970, 1980, 1990, and 2000. Changwat Level Summary Tables. The National Statistics Office, Kingdom of Thailand. Data not available at Changwat Level prior to 1960.

Brazil

Our Brazil sites represent varied settlement histories, livelihoods, and levels of human health and well-being. Altamira lies on the Trans-Amazonian Highway (BR-230), and was an early demonstration settlement area. Colonists brought from other regions of Brazil were granted land for agriculture, provided they cleared a portion, improved it through planting crops and adding buildings, and paid a nominal sum to the government (Moran 1981). Early settlement was characterized by high rates of deforestation, malaria, farm failure, and onward or return migration of some initial settlers. These failures in part resulted from settlers lack of knowledge about biophysical conditions of farms. Today, Altamira retains a smallholder character with small- to mid-scale ranching and agriculture, and an increasingly urban orientation of landholders (VanWey *et al.* 2008; Walker *et al.* 2000). For a discussion on frontier settlement and its implications for human health in a similar frontier, see Chapter 7.

Santarém sits at the confluence of the Amazon and Tapajós rivers, at the northern terminus of the Cuiabá-Santarém Highway—the "Soy Highway." A traditional port city, Santarém has long functioned as a trade center between the upper Amazon and the estuary. A deep-water port was constructed in 2003 to permit direct export of soy to international markets (Steward 2007). Agricultural settlement of surrounding regions began in the early twentieth century, but Santarém's low-fertility soils supported low crop yields, leading much of the region to be fallowed again (D'Antona *et al.* 2006; Futemma and Brondízio 2003). Unlike the government-sponsored colonization of Altamira, much of the settlement of Santarém was spontaneous, and colonists came from nearby, suggesting that they had greater understanding of the region and its limitations. Settled properties here were, and remain small and precarious today.

Two sites in Mato Grosso represent yet another history, centered on the arrival of larger (but still familial) farmers. Both were colonized around the 1980s and experienced rapid land clearing for production. Both also benefited from more favorable soil and climatic conditions and from more similarity to the origin areas of settlers. The first site, near Canarana in the east of the state, was colonized by cooperatives of large owners from the south and southeast of Brazil (Jepson 2006). This tradition of large owners and hierarchical labor relations among owners, managers, and farm workers persists today even while the original ranching is giving way to some row crop agriculture. The second site, Lucas do Rio Verde, in the center of the state, was also settled by larger landowners and large row crop farms continue here today, but are supported by verticalized industrial agribusiness. The county houses a processing facility converting raw soy into oil and meal, an incubating operation for chicken eggs, and poultry and swine slaughterhouses. Recent investment in the region by Brasil Foods Corporation (formerly Perdigão) and Sadia has led to the rapid expansion of confined poultry and swine operations that take advantage of local production to feed their stock. The agribusinesses in town coexist with strong local family involvement in farms and little absentee ownership.

Thailand

Nang Rong District, located in northeast Thailand, experienced rapid demographic, economic, and environmental change beginning in the mid-twentieth century (Entwisle *et al.* 2008; Walsh *et al.* 2005). Rapid in-migration and land clearing ensued, reducing forest cover from 50 percent in the 1950s to less than 20 percent by century's end (Entwisle *et al.* 2008). As in Brazil, settlers took advantage of land tenure laws to claim lands and gain informal rights through clearing the land, in this case along rivers and floodplains, and by making improvements supporting paddy rice cultivation (Keyes 1976; Entwisle *et al.* 2008; Phongphit and Hewison 2001). Gradually, rice production has expanded upward from alluvial plains to less suitable environs. When demand for first jute and then cassava (for European livestock) spiked in the 1970s and 1980s, agricultural production expanded further into the uplands, driving deforestation (Entwisle *et al.* 2005; Rigg 1987). Agricultural mechanization first begun in the 1980s was not fully incorporated for another two decades, leaving many agricultural activities to be performed by hand during the period examined here (Hull 2008). Today, small manufacturing sponsored by private development initiatives supports limited local off-farm employment in higher skill jobs (Alva and Entwisle 2002).

Health through the lens of a livelihoods-context capitals framework

What follows is a condensed introduction to our framework, which theorizes the linkages between development, land-use and land-cover change, and household livelihood transition across smallholder frontiers. It builds upon the conceptual frameworks of Bebbington (1999), Lambin *et al.* (2003), Curran and DeSherbinin (2004), Axinn *et al.* (2010), and McCusker and Carr (2006). A response to calls for an integrative theory of coupled human and natural systems, this framework speaks to multiple substantive domains, including health, deforestation, outmigration, and changing economic exchange relations. For a related discussion on how to conceptualize human health, see Chapter 2.

Common to all of these processes is the interdependency of household-level decision-making and evolving endogenous context. Our framework, depicted in Figure 3.2, has five principal elements. In the lower left are capital portfolios, consisting of stocks of each of five capitals—natural, physical, social, human, and financial—which constrain and enable livelihoods. In the upper left are returns to each class of capital, which also shape livelihoods and are impacted by the endogenous evolution of context. Context, represented in the upper right, consists of six major dimensions—accessibility, demographic, social, economic, political, and environmental—which respond to the livelihood decisions of actors, while also influencing future decisions. In the lower right we depict key outcomes of interest to researchers and policymakers, particularly health, which are determined by livelihood decisions. We now describe each of these elements, first describing the articulation between health outcomes, well-being, and livelihoods.

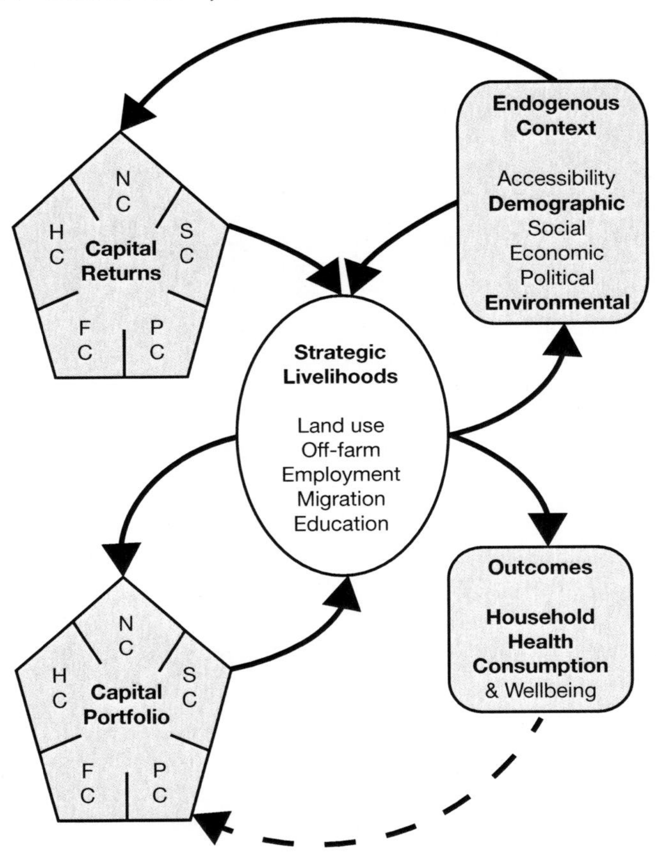

Figure 3.2 Human and ecological well-being co-evolve over time.

Note: Health intersections are emphasized in bold.

Health, multidimensional well-being, and livelihoods

In smallholder frontiers households are the locus of income pooling and joint consumption, including joint decisions about investments in various forms of capital and about key health-promoting behaviors. The pursuit and maintenance of health is an important and distinct goal of household decision-making calculus. Ill-health decreases income and increases costs, and may be conceptualized as a factor competing with other dimensions of well-being for scarce household resources, thus exerting unique influence on livelihood decisions (Hampshire *et al.* 2009; for a related discussion on disease as shock to livelihood systems, see Chapter 14 of the present volume). Simultaneously, good health constitutes a basic element of human capital, linking our framework to early work on human capital showing health as equivalent to education and knowledge (Becker 2007; Grossman 1972). Health is connected to household well-being by an endogenous system of *health status*, *nutritional intake* and *labor productivity* (Joffe 2007). Ill-health events can drive impoverished households into a vicious cycle of depleting multiple forms of capital to cover costs of the illness (including treatment, food, medication, transportation, but also the shadow price of an absent labor). This cycle affects the ability of households to maintain optimal nutrition, leading to productivity declines and increasing vulnerability to further ill-health shocks (Sauerborn *et al.* 1996; Joffe 2007). While this is an endogenous system at the household level, decisions about health investments are not made in a vacuum. The changing contexts of healthcare institutions and accessibility influence the costs of health.

Capital portfolios

We argue that households pursue livelihoods in order to maximize stocks of preferred capitals, where subjective preferences for particular types of capitals evolve along with the frontier. In turn, households' stocks of capitals (collectively "portfolios") simultaneously constrain and enable various livelihood decisions. We consider five classes of capital: natural capital (e.g. access to water, forest products), physical capital (e.g. machinery, buildings), financial capital (e.g. wages, remittances, public transfers), human capital (e.g. health, education, on-the-job and on-the-farm training), and social capital (e.g. familial and social networks, associations/unions, generalized social trust). The relative share of a household's total portfolio constituted by each type of capital is altered as the household chooses livelihood strategies that draw on existing forms of capital in the hopes of obtaining more of the same or converting lower-return capitals into forms of capital with higher returns. The strategic use of capitals based on returns is key to understanding health impacts on capital portfolios. If a health shock induces a household with limited capital diversity to capital depletion, the ability to overcome health shocks is further compromised, since the absence of specific capitals prevents households from taking advantage of potential higher returns (Joffe 2007). This iterative process of capital accumulation, depletion, and conversion to alternative forms is summarized by the arrows flowing from capital portfolio to livelihoods and back in Figure 3.2.

Capital returns

It is useful to distinguish the amounts of various capitals in a portfolio from the advantages and disadvantages of possessing them within a specific context. Returns to capital are dynamically shaped by context, subsequently influencing livelihood strategies in the same manner as capital portfolios themselves (depicted by the arrow from endogenous context to capital returns in Figure 3.2). Considering either in isolation provides an incomplete explanation of household behavior. These two major elements of the framework do not interact directly, however. Instead, they impact one another indirectly through livelihoods and endogenous context.

Figure 3.3 shows the stylized empirical facts underlying our theory of changes to capital portfolios and capital returns over time in smallholder frontiers. Returns to each form of capital over frontier development are depicted in Figure 3.3a, starting with settlement by smallholders and ending with the integration of the frontier into national and global systems. Figure 3.3b shows how distributions of capitals in household portfolios change over the same span. The net effect of frontier development is conversion of a substantial amount of natural capital into financial, physical, and human capitals which begin providing greater returns by the later stages of development. Figure 3.3b can be conceptualized as the static result of the dynamic transitions depicted in Figure 3.3a. We briefly describe these shifting returns and stocks.

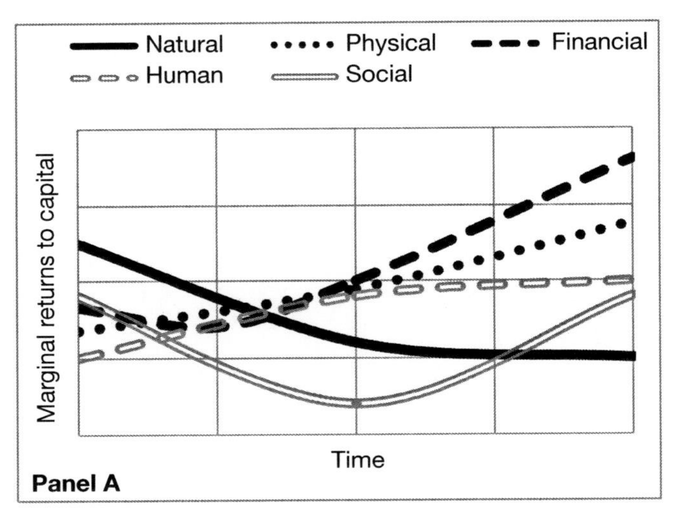

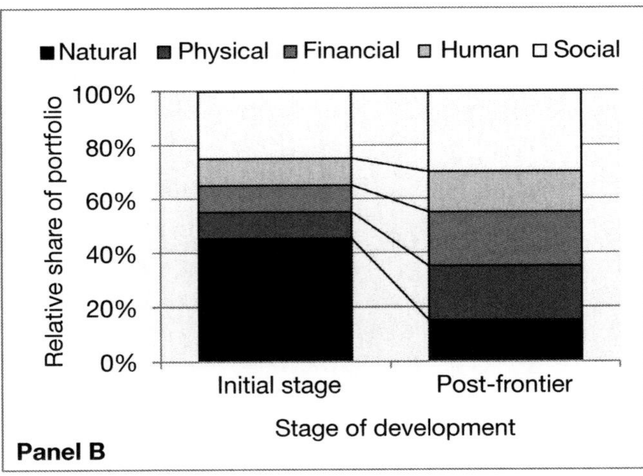

Figure 3.3 Capital returns and portfolios change as smallholder frontiers transition.

Natural capital

Natural capital (NC) is especially important during the initial phase of frontier development because economic institutions are limited. At this point, subsistence is largely dependent on ecological services derived from NC. Off-farm employment and markets for the purchase of goods both expand as the frontier develops, reducing households' dependence on natural resources to meet consumption needs. Returns to NC likewise decrease as households adopt newer and more efficient technologies for agricultural production (Figure 3.3a). The relative representation of NC in total household portfolios declines as households diversify into other capitals, particularly human (Figure 3.3b). Declining returns remove incentives for households to reinvest in NC through practices like fertilizer application or erosion control. In Altamira, rather than investing in improvements to common poor soil types, settlers clear more extensive areas and plant hardy pasture grasses that will survive with few inputs (Moran *et al.* 2002). In Santarém, the depreciation of NC over time is evident in the belief that forest is a nuisance impeding other property uses, leading buyers to pay a premium for cleared land (Adams 2008).

Social capital

A distinctive U-shaped curve characterizes returns to social capital (SC). This reflects the synthesis of monotonically decreasing returns to bonding SC (strong bonds among close associates) and the monotonically increasing value of bridging social capital (relations that connect people across much larger social distances; Woolcock and Narayan 2000). The value of bonding SC decreases as population and markets grow and households become either internally diversified or possess enough FC to purchase goods to meet needs. With further development, social networks connect rural households to larger networks of goods, information, and opportunities, but these networks necessarily embody bridging SC rather than bonding SC. Early in frontier evolution, social networks enable settlers to overcome poor or nonexistent labor and credit markets to achieve production essential to well-being. Networks also promote investment in human capital in the form of health by putting settlers in touch with information and access to improved healthcare, and providing an important source of care-giving during ill-health events (Smith and Christakis 2008). But while current theories about social network impacts on health implicitly assume a static social structure, our theory explicitly argues that the content and form of social networks, and the overall returns to social capital vary endogenously over time, with corresponding variations in the health implications of network structure.

In Brazil, for example, when returns to NC are high, bridging connections to business owners and government officials in São Paulo or Belém are of little help in clearing, planting, and harvesting, and therefore provide little in the way of assistance with health investments. With improving accessibility, such connections yield higher returns, helping households secure access to distant markets, documentation for credit, and favorable access to education and quality healthcare, while bonding ties continue to provide access to the same opportunities to which a household already has access. The competing effects of declining investment in bonding SC and increasing investment in bridging SC lead to a small net increase in the SC component of household portfolios (Figure 3.3b), but large potential changes in the ability of a household's network to respond to ill-health episodes.

Physical capital

Settlers initially have little physical capital (PC): machinery, buildings, or landesque capital such as bunds, fences, or irrigation systems (Blaikie and Brookfield 1987). Households are often PC-poor before relocating to the frontier, and are further limited to what can be easily transported. PC increases steadily over time (Figure 3.3b) through conversion of NC and FC. Early increases in PC are essential for the health dimension of household well-being; improvements to housing stock reduce disease risk through improving water access and reducing exposure to disease vectors. Returns to PC (Figure 3.3a) are initially low, but increase as markets develop for local sale and export of surplus production, incentivizing investment in machinery such as tractors or trucks. Early on in our Brazilian sites, households extracted NC such as timber to construct houses, furniture and other PC such as livestock fences. In Nang Rong, the early years saw construction of PC such as ponds, fish weirs, and elevated buildings for keeping equipment dry. As incomes increase, households also invest in house improvement (Rindfuss *et al.* 2007). Later still, some acquire enough cash income from production or employment to purchase chainsaws, refrigerators, radios, vehicles, and other durable goods.

Human capital

Human capital (HC) levels are low at frontier settlement (Figure 3.3b), but increasing returns over time (Figure 3.3a) lead households to seek more and different HC. Early on, site-specific knowledge and pure labor power drive agricultural success (Moran 1981) while at later stages returns are highest to formal HC like education and work experience. Upon arrival, migrants have low site-specific HC and may experiment with the new biophysical environment as an investment in HC. As noted above, they also make investments in PC that are simultaneously investments in HC as the health of family members improves. In early years in Altamira and Santarém, we see experimentation and heterogeneity at the property level in land use, while later years are characterized by specialization in the highest return crop (given soils, water and accessibility; McCracken *et al.* 2002; VanWey *et al.* 2007). Early villages in Nang Rong were situated on the margins of river flood plains where rice could be reliably cultivated using the traditional paddy system, with which most settlers were already familiar, only expanding in other agroecological regions once settlers developed the site-specific HC needed to exploit them (Entwisle *et al.* 2008). As further investment in site-specific HC yields lower marginal returns, families adopt new strategies of investing in formal HC, requiring strategic decisions about which members receive additional education and which stay "down on the farm" (Curran *et al.* 2004).

Financial capital

Returns to financial capital (FC) also follow a U-shaped curve (see Figure 3.3a), but the upturn occurs before that for SC. FC returns are high when establishing a farm, but decline again until the frontier becomes well-integrated with broader markets. In the interim, even households with high levels of FC experience low returns. In our study areas, only the construction of better roads facilitating exports and the development of labor markets allowed some households to invest FC into production through purchases of PC and labor. With improvement in infrastructure and development of urban markets, returns to FC rapidly increase (the upward turn in Figure 3.3a). In the context of limited credit markets or higher rates for credit, characteristic of our study sites (Ludewigs 2002), FC returns respond even more to these improvements. Once the rate of return to investments in FC is reliably increasing, households seek to rapidly expand their potential sources of cash income. Migrant remittances, profits from the sale of cash crops, off-farm employment, rental of land and physical capital, and the rapid development of a monetized labor market are paired with "public" strategies for FC accumulation such as the rural retirement and Bolsa Família programs in our Brazil study sites, subsidized credit (as we see in the rural Brazilian Amazon), and monies for community development from NGOs in Nang Rong (David and Viravaidya 1986).

Returns to each of these forms of capital are altered regularly by changes to multiple dimensions of context (upper right box of Figure 3.2). We discuss these dimensions in the next section, highlighting the way in which they change endogenously as household investment decisions change the context in which future returns are determined.

Endogenous context

Understanding the endogenous evolution of context is vital to explaining how and why livelihood strategies change over time. Our approach goes beyond existing work on the political economic context for decision-making by characterizing context as dynamic, endogenous, and multidimensional. While we do not discount the importance of exogenous changes in determining returns to capitals, our goal is rather to highlight the predictable ways in which context changes endogenously. Frontiers are an ideal natural laboratory for the study of endogenous changes in each of these domains.

Livelihood strategies are determined in a given time period by capital portfolios and returns, as well as by context (Figure 3.2, center). Livelihood strategies also modify the multiple elements of context. These factors influence in turn the returns to various forms of capital in the future (arrow at top of Figure 3.2). In considering the temporal scale of changes, we distinguish slow, medium, and fast endogenous feedbacks observable empirically (cf. Lambin *et al.* 2003). This distinction parallels recent interdisciplinary health systems frameworks that classify endogenous system processes as either proximal or distal (Eisenberg *et al.* 2007, Batterman *et al.* 2009). At the finest temporal scale, we consider endogenous change from year to year, comprising, for example, this year's crop choice in response to last year's yield or prices. At the medium scale, we consider several years to approximately 15–20 years, the scale over which, for example, environmentally-oriented social movements may form, lobby for changes, and achieve policy changes. We limit the upper end of the continuum to changes that occur on the scale of approximately 20–40 years, the scale of a human generation, over which, for example, childbearing decisions by parents modify regional populations and feed back into the fertility choices of the next generation. Figure 3.4 summarizes the time-scale dependencies for representatives of each dimension of endogenous contextual change.

Accessibility

The timing of accessibility improvements determines the speed at which frontiers move along the curves in Figure 3.3a, raising returns to PC and the conversion of NC through agriculture and influencing the speed at which households move toward more diversified and monetized livelihoods. A dramatic exogenous change in accessibility marked the opening of the frontier in each of our study areas with externally initiated road-building projects. But we focus here on the ways in which accessibility subsequently changes endogenously across multiple spatial scales. As farmers move across properties and travel to local urban areas and other destinations, paths become small and then larger roads, especially as greater FC permits the purchase of motorcycles and trucks. The utility and permanence of paths and small roads increase as a function of land use and other decisions within a year while the need for and construction of all-weather roads happens over years to decades. The construction or paving of state or federal highways is a more complicated and long-term process, but it too depends on the ingenuity, labor, and desires of local populations, and on their ability to make claims on non-local actors (Figure 3.4). State or federal roads will at times follow existing roads because they are the path of least resistance (physically and socially), a phenomenon that has been well documented in regions proximate to our Altamira study area, where informal roads, most built for logging, became formalized all-weather roads and even state highways (Arima *et al.* 2005; Perz *et al.* 2007). Accessibility also modifies household investment in HC, including healthcare, by changing the costs of such investments. The nominal and lost-labor costs of a visit to a clinic or to the city to buy medicine decline dramatically as transportation infrastructure develops. As households invest more in this form of HC, demand increases, and local social leaders and politicians can make added demands on the central government for new clinics or more funding for existing clinics (in settings such as ours where healthcare is publicly funded).

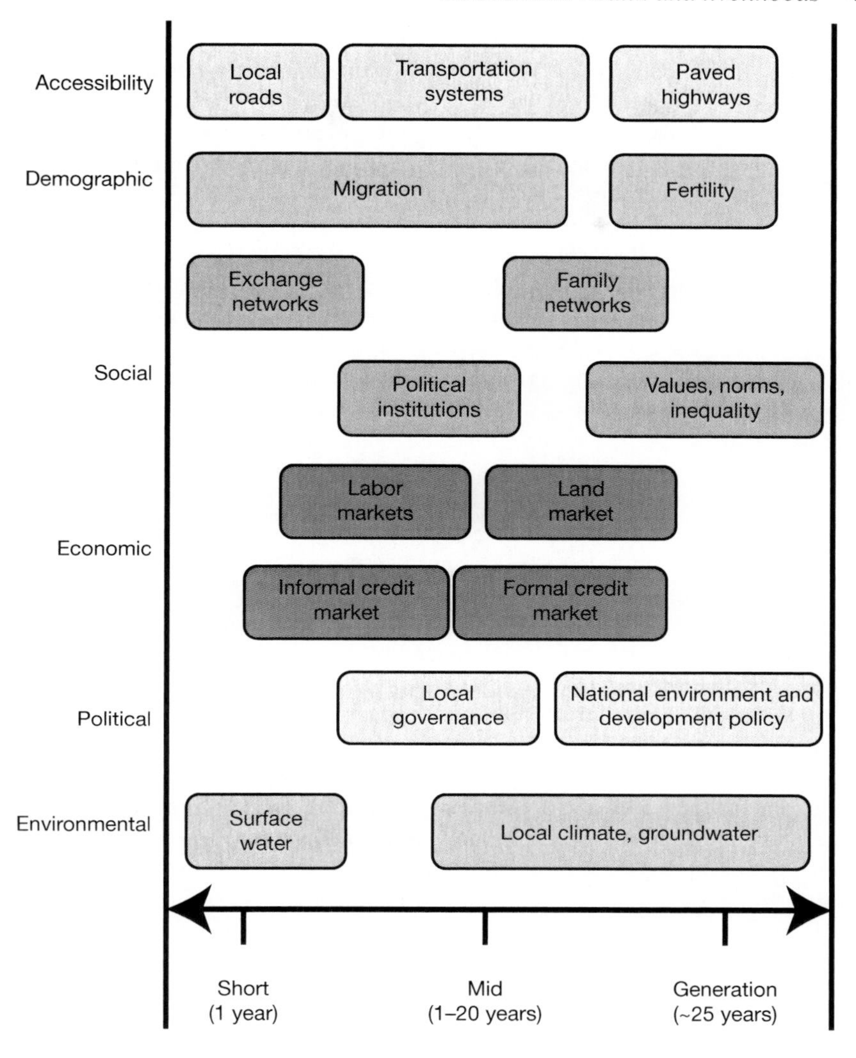

Figure 3.4 Time scale of key endogenous feedbacks.

Demographic factors

Across generations, fertility decisions and health investments that alter mortality rates change the demographic context endogenously. These actions are particularly relevant in frontiers, where age distributions are weighted heavily toward the child-bearing years. As these large cohorts of young adults age, the characteristic bulge moves through the age structure, followed by an "echo" bulge of their children, with the size of this echo influenced by infectious disease mortality and a household's ability to invest in health (Joffe 2007; Russell 2004). Between these bulges lie smaller cohorts (in the absence of continuous in-migration). The size of cohorts can influence labor availability, with small cohorts creating labor scarcity and changing the context in which agricultural and HC investment decisions are made. Migration also endogenously changes the context of future household decisions. As connectivity with other regions increases and population exceeds the capacity of local labor markets to absorb it, temporary out-migration on a seasonal or multiyear basis becomes an alternative path to meeting consumption needs and accessing scarce FC (VanWey *et al.* 2009). In our Brazilian sites, daily or weekly commuting between cities and urban hinterlands occurs for work or leisure. In Nang Rong, we observe seasonal migration for farm work in the region combined with longer-term migration to urban centers for service, factory, and construction jobs (Korinek *et al.* 2005). These population movements further alter livelihood strategies as some migrants leave children in their parents' care (Piotrowski 2009).

Social factors

Formal and informal social institutions develop over long human histories. Some shared institutions arrive with settlers. Others, such as cooperative agreements about labor sharing and local governance arrangements, must develop rapidly in frontiers. Settlers in a new frontier may share national or regional histories, but often have no shared local history, forcing them to devise new institutions governing collective behavior. Simple collective governance systems may develop rapidly (less than a decade) to address inadequate or non-existent institutions on frontiers, but more complex governance such as organized parties and political systems takes considerably longer (Figure 3.4). Another important set of institutions regulate how land is intrinsically valued, and these develop on the scale of generations. Since settlement, norms have emerged in Altamira that identify "clean" land as preferable to "dirty" land. "Clean" in this context is land cleared of trees and shrubby vegetation, with fences and views to the horizon, and is interpreted as the mark of a superior farmer. The normative land-use pattern in much of Santarém, meanwhile, would be viewed in Altamira as the hallmark of the lazy farmer, with its mixed areas of fallow, fruit trees, intercropped annuals, and native vegetation, which obstruct clear views of the horizon (Adams 2008). Such emerging value systems have important impacts on livelihood and land-use decisions in critical ecosystems throughout the globe. As frontiers age, the social networks linking various social entities ranging from individuals to state actors change. As Figure 3.3 suggests, such connections have high returns initially, and many sorts of new networks emerge in the space of a few years following frontier opening. By investing in production for sale, farmers then form new, longer-distance networks. Similarly, migration for employment creates longer-distance networks. In Nang Rong these pass information about family planning options (Entwisle *et al.* 1996), information about job prospects (Curran *et al.* 2005), and FC in the form of remittances (Piotrowski 2006).

Economic factors

Development of markets for labor, credit, and land directly impact agricultural production, land use, and ecological change, as well as household well-being. Equally important, development of these markets permits households to pursue more diverse livelihoods and capitals. Labor markets develop slowly in most frontiers for two reasons. First, scarce access to monetary currency in many frontiers constrains the development of monetized labor systems. Second, abundant land resources and low efficiency require that most of a household's labor be invested in its own production. It takes time for population dynamics (in-migration and natural increase) and income growth (external investment or internal accumulation) to reach a relatively stable equilibrium in which a labor market can be sustained (Figure 3.4). Credit markets also take time to develop, making room for informal credit markets and mutual aid societies to develop for the smoothing of consumption and to support investments in new activities (Alvi and Dendir 2009; Boucher and Guirkinger 2007). Such groups may take a decade or more to develop (Figure 3.4), both because the trust that underlies them is slow to solidify and because some monetary accumulation is requisite. The process can be accelerated if demand for credit and frontier accessibility are high enough that outside actors have incentives to provide credit, as has occurred in much of rural Brazil, where demands (and qualifications) for credit are artificially inflated by public provision of credit.

Political factors

Political factors consist of formal laws governing, for example, labor relations, forest exploitation, taxes, and education, as well as informal but mutually recognized organizations that allocate power and resources among people. Traditional (post-colonial) Brazilian power relations follow a patron–client model, with a small group of individuals controlling access to resources among their constituents (Houtzager 2000). In contrast, Nang Rong exhibits a traditional form of relatively flat political relations, vesting little power in village headmen. The hierarchical relations that exist are based instead on age and gender (Phongphit and Hewison 2001). These are among the slowest of all the endogenous processes we consider (Figure 3.4), and among the most important for health. While the implementation of a particular policy or government agency may occur rapidly, the process of garnering support for, lobbying, and designing it is usually slow. Even local governance structures, formal or informal, change only over the span of decades. We have yet to study the development of local basic healthcare provision in our study areas, in part because of the slowness of the process. We can point to healthcare as one of the top two (along with education) reasons for moving among migrants from the countryside to the city in Altamira and Santarém. Anecdotally, we can point to the reported importance of healthcare facilities (among other types) in the success of Lucas do Rio Verde in attracting businesses and workers, and to the reported struggles of local healthcare workers in getting more supplies, personnel and equipment in Santarém. These processes, however, are essential for the ability of a household to invest in both preventative medicine and treatment, and therefore the HC of its members.

Environmental factors

Like political factors, environmental factors may change quickly following a long process of endogenous change. Global changes in greenhouse gases may trigger rapid climate change, but the buildup to that point may take decades or longer. At the local level, deforestation and withdrawals of water for domestic and agricultural use can change regional climate on a similarly long time scale. In Altamira, deforestation near the city led in less than a generation to a warmer and drier climate in that area than in regions just 100–200 kilometers west. Farmers attribute drying up of favorite fishing holes, for example, to increased frequency of flash precipitation that exceeds percolation capacity and simply runs off the landscape. Collective agreements to protect forest in the interest of protecting rainfall (and also to strengthen indigenous groups) led to the so-called Terra do Meio south of Altamira remaining forested from the 1970s until very recently, without formal protected status. Within the past decade, loggers and settlers have encroached on this intact forest, generating frequent conflicts, a strong forest protection movement, and formal designation of most of the region as protected or indigenous lands. This history provides a clear example of the circular relation between livelihood decisions and environmental and political change. Over shorter time-scales, the land-use component of the livelihood strategies can affect surface water quality and availability for other residents on the frontier. In extreme cases, the construction of bunds or other water diversion systems for rice production can leave downslope neighbors with insufficient or polluted water for their own fields, interactively driving the development of new political and social institutions. Each development alters inputs to the decision-making of frontier residents about land use and the benefits of on-farm or off-farm activities.

Discussion

Interactions among health, capitals, and livelihoods

We describe above investments in health as a form of human capital, and turn now to the role of health shocks in our framework. It is well-known that greater household capacity to invest in health results from (usually) growing stocks of capital while resilience to shocks, including health problems, may be attained through more diversified capital portfolios (Ellis 1993). In some situations, however, a health shock or long-term illness can create a downward "vicious circle" (White *et al.* 2005). Households may deplete household assets to cope with costs of treatment (e.g. food, medication, transportation, and the shadow price of absent labor), which reduces their ability to maintain the health of other members and of the overall household unit. Depletion can come both in terms of the absolute reduction of capitals in various classes and in terms of the diversity of a household's portfolio, which when reduced may further increase that household's overall vulnerability.

Our approach is distinguished from previous literature in this area by a focus on the interplay between household strategies and evolving context in frontiers, which leads to stage-specific predictions about the order in which the five capitals will be depleted in response to an ill-health event. Our predictions are summarized in Figure 3.5 and compared with predictions made by Sauerborn *et al.* (1996). These draw on the changing relative returns to, and therefore investment preferences for, each form of capital (shown in Figure 3.3), the representation of the capital in portfolios, and the exchangeability of the different forms of capital. Representation in portfolios and exchangeability determine whether a capital can be used to meet needs during an ill-health event while investment preferences determine the results of foregone investment opportunities for a household. FC is the first depleted in any illness because of its exchangeability for healthcare costs both direct and indirect.

While the FC is being depleted, households will forego investments in other forms of capital (rather than using up other capitals). In the early stages of frontier development, households are investing in PC, SC and HC, and foregoing these investments will leave them with little to draw on as contexts change. PC may be depleted if it can be exchanged (e.g. chainsaws, tractors), and HC will then be converted to money for care through negative investments in HC (healthy family members working until ill). Households are then left with the NC and SC they arrived with. The NC and SC (mostly bonding) are both declining in returns. These may be used to obtain FC if possible, but this action usually signals the departure of the household from the area as they sell the farm and overdraw their social ties. Thus, early in frontier development, we posit a strong pattern of preserving social and natural capital. These predictions for the initial phases

Order of depletion	**Sauerborn *et al.* 1996**	**Initial stage**	**Post-frontier**
First depletion	Financial capital	Financial capital	Financial capital
	Physical capital	Physical capital	Physical capital
	Human capital	Human capital	Natural capital
	Social capital	Social capital	Human capital
Last depletion	Natural capital	Natural capital	Social capital

Note: Cells of the same color indicate roughly equivalent degrees of hardship in depleting each type of capital. Darker cells indicate increasing hardship.

Figure 3.5 Order of capital depletion during illness episode as frontiers evolve.

of frontier evolution, when natural capital is plentiful and essential to guaranteeing household well-being, map closely onto the empirical observations made in previous studies examining the African Sahel (Adams *et al.* 1998).

In post-frontier contexts, households initially are more resilient because they have diversified capital portfolios and larger stocks of capital. They are heavily investing in PC, HC, and SC for the acquisition of the most exchangeable FC. As in other settings, illness will initially lead to a depletion of FC because of its ready acceptance as payment. It will also result in households foregoing investments in PC, HC, and SC. Next to be depleted will be PC, because of its higher level of exchangeability in the new context. With the presence of land, credit and product markets, higher stocks of PC such as vehicles, machinery, housing, and livestock can be converted into cash for healthcare through sale, mortgage, or rental. Because of the much lower returns to NC and the presence of these markets, NC is next to be depleted, leaving households with HC and SC. In post-frontier contexts, sprawling networks of bridging ties link households to temporary housing in urban areas (near health centers), migrant remittances, and political officials, all of which become important to strategies for combating illness (Padoch *et al.* 2008). Depending on the length of illness, HC and SC may be enough to regain capital stocks after an illness, as education and social connections will help household members get good jobs, pay the rent, and save money for purchase of land, or other NC or PC. With long-standing illness, or illness that affects households during critical years of investment in HC or SC, foregone investments may limit the accumulation of HC and SC and lead to a downward spiral. Successive ill-health events not only deplete household resources to dangerous levels, but tend to drive household capital portfolios closer to the characteristics they displayed earlier in the frontier. With the depletion of FC and PC, and sometimes NC, households' capital portfolios resemble (though with different forms of HC and SC) those of frontier settlers, and they may become settlers in new frontiers elsewhere.

Conclusions

In frontier settings, as well as low-income settings generally, avoiding a vicious circle may depend on interventions, such as provision of health services or subsidized transportation to reduce transportation cost (Russell and Gilson 2006). In our Brazilian study areas, multi-local households (those maintaining multiple residences in different regions) attempt to benefit from proximity to health services in urban areas, while maintaining control over NC and SC on the frontier (Padoch *et al.* 2008). This duality is facilitated by effective and affordable (subsidized for the elderly) bus transport. In this sense, free health services leverage households' productivity by providing low cost investments in human capital (Grossman 1998) and by removing the need for households to finance health expenses through depleting other capitals.

As a result, the provision of free or low-cost healthcare in poorly developed settings like the smallholder frontiers we describe can be seen to reduce necessary expenditures and free up household capitals for other investments (Russell and Gilson 2006). With this application of our theory, we have described the sorts of capital likely to be available to households in a frontier and what sorts of investments are likely in response to decreasing the cost of healthcare (as well as likely patterns of depletion from increasing cost of care). In early frontier settings, few capitals are available and the lack of affordable and available healthcare leads to out-migration as households must spend the SC and NC available to them, a lesson learned in Altamira with early high rates of malaria and out-migration (Moran *et al.* 2005). In post-frontiers, investments in subsidized transport for healthcare and in inexpensive care facilitate more rapid increases in HC and SC, and the failure to provide such care leads to stalled development (households reverting to early frontier capital portfolios) and rapid turnover in land ownership if markets exist for land.

The politics of health have traditionally focused on equitable health supply, based on technical and scientific intervention in unassisted or under-covered areas, which are key, given the importance of accessibility in household decision-making. Yet the development of markets and increased accessibility independent of directly targeted healthcare policy can have the same impact (as in the dual-residence households). This mixed approach is likely to benefit a wider group of households, addressing issues of equity and human rights in the process. We argue that a sustainable approach to long-term improvement of productive capacity of households can facilitate both health investments and the diversification of capital portfolios, ultimately improving both health and overall well-being (Hampshire *et al.* 2009).

References

Adams, R.T. (2008) Large-scale mechanized soybean farmers in Amazônia: new ways of experiencing land. *Culture and Agriculture* 30: 32–37.

Adams, A., Cekan, J., and Sauerborn, R. (1998) Towards a conceptual framework of household coping: reflections from rural west Africa. *Africa: Journal of the International African Institute* 68: 263–283.

Alva, S. and Entwisle, B. (2002) Employment transitions in an era of change in Thailand. *Journal of Southeast Asian Studies* 40: 303–326.

Alvi, E. and Dendir, S. (2009) Private transfers, informal loans and risk sharing among poor urban households in Ethiopia. *Journal of Development Studies* 45: 1325–1343.

Arima, E.Y., Walker, R.T., Perz, S.G., and Caldas, M. (2005) Loggers and forest fragmentation: behavioral models of road building in the Amazon basin. *Annals of the Association of American Geographers* 95: 525–541.

Axinn, W.G., Barber, J.S., and Biddlecom, A.E. (2010) Social organization and the transition from direct to indirect consumption. *Social Science Research* 39: 357–368.

Batterman, S., Eisenberg, J., Hardin, R., Kruk, M.E., Lemos, M.C., Michalak, A.M., Mukherjee, B., Renne, E., Stein, H., Watkins, C., and Wilson, M.L. (2009) Sustainable control of water-related infectious diseases: a review and proposal for interdisciplinary health-based systems research. *Environmental Health Perspectives* 117: 1023–1032.

Bebbington, A. (1999) Capitals and capabilities: a framework for analyzing peasant viability, rural livelihoods and poverty. *World Development* 27: 2021–2044.

Becker, G.S. (2007) Health as human capital: synthesis and extensions. *Oxford Economic Papers* 59: 379–410.

Berman, E., Bound, J., and Griliches, Z. (1994) Changes in the demand for skilled labor within US manufacturing industries: evidence from the annual survey of manufacturing. *Quarterly Journal of Economics* 2: 367–397.

Bilsborrow, R.E. (2002) Migration, population change and the rural environment. *Environmental Change and Security Report* 8: 69–94.

Blaikie, P. and Brookfield, H. (1987) *Land Degradation and Society*. London, UK: Methuen.

Boucher, S. and Guirkinger, C. (2007) Risk, wealth and sectoral choice in rural credit market. *American Journal of Agricultural Economics* 89: 991–1004.

Curran, S.R. and de Sherbinin, A. (2004) Completing the picture: the challenges of bringing "consumption" into the population–environment equation. *Population and Environment* 26: 107–131.

Curran, S.R., Chung, C.Y., Cadge, W., and Varangrat, A. (2004) Educational opportunities for boys and girls in Thailand. *Review of Sociology of Education* 14: 59–102.

Curran, S.R., Garip, F. Chung, C.Y., and Tangchonlatip, K. (2005) Gendered migrant social capital: evidence from Thailand. *Social Forces* 84: 225–255.

D'Antona, A.O. and VanWey, L.K. (2007) Estratégia para amostragem da população e da paisagem em pesquisas sobre uso e cobertura da terra. [A strategy for sampling the population and the landscape in research on land use and land cover.] *REBEP (Revista Brasileira De Estudos De População)* 24: 263–275.

D'Antona, A.O., VanWey, L.K., and Hayashi, C.M. (2006) Property size and land cover change in the Brazilian Amazon. *Population and Environment* 27: 373–396.

David, H.P. and Viravaidya, M. (1986) Community development and fertility management in rural Thailand. *International Family Planning Perspectives* 12: 8–11.

de Sherbinin, A., VanWey, L., McSweeney, K., Aggarwal, R., Barbieri, A., Henry, S., Hunter, L.M., and Twine, W. (2008) Rural household demographics, livelihoods and the environment. *Global Environmental Change—Human and Policy Dimensions* 18: 38–53.

Eisenberg, J.N.S., Desai, M.A., Levy, K., Bates, S.J., Liang, S., Naumoff, K., and Scott, J.C. (2007) Environmental determinants of infectious disease: a framework for tracking causal links and guiding public health research. *Environmental Health Perspectives* 115: 1216–1223.

Ellis, F. (1993) *Peasant Economics: Farm households and agrarian development*, 2nd edition. Cambridge, UK: Cambridge University Press.

Ellis, F. (1998) Household strategies and rural livelihood diversification. *Journal of Development Studies* 35: 1–38.

Entwisle, B., Rindfuss, R.R, Guilkey D.K., Chamratrithirong, A., Curran, S.R., and Sawangdee, Y. (1996) Community and contraceptive choice in rural Thailand: a case study of Nang Rong. *Demography* 33: 1–11.

Entwisle, B., Walsh, S.J., Rindfuss, R.R., and VanWey, L.K. (2005) Population and upland crop production in Nang Rong, Thailand. *Population and Environment* 26: 449–470.

Entwisle, B., Rindfuss, R.R., Walsh, S.J., and Page, P.H. (2008) Population growth and its spatial distribution as factors in the deforestation of Nang Rong, Thailand. *Geoforum* 39: 879–897.

Futemma, C. and Brondizio, E.S. (2003) Land reform and land-use changes in the lower Amazon: implications for agricultural intensification. *Human Ecology* 31: 369–402.

Grossman, M. (1972) On the concept of health capital and the demand for health. *Journal of Political Economy* 80: 223–255.

Grossman, M. (1999) *The Human Capital Model and the Demand for Health*. NBER Working Paper 7078. Cambridge, MA: National Bureau of Economic Research.

Hampshire, K.R., Panter-Brick, C., Kilpatrick, K., and Casiday, R.E. (2009) Saving lives, preserving livelihoods: understanding risk, decision-making and child health in a food crisis. *Social Science and Medicine* 68: 758–765.

Hayward, M.D. and Gorman, B.K. (2004) The long arm of childhood: the influence of early-life social conditions on men's mortality. *Demography* 41: 87–107.

Houtzager, P.P. (2000) Social movements amidst democratic transitions: lessons from the Brazilian countryside. *Journal of Development Studies* 36: 59–88.

Hull, J.R. (2008) Migration, remittances, and monetization of farm labor in subsistence sending areas. *Asian and Pacific Migration Journal* 16: 451–484.

Jepson, W. (2006) Private agricultural colonization on a Brazilian frontier, 1970–1980. *Journal of Historical Geography* 32: 839–863.

Joffe, M. (2007) Health, livelihoods, and nutrition in low-income rural systems. *Food and Nutrition Bulletin* 28: S227–S236.

Keyes, C. (1976) In search of land: village formation in the central chi river valley, northeastern Thailand. *Contributions to Asian Studies* 9: 45–63.

Korinek, K., Entwisle, B., and Jampaklay, A. (2005) Through thick and thin: layers of social ties and urban settlement among Thai migrants. *American Sociological Review* 70: 779–800.

Lambin, E.F., Geist, H.J., and Lepers, E. (2003) Dynamics of land-use and land-cover change in tropical regions. *Annual Review of Environment and Resources* 28: 205–241.

Ludewigs, T. (2002) Agricultural credit and the build-up of social capital in the Brazilian Amazon Frontier. Institutional Analysis and Development Mini-Conference, 2002. See www.indiana.edu/~workshop/seminars/papers/y673_fall_2002_ludewigs.pdf (accessed 7 July 2012).

McCracken, S.D., Siqueira, A.D., Moran, E.F., and Brondízio, E.S. (2002) Land use patterns on an agricultural frontier in Brazil: insights and examples from a demographic perspective. In *Deforestation and Land Use in the Amazon*, C.H. Wood and R. Porro (eds). Gainsville, FL: University Press of Florida, 162–192.

McCusker, B. and Carr, E.R. (2006) The co-production of livelihoods and land use change: case studies from south Africa and Ghana. *Geoforum* 37: 790–804.

Moran, E.F. (1981) *Developing the Amazon*. Bloomington, IN: Indiana University Press.

Moran, E.F. (2010) *Environmental Social Science: Human Environment Interactions and Sustainability*. Malden, MA: Wiley-Blackwell.

Moran, E.F., Brondizio, E.S., and McCracken, S.D. (2002) Trajectories of land use: soils, succession, and crop choice. In *Deforestation and Land Use in the Amazon*, C.H. Wood and R. Porro (eds). Gainesville: University of Florida Press, 193–217.

Moran, E.F., Brondizio, E.S., and VanWey, L.K. (2005) Population and environment in Amazonia: landscape and household dynamics. In *Population, Land Use, and Environment: Research Directions*, B. Entwisle and P. Stern (eds). Washington, DC: National Academies Press, 106–134.

Padoch, C., Brondizio, E., Costa, S., Pinedo-Vasquez, M., Sears, R.R., and Siqueira, A. (2008) Urban forest and rural cities: multi-sited households, consumption patterns, and forest resources in Amazonia. *Ecology and Society* 13: 2.

Perz, S.G., Caldas, M.M., Arima, E., and Walker, R.J. (2007) Unofficial road building in the Amazon: socio-economic and biophysical explanations. *Development and Change* 38: 529–551.

Phongphit, S. and Hewison, K. (2001) *Village Life: Culture and Transition in Thailand's Northeast*. Bangkok, Thailand: White Lotus.

Piotrowski, M. (2006) The effect of social networks at origin communities on migrant remittances: evidence from Nang Rong District. *European Journal of Population* 22: 67–94.

Piotrowski, M. (2009) Migrant remittances and skipped generation households: investigating the exchange motive using evidence from Nang Rong, Thailand. *Asian and Pacific Migration Journal* 18: 163–196.

Reardon, T. and Vosti, S.A. (1995) Links between rural poverty and the environment in developing countries: asset categories and investment poverty. *World Development* 23: 1495–1506.

Rigg, J. (1987) Forces and influences behind the development of upland cash cropping in Northeast Thailand. *Geographical Journal* 153: 370–382.

Rindfuss, R.R., Piotrowski, M., Thongthai, V., and Prasartkul, P. (2007) Measuring housing quality in the absence of a monetized real estate market. *Population Studies* 61: 35–52.

Rindfuss, R., Entwisle, B., and Walsh, S. (2009) *Nang Rong Projects [Thailand]*. Computer file. Ann Arbor, MI: Inter-University Consortium for Political and Social Research.

Russell, S. (2004) The economic burden of illness for households in developing countries: a review of studies focusing on malaria, tuberculosis, and human immunodeficiency virus/acquired immunodeficiency syndrome. *American Journal of Tropical Medicine and Hygiene* 71: 147–155.

Russell, S. and Gilson, L. (2006) Are health services protecting the livelihoods of the urban poor in Sri Lanka? Findings from two low-income areas of Colombo. *Social Science and Medicine* 63: 1732–1744.

Sauerborn, R., Adams, A., and Hien, M. (1996) Household strategies to cope with the economic costs of illness. *Social Science and Medicine* 43: 291–301.

Saxena, N.C., Speich, N., and Steele, P. (2005) *Review of the Poverty—Environment Links Relevant to the IUCN Programme*. Gland, Switzerland: IUCN.

Smith, K.P. and Christakis, N. (2008). Social networks and health. *Annual Review of Sociology* 34: 405–429.

Steward, C. (2007) From colonization to “environmental soy”: A case study of environmental and socio-economic valuation in the Amazon soy frontier. *Agriculture and Human Values* 24: 107–122.

VanWey, L.K., D’Antona, A.O., and Brondizio, E.S. (2007) Household demographic change and land use. *Population and Environment* 28: 163–185.

VanWey, L.K., D’Antona, A.O., and Guedes, G.R. (2008) Land use trajectories after migration and land turnover, New Orleans, LA. Annual Meeting of the Population Association of America, New Orleans, LA, 17–19 April.

VanWey, L.K., Guedes, G.R., and D’Antona, A.O. (2009) Out-migration and household land use change in Altamira, Pará, Marrakech, Morocco. Proceedings of the XXVI International Population Conference, Marrakech, Morocco, 27 September–2 October. See http://iussp2009.princeton.edu/download.aspx?submissionId=93059 (accessed 7 July 2012).

Walker, R., Moran, E., and Anselin, L. (2000) Deforestation and cattle ranching in the Brazilian Amazon: external capital and household processes. *World Development* 28: 683–699.

Walsh, S.J., Rindfuss, R.R., Prasartkul, P., Entwisle, B., and Chamratrithirong, A. (2005) Population change and landscape dynamics: Nang Rong studies. In *Population, Land Use, and Environment: Research Directions*. B. Entwisle and P.C. Stern (eds). Washington, DC: National Academy Press, 135–162.

White, P.J., Ward, H., Cassell, J.A., Mercer, C.H., and Garnett, G.P. (2005) Vicious and virtuous circles in the dynamics of infectious disease and the provision of health care: gonorrhea in Britain as an example. *Journal of Infectious Disease* 192: 824–836.

Woolcock, M. and Narayan, D. (2000) Social capital: implications for development theory, research, and policy. *World Bank Research Observer* 15: 225–249.

4 Change in tropical landscapes

Implications for health and livelihoods

Kenneth R. Young

Introduction

The original land covers of tropical landscapes are preconditioned by climate, topography, and soils. Human land use is partly in response to the opportunities from the land resources, but is mostly a function of people's needs and abilities to alter land use and land cover for their benefit. People who depend on natural resources utilize livelihood strategies to produce or acquire the food, fiber, and shelter they need. Health and nutrition are affected in many ways by these kinds of dynamics; for examples of these types of health patterns see Chapters 3 and 7. While the lands and landscapes in tropical latitudes are characterized by different biophysical environments and multiple land use types, they are all affected in fundamentally asymmetrical ways by current and future global environmental changes (Young 2007). In terms of human health, King and Bertino (2008) described the "asymmetries of poverty," specifically highlighting the consequences of overlooking the effects of chronic infections (helminthiasis and infections caused by protozoa, bacteria, fungi), and other neglected tropical diseases.

Health concerns in tropical landscapes arise at the interfaces of biophysical environment, land use/land cover systems, and their respective drivers of change and feedbacks. The reciprocal human–environment interactions involved are common within socio-economic systems that utilize natural resources (Chapin *et al.* 2009). Making the coupling(s) of the natural–human system a central investigative focus opens up many new ways to look at system behavior, including evaluating consequences for human health and livelihoods. An overview by Liu *et al.* (2007) examined several agricultural, forestry, and ecotourism-based systems, showing that they shared emergent features such as feedbacks, non-linearities, and legacy effects. There were also identifiable similarities in how resilience and sustainable resource use could be characterized for the different systems. At the same time, all the studied landscapes showed heterogeneity, both temporally as processes wax and wane, and spatially as the resource base (and hence land-use strategies) rearrange from place to place.

People in tropical countries are more vulnerable as they have greater exposure to health risks at the same time as fewer health services and less technology is available for them to ameliorate conditions. Hotez and Kamath (2009) evaluated these concerns for sub-Saharan Africa and for the 500 million poor people who live there. They estimate that those inhabitants are infected with from 14 percent to 100 percent of the world's disease burden, for example with 93 percent of the cases of schistomiasis, 99 percent of those of onchocerciasis, 90 percent of those of yellow fever, and all of the cases of dracunculiasis and human African trypanosomiasis. Because of ongoing socio-economic changes that encourage livelihoods that in turn increase fire frequencies, result in deforestation, foment conversion for agriculture, add to growth of urban areas, and increase the influence of market forces on many decisions made by rural inhabitants, it appears likely that disparities globally and within the respective countries, will continue to increase.

Observers have noted that the current confluence of environmental and interacting social changes represent a fundamental shift in human history. Harper and Armelagos (2010) point to epidemiological and demographic transitions that are combining to produce large human populations with many young people, but also with elderly exposed to delayed chronic diseases (for a related discussion, see Chapter 12 of the present volume). These changes are accompanied by increased failures of antimicrobials and faster spread of novel infections. They direct attention to countries such as Guatemala, El Salvador, Kenya, and the Democratic Republic of Congo, which are particularly stressed by prevailing social inequalities that can perversely increase negative health effects of both under-nutrition and over-nutrition in different parts of society.

This chapter delineates major sociopolitical and ecological concerns for tropical landscapes and their associated land use systems found in the lowland, highland, and seasonally dry tropics, plus those changed by rural–urban connections. The coupled systems of tropical livelihoods are conceptualized in relation to likely future changes, especially for those with implications for vulnerability, well-being, and inequality. A brief overview of coupled natural–human systems begins the chapter, which is then followed with commentary on the major expectations for the humid, upland, and seasonal tropics. A final part highlights how future changes in health and livelihoods will be linked not only to environmental influences on biophysical controls, but will also be exacerbated by increased asymmetries among urban and rural populations of tropical countries.

Coupled natural–human systems in the tropics

Global environmental change will have many direct and indirect effects on health and livelihoods, for example through altering water flows and other landscape-scale or ecosystem processes. Patz *et al.* (2005) identified likely effects due to warmer temperatures and greater climatic variability acting (1) on crop yields and food supplies, (2) through increased vulnerability to ENSO (El Niño Southern Oscillation) and other climate system effects, (3) through possible (although not fully demonstrated) effects on infectious diseases through temperature controls on vectors and infectious agents, and (4) on urban areas with augmented heat island effects.

Comrie (2007) asserts the need to look at the coupled natural–human interactions involved, given that vulnerability is expressed through the adaptability of the social group affected and the sensitivity and exposure of the place involved. As an intellectual exercise to consider the consequences, imagine a simple coupled system of land use on a high tropical mountain landscape, as represented in Figure 4.1 (see also Young 2009). Further, consider it as shaped by people who affect the land cover consisting of open tropical alpine vegetation, sometimes with an admixture of low shrubs or wetlands, through land use that includes the grazing of livestock. Part of the coupling of concern would include the effects of those herbivores on plant species composition, with relative species abundance affected by the differential consumption of palatable plants and an increase in less-desired forage species. This particular coupling could lead to overgrazing, followed by an adaptive shift in land management through a change in stocking rates, in pasture rotation intervals or in the species of livestock utilized. Or the coupling could even lead the people to cease grazing the site, followed by a subsequent recovery of the original land cover. Perhaps outside income sources from wage labor or remittances would be used to replace the earnings once derived from the livestock.

Figure 4.1 Pastoralist choosing grazing areas for a mixed flock of llamas, alpacas, and sheep in high elevations of the Peruvian Andes. Use of particular grazing areas is mediated through family and community interactions, while the ecological productivity of those areas is increasingly affected by climate extremes and reduced glacial meltwater. Often the power for legal and political decision-making over lands and natural resources is located in distant lowland cities.

Note that one implication of this thought experiment is that even a simple coupled land-use/land-cover system could require the assessment of multiple possible trajectories, and numerous social, ecological, and physical parameters and feedbacks. Prediction is difficult and complexity is inherent. In addition, all the connections of the landscape back to livelihoods could also have additional feedbacks or complications for human nutrition or for illnesses at household and community levels. Particular households with less livestock would be more exposed to financial strains should it be necessary to pay for medical expenses. Communities unable to manage land resources held in common may also be not capable of providing other aspects of social capital.

Further imagine in relation to the landscape shown in Figure 4.1 that directional climate change is affecting the system, perhaps as mediated locally through an increasingly longer plant-growing season. In this hypothetical case the result could be that some plant species become more productive and able to tolerate more grazing pressure. As a result, a biophysical shift has altered the possible sustainable use of the land cover, in turn altering prospective land use goals and means. This shift would then affect perceptions of incomes, agricultural productivity, household dynamics, and family welfare, and hence resiliency to health concerns. However, if the directional climate change continued the characteristics of the vegetation and biological interactions would also continually change, implying that ecological conditions of the land will continue to evolve and that land use decision-making must similarly be adaptive and constantly reevaluated. These socio-economic and managerial shifts presume the needed land and labor resources exist, that the requisite knowledge and wisdom are present among the people, and that land tenure and water rights are embedded in systems with the required flexibility.

Thus, the behavior of a simple coupled system becomes even more complex when global environmental change affects particular places. Coincident change in people's values, objectives, and technologies can produce multitudinous scenarios of change. These dilemmas are often discussed as generalities in global overviews (Millennium Ecosystem Assessment 2005; see also the website of the Intergovernmental Panel on Climate Change at www.ipcc.ch), or as evaluated in terms of products, nation states, and economic development, but are less commonly evaluated with reference to the landscape scale where people and environment interact through land use, as is done in this chapter and elsewhere in this book.

Public health responses to these kinds of issues would be well served by incorporating the findings of land-use/land-cover and infrastructure change research. To do this may require new institutions and social networks, as, for example, experts on wildlife and domesticated animal diseases do not necessarily know about or interact with researchers, managers, administrators, remote sensing experts, or NGOs involved with either human diseases or ecosystem functioning. Climate and health modelers will need to form integrated research teams. For example, it is sand flies that connect people to rodent reservoir species and American cutaneous leishmaniasis. Chaves *et al.* (2008) suggest that a threshold shift best explained human infection by *Leishmania* because forest cover (spatial patterns of which would control rodent populations) was statistically significant as an explanatory variable only at small spatial scales (i.e. for people living within 5 km of forest) and otherwise was controlled by measures of social marginality. González *et al.* (2010) used predictive species distribution modeling to assess possible shifts in both the sand fly vector species and the rodent reservoir species due to climate change in Mexico, the US, and Canada. Similarly, human plague in east Africa can be modeled in terms of risk associated with elevation, rainfall, and number of land cover types (Winters *et al.* 2009). It is likely that the reservoir rodent populations involved and their respective pathogens are at least partially tracking environmental shifts that are represented in remotely sensed data, permitting monitoring. In all these cases, multiple academic disciplines need to be involved to understand the effects of landscape dynamics, and team-based approaches to prediction and response would be necessary.

Future scenarios for coupled tropical systems

Global change causes disparate shifts in the coupled natural–human systems associated with land use in tropical landscapes. Continued future shifts in those drivers, may well force land-use systems over thresholds, with contingent influences on migration, land tenure, and public health. The rural sites where small-scale agriculture will become more marginalized could act as sources of environmental refugees, with people moving to urban-fringe settlements. In this section, a selection of health concerns from the Global Tropics is discussed in terms of socio-ecological changes expected in the humid lowlands, tropical highlands, tropical biomes with long dry seasons, and urbanized areas.

Humid lowlands

For the tropical rainforest areas of the Earth, there is much concern that higher carbon dioxide levels, plus additional climate system responses, have already begun to alter growing conditions for plants (Körner 2009). Data from monitored forest inventory plots is suggestive, with some indications of increased biomass and faster tree growth rates (Chave *et al.* 2008). Lewis *et al.* (2009: 529) found a number of lines of evidence of "large-scale, directional changes." The shifts can be manifested in altered forest dynamics and tree regeneration patterns (Feeley *et al.* 2007), increased litterfall that once decomposed introduces more carbon dioxide to the atmosphere (Sayer *et al.* 2007), and a proposed "threshold response" of tropical plants switching their dominant nitrogen sources following climate change (Houlton *et al.* 2007).

The humid lowlands of the tropical Americas, Africa, and Asia by definition have more rainfall over the course of the year than the potential evapotranspiration that would otherwise act to return that moisture to the atmosphere. As a generalization, this results in much weathering as water percolates down through regoliths, poor and acidic soils, large river systems, great quantities of biomass and primary production, and high biodiversity. Land use must cope by using adaptive systems such as shifting cultivation (Kellman and Tackaberry 1997), which acts to limit pests and recuperate soil fertility, or people may depend on forest product extraction and fishing. Much recent land cover change in the lowland tropics is associated with commercial ventures, logging of tropical timbers, and the installation of livestock grazing in sites once covered by forest (e.g., Fox *et al.* 2009).

For indigenous forest-dwelling people, the persistence of forest is crucial. The Achuar, for example, make small openings in the Amazon rainforest for their settlements and gardens (Figure 4.2). Descola (1994) described how this system had functioned for many centuries, but also pointed to recent demographic changes inspired by the arrival of the outside world. On a trip to this area in 2006, I was struck both by the continuity of their experiences as manifested in cultural richness and sustainable use of the forests, but at the same time by the fragility of those land use and social systems given an increased preference for long-term use of the same areas. This was due to the construction of numerous air strips for use by small planes that permitted some limited economic exchange and also served for emergency evacuations for sick people to towns located more than a one-hour flight away. In subsequent research, López and Sierra (2010) used land-use modeling to show that despite overall low population densities and relatively large extensions of rainforest, lands suitable for agriculture were becoming a limiting resource given increasingly sedentary settlement patterns.

Figure 4.2 Lowland rainforest as used for swidden cultivation in an Achuar indigenous community in the Ecuadorian Amazon. In the past, village sites were changed as soils became more impoverished. However, increasingly perceived needs for airplane strips that allow for emergency health flights and other amenities have made for more permanent settlements in this isolated corner of the Amazon basin.

Modeling consistently finds worrisome future scenarios for plant growth and ecosystem processes in humid tropical forest areas such as these: increased drought is of particular concern, with feedbacks increasing return of carbon dioxide to the atmosphere, thus amplifying the forcing (Sitch *et al.* 2008). Observations of an additional feedback connecting drier conditions to increased use and influence of fire (Nepstad *et al.* 2008), suggest that forested tropical ecosystems will tend to switch to non-forest states, either due to fire-caused conversion to scrub or savanna, or else by the expansion of pasturelands and agricultural fields. Households in the humid tropics often connect rural areas to cities through production, extraction, and trade of natural resources. Connecting national and regional land use to global demands for commodities and forest products would likely continue to provide increased motivation for land-use/land-cover change and contingent effects on the atmosphere.

Malaria distribution in future tropical lowlands will presumably be a function of the shifting distributions of land cover changes, mosquito vectors, the biologies of the *Plasmodium* parasite species, and interactions among human exposure and mosquito behavior. For example, subdecadal variations from ENSO are known to alter malaria–climate associations (Poveda *et al.* 2001). Simultaneously, genetic and population studies permit reconstructions of past distributions of particular genotypes that may vary in ways that have important spatial implications for public health responses. Levine *et al.* (2004) used the current distribution of *Anopheles gambiae* and closely related species in Africa, plus climatic, topographic, and land-cover data, to model the expected distribution of the vectors, which revealed some anomalous gaps in species ranges. This same general approach allowed them to explore implications of historic and possible future colonization and range expansions in the New World tropics. Similar analyses linked to climate change scenarios would permit assessments of possible future distributions of vector species within the context of shifting land covers.

These shifts are also to be expected for the *Plasmodium* parasite species, which include regional genotypes resistant to some malaria drugs (Roshanravan *et al.* 2003). In addition to these spatio-temporal phenomena, it is well known that many local and regional land-use and land-cover changes affect the prevalence of malaria, from the installation of fish ponds (Maheu-Giroux *et al.* 2010) to the building of roads and the cutting of forests (Vittor *et al.* 2006). However, integrated predictive assessments will also require better understanding of the role of natural environmental fluctuations in rainfall, surface water, and soil moisture, plus concurrent socio-economic changes (Wandiga *et al.* 2010).

More broadly, Gage *et al.* (2008: 436) pointed out that the climate influence on vector-borne diseases will be mediated through additional feedbacks with other drivers of change such as "increased trade and travel, demographic shifts, civil unrest, changes in land use, water availability, and other issues." Costello *et al.* (2009) acknowledged the likely future increase in vector-borne diseases, but concluded that indirect effects on water sources, food supplies, and the effects of extreme climate events will have even larger influences on public safety and welfare.

Other health concerns associated with mosquitoes include dengue, with epidemics that may be predictable given more knowledge on the links among insect populations, climatic variation, and disease transmission (Degallier *et al.* 2010). Monitoring necessarily includes the genetics and population dynamics of multiple species in addition to the mosquitoes. For example, numerous wild bird species serve as reservoirs for the West Nile virus (Ezenwa *et al.* 2006). Yellow fever, maintained in wild primate populations, shows distribution patterns of genetic variation that appear to be related to limitations on the dispersal of the monkeys, the mosquito species, or both (Bryant *et al.* 2003). These multiple species assemblages are affected in complex ways by environmental dynamism, in addition to responding to feedbacks from social dimensions including individual aspirations and behaviors of people, along with local to national-scale infrastructure development.

As an example of the kind of study needed, Eisenberg *et al.* (2006) evaluated the role of physical accessibility on the health of villages in western Ecuador as mediated through proximity and remoteness, functions of travel time and distance. By controlling for various social and biophysical parameters, they showed that diarrhea-causing pathogens had more negative effects on well-connected villages, after taking village size and sanitation level into account. It was children especially who were susceptible to *Escherichia coli*, *Giardia*, and rotavirus. And it was the development of the transportation infrastructure, and associated social and ecological changes, that altered vulnerability and disease transmission. The social factors proposed to be critical were demographic (in-migration, out-migration, short-term visits), degree of connectedness through social networks, and sanitation infrastructure. The system defined in this fashion could then be used to assess the spread and persistence of different pathogens.

Highlands

Tropical highlands are already changing due to shifts in climate. These mountain areas include the centers of origin of many of the world's crops, along with very early agricultural systems; as a result, they include some of the world's first humanized landscapes. Directional biophysical changes, such as those that marked the end of the Pleistocene or major mid-Holocene shifts, reshuffled possibilities and opened up new lands for settlement. Similarly, the last several decades of current times have seen directional declines in the size of tropical glaciers (Thompson *et al.* 2006), primarily due to warmer temperatures in tropical highlands, hence altering ecological conditions and agronomic potentials.

An integrated evaluation of current trends was done in the central Andean range of Colombia by Ruiz *et al.* (2008). They report not only warmer temperatures overall, but higher maximum temperatures, larger annual and daily temperature ranges, more sunny days, less relative humidity, and increased variability in occurrences of heavy rains, but generally lower rainfall totals. As a result, the high mountains there have smaller icecaps and dried-up lakes, as well as more frequent fires in the vegetation. They predict important reductions in water supply for people downslope, with alterations in water cycles (e.g, Mark *et al.* 2010).

More generally, Andean land use will need to change, mediated through where production zones are located and what useful plants and animals are cultivated and managed.

In earlier studies (Young 2008, 2009) I put these current changes in the tropical Andes into perspective by pointing to evidence of rapid ongoing changes in land cover with small holders often responding within weeks or months to perceptions of socio-ecological change. But those drivers of change are acting upon anciently inhabited landscapes that experienced at least two previous bouts of deforestation/reforestation and land cover conversion, with European colonization in the sixteenth century and with previous changes dating back 500 to over 3500 years, depending on the location. The past resiliency of Andean livelihood systems required either extra land to where agriculture could be shifted, or else much time needed to pass while human populations rebuilt.

Given current constraints acting on the kinds of landscapes shown in Figure 4.1, the ability of farmers and pastoralists to shift production zones in response to unusual climatic events or with directional climatic change may be limited. It seems like recent economic development in the tropical Andes often has consisted of programs and policies designed to make land tenure private rather than communal. Mining and industrial farming is on the increase, incentivized by global commodity prices and government subsidies. Related programs act to place monetary values on water use and to prioritize private property regimes. Family health is shaped by these changes, with many previously adaptive responses tied to reciprocity at local levels or acting through familial bonds maintained despite out-migration (Young and Lipton 2006). Increasingly, both households and rural communities are connected to major urban areas through migration, permitting the return of remittances and other subsidies that may buffer the exigencies of rural life (Adams and Page 2005).

Savannas/dry forests

The seasonal tropical lands have natural land covers of dry forests, savannas, or shrublands (e.g., Figure 4.3). Of concern for the future of the seasonal tropics are combinations and interactions of global environmental change and socio-economic processes that would force large areas of humid–dry forests into more open land cover types, hence also changing livelihoods. Pueyo *et al.* (2010) find that the "self-organized criticality" of these ecosystems is affected by fire regime; drought conditions and burns could switch forested areas into savannas. These and other non-linear shifts would affect native biodiversity (Biggs *et al.* 2008). The balance among herbaceous versus woody plant dominances in particular landscapes is often set by fire disturbance dynamics, with grasses favored by fire, but trees and shrubs persisting through fire-proof bark, resprouting, and post-fire colonization. Human land use frequently incorporates burns into landscape management, increasing the productivity of grasslands for livestock.

Spatial heterogeneity defines these seasonal tropical ecosystems. Pringle *et al.* (2010) provide an example of how termite mounds form hotspots of plant growth and animal abundance, and thereby promote much place-to-place variation in

Figure 4.3 Savanna and elephant along the Boteti River in Botswana. Note that the photograph is taken looking into a national park; a photograph taken at the same site but in the opposite direction would show people and livestock grazing just outside the park amid a transformed landscape. This park–people interface is a major zone of conflict in rural Africa; in addition, it is a zone of the intermingling of diseases and parasites shared between domesticated and wild animals.

land cover, with higher overall productivity of savanna at landscape scales. They propose that termites and their couplings to other parts of the system create emergent patterns with consequences for ecosystem functioning. Earlier research (McNaughton 1983) had ascribed causality of savanna spatial heterogeneity to the effects of different kinds of herbivores on vegetation (Figure 4.3), combined with differences from place to place in soil texture. Anderson *et al.* (2010) found that it is not only soil fertility or other edaphic properties, but also the risk of predation that affects the behavior of native herbivores, and hence eventually affects the patterning of the landscape. They point out that both bottom-up (e.g., resource and edaphic related factors) and top-down (predator behavior and abundance) processes affect the heterogeneity of tropical savannas.

In utilized landscapes, humans take over aspects of both bottom-up and top-down processes through a simplification of landscape configuration and by control of wild species deemed to be undesirable. Figure 4.3 shows a common but more complex situation where the landscape in the photograph is protected for biodiversity conservation but is located adjacent to landscapes used by local people for livestock raising and dryland farming; livelihoods are negatively affected by nuisance wild animals, but ongoing development efforts also try to harness income from ecotourism to local needs. Outcomes for particular families and communities will vary markedly.

The ecological side of interconnectedness and nestedness is being studied. For example, Graham *et al.* (2009) provided an assessment of the kinds of networks that inter-relate vertebrates and their ectoparasites, including ticks, mosquitos, and tsetse flies. They develop means to quantify the nestedness of 29 different systems. They contrast these measures with hypothetical networks with random interactions or that show perfect nestedness. The implications for disease transmission in general are significant, for example with some network structures that would spread novel diseases among many vertebrate host species; species-rich environments are probably more susceptible and vulnerable. Holdo *et al.* (2009) examined the hierarchical structure of food webs and trophic chains on diseases. They show that disease (in this case rinderpest) can at times cause novel feedbacks, resulting in shifts in ecosystem regimes, illustrated in their study with an example from an African savanna system of vegetation, elephants, and wildebeests, as affected by fire and rainfall. The increased dominance of woody plants following wildlife die-offs from rinderpest caused shifts in biomass, carbon structure, and landscape mosaics.

These ecological processes will interact in complicated ways with socio-economic needs and activities of pastoralists and farmers using seasonally dry landscapes. People affect wildlife and vegetation with the grazing of cattle and goats, and through the establishment of settlements and agricultural fields; in turn, wildlife can alter vegetation structure, they may host diseases and parasites that affect domesticated animals, and some wildlife species cause direct economic loss to crops and livestock. Ogutu *et al.* (2010) developed a model meant to help with sustainable management strategies. Densities of wildlife and livestock were set along two environmental gradients: distances to and among water sources, and to and among pastoral settlements. Indeed, although the length of the dry season and occasional decadal and subdecadal occurrences of droughts are thought to be fundamental organizers of savanna and other semi-arid landscapes, local people are not limited to transhumance with their livestock. Nielsen and Reenberg (2010) point out that this essentially makes them "beyond climate," as they are not altering their land use in a monotypic fashion with climatic variation. Coping strategies would include the switching of cropping systems and the development of more efficient irrigation methods, all of which would connect to health outcomes at household and village levels.

Cities and settlements

Urban environments have structures that in part derive from environmental and historical legacies, and in part are the result of socio-economic processes (Pickett *et al.* 2005), including the networks and clusters created by transportation and economic processes connecting cities to rural areas. Urban ecosystems increasingly alter not only themselves but all the other ecosystems of the Earth through altered biogeochemical cycles (Grimm *et al.* 2008). Their growth is complex, affected by bottom-up and top-down processes (sensu Irwin *et al.* 2009), with important effects on environmental quality for residents and for the hinterland.

The physical structure of settlements and cities can create spatial patterning of health. Hemme *et al.* (2010) found that a highway in Trinidad split *Aedes aegypti* into two genetically separated populations, with possible implications for the efficacy of mosquito monitoring and suppression programs on either side of the road. In this case, the differences have consequences at the scale of the dispersal distance of individual insects, on the order of several kilometers. Barton (2009) examined health issues in planning of urban morphology, making it clear that a broader perspective on health is crucial, including not only pollution and inequities in housing, labor opportunities, and transport, but also aspects that affect well-being such as recreational choices and mental health.

Tong *et al.* (2010) recently evaluated the state of knowledge concerning vulnerability, which they classified in relation to: (1) weather extremes, including heatwaves, floods, and droughts; (2) food security and safe drinking water; (3) infectious diseases, especially the potentially climate-sensitive diseases such as malaria, dengue fever, and schistomiasis; and (4) climate-related human migration. The urban poor are particularly vulnerable to these alterations. Dye (2008) found that present-day health in urban areas is on average better than in rural places, as seen for example in lower fertility and infant mortality rates. These differences are conditioned by public services and opportunities for jobs and education. Most of the world's people are now or shortly will be city-dwellers, including a predicted 5.26 billion urbanites in developing countries by 2050 (Montgomery 2008). In the future, there will be hundreds of large cities in Asia and Latin America.

As one example of urban vulnerability and rural–urban connections, Chagas disease infects humans through contact with about 70 species of blood-consuming triatomine bugs (Abad-Franch *et al.* 2010). It has been associated with mammalian reservoir species (armadillos, marsupials, rodents, and skunks) for millions of years. The current household context wherein most people are infected was evaluated with a mathematical model by Cohen and Gürtler (2001), based on data from subtropical northern Argentina. Chickens are not infected by the *Trypanosoma cruzi* pathogen, but the bugs do feed on them (and they on the bugs). Dogs are not only a target for the bugs, but also can carry the infection. A series of model outcomes were produced in that study for seasonal variation in disease exposure to humans as conditioned by the number of chicken and dogs kept by the family. Although a complex system is modeled, the researchers conclude that the relatively simple suggestion of restricting the number and presence of dogs (and other infected vertebrates) from the sleeping areas of the house could dramatically lower infections for people. Bayer *et al.* (2009) showed that the movement of rural people into a rapidly expanding urban area in southern Peru (Polk *et al.* 2005) also brought Chagas disease along with the in-migration to the shanty towns. Interestingly, it was not the original rural–urban migration itself that conveyed the disease, but recurrent trips originating from the need of the new residents to do seasonal agricultural jobs in outlying areas. The presence of many domesticated animals (guinea pigs, dogs) in the shantytown settlements then acted to maintain vertebrate reservoirs of the pathogen.

Criteria for research priorities on rural–urban interactions would need to consider equity, the severity of potential impact, the need for intervention, and possibilities for addressing multiple needs simultaneously. Thus, tropical cities should prepare for public health concerns for marginalized human populations exposed to extreme conditions, malnutrition, and limited social and economic mobility. New settlements in frontier areas of the tropics or on the fringes of cities will have those same pressures, as out-migrating rural dwellers often move first to them. Keeping rural inhabitants in place means that the support systems required for adaptation, health services, educational advancement, and land/food security must reach out to remote locations in socially acceptable and economically sustainable manners.

Conclusions

Patz *et al.* (2000) stated that feedbacks coming from land-use change would increase future global health concerns due to road construction, water control and diversion projects, forest cutting, and new settlements. These drivers would be exacerbated by warmer temperatures and alterations in the hydrological cycle originating with changes in rainfall intensity, length of the rainy season, and the strength of phenomena such as ENSO. In terms of parasitic diseases, they predict increases in the effects of malaria, leishmaniasis, cryptosporidiosis, giardiasis, tyrpanosomiasis, trematode-caused diseases (schistosomiasis), and those resulting from tissue nematodes (filariasis, onchocerciasis, loiasis). It is the developing countries that are particularly susceptible to increased infections (e.g., malaria, dengue), malnutrition, respiratory ailments (from increased fires), and diarrhea-causing diseases such as cholera.

Tropical countries will continue to urbanize and globalize over the next several decades. There has been much internal migration to urban areas, making cities into decision-making entities with considerable political and economic power, which will continue to increase. The demand for commodities and for labor reaches out into tropical landscapes, creating disparities and altering decision making by almost all farmers and other rural residents. More than a decade ago, researchers described how Amazon cities were increasingly structuring rural landscapes and livelihoods in the entire Amazon Basin, and this trend has continued there (Guedes *et al.* 2009) and elsewhere.

At the same time, many lowland tropical landscapes will become drier or more seasonal, agricultural zones in the highlands will shift upward, and novel ecosystems will form with different species as dominants and new assemblages of species. The forested humid lowlands will be influenced by city growth, with rural land uses also constrained by the increased rainfall seasonality predicted for many areas. Future scenarios include the tropical highlands becoming even more important for extraction of water and mineral resources, but also less intensively used by local people due to out-migration and increased control by often distant urban areas. Most predictions for tropical savannas make them more prone to large-scale alterations by burning and industrial agriculture. All of these trends lead to increased vulnerability, especially of poor people.

Some middle elevational sites in tropical highlands might become more productive if soil moisture is adequate or if irrigation is feasible; the same may be true for seasonal tropical areas, especially if they are located near infrastructure. However, these positive aspects of global change would require technology, finance, and transportation to be utilized to their maxima. Overall, the negatives and increased difficulties for many rural smallholders would appear to be greater in number and degree of influence. In those cases, increased rural–urban demographic flows and economic interconnections would be responses and adaptations to be expected; being able to predict these population and financial shifts would allow for better planning for health and development interventions.

Rural livelihoods and health outcomes are affected by global commodity prices, which can send discouraging market signals to farmers (Clapp 2009), especially if national governments are politically beholden to urban-based majorities. In some cases, food standards in overseas markets may act to create opportunities for small-scale producers in the tropics, depending on partnerships that reach farmers. A worrisome counterexample comes from research by McKey *et al.* (2010), who described the agronomic and technical constraints acting upon how manioc is grown and used in Africa and South America; they worry that changing biophysical and economic conditions may make current farmer strategies non-adaptive in the future. Stresses acting upon household welfare imply greater exposure to health risks.

Biophysical and socio-economic drivers that force land-use systems over thresholds will have contingent influences on millions of individuals and their families, causing demographic shifts, difficulties with land tenure and security, and increased infectious diseases, in addition to acting through indirect processes that augment other health-related concerns. Global change will tend to exacerbate inequalities at local, regional, and national levels, and in both rural and urban environments. The health and welfare concerns of the rural and urban poor will increase under these scenarios, requiring greater multidisciplinary and multisector commitments to addressing the implications of environmental and social inequities.

References

Abad-Franch, F., Santos, W.S., and Schofield, C.J. (2010) Research needs for Chagas disease prevention. *Acta Tropica* 115: 44–54.

Adams, Jr., R.H. and Page, J. (2005) Do international migration and remittances reduce poverty in developing countries? *World Development* 33: 1645–1669.

Anderson, T.M., Hopcraft, J.G.C., Eby, S., Ritchie, M., Grace, J.B., and Olff, H. (2010) Landscape-scale analyses suggest both nutrient and antipredator advantages to Serengeti herbivore hotspots. *Ecology* 91: 1519–1529.

Barton, H. (2009) Land use planning and health and well-being. *Land Use Policy* 26S: S115–S123.

Bayer, A.M., Hunter, G.C., Gilman, R.H., Cornejo del Carpio, J.G., Naquira, C., Bern, C., and Levy, M.Z. (2009) Chagas disease, migration and community settlement patterns in Arequipa, Peru. *PLOS Neglected Tropical Diseases* 3(12): e567 (doi:10.1371/journal/pntd.0000567).

Biggs, R., Simons, H., Bakkenes, M., Scholes, R.J., Eickhout, B., van Vuuren D., and Alkemade, R. (2008) Scenarios of biodiversity loss in southern Africa in the 21st century. *Global Environmental Change* 18: 296–309.

Bryant, J., Wang, H., Cabezas, C., Ramirez, G., Watts, D., Russell, K., and Barrett, A. (2003) Enzootic transmission of yellow fever virus in Peru. *Emerging Infectious Diseases* 9: 926–933.

Chapin, F.S., Kofinas, G.P., and Folke, C. (eds). (2009) *Principles of Ecosystem Stewardship: Resilience-Based Natural Resource Management in a Changing World.* New York, NY: Springer.

Chave, J., Condit, R., Muller-Landau, H.C., Thomas, S.C., Ashton, P.S., Bunyavejchewin, S., Co, L.L., Dattaraja, H.S., Davies, S.J., Esufali, S., Ewango, C.E., Feeley, K.J., Foster, R.B., Gunatilleke, N., Gunatilleke, S., Hall, P., Hart, T.B., Hernández, C., Hubbell, S.P., Itoh, A., Kiratiprayoon, S., Lafrankie, J.V., Loo de Lao, S., Makana, J.R., Noor, M.N., Kassim, A.R., Samper, C., Sukumar, R., Suresh, H.S., Tan, S., Thompson, J., Tongco, M.D., Valencia, R., Vallejo, M., Villa, G., Yamakura, T., Zimmerman, J.K., and Losos, E.C. (2008) Assessing evidence for a pervasive alteration in tropical tree communities. *PLOS Biology* 6(3): e45 (doi:10.1371/journal.pbio.0060045).

Chaves, L.F., Cohen, J.M., Pascual, M., and Wilson, M.L. (2008) Social exclusion modifies climate and deforestation impacts on a vector-borne disease. *PLOS Neglected Tropical Diseases* 2(2): e176 (doi:10.1371/journal.pntd.0000176).

Clapp, J. (2009) Food price volatility and vulnerability in the Global South: considering the global economic context. *Third World Quarterly* 30: 1183–1196.

Cohen, J.E. and Gürtler, R.E. (2001) Modeling household transmission of American trypanosomiasis. *Science* 293: 694–698.

Comrie, A. (2007) Cimate change and human health. *Geography Compass* 1: 325–339.

Costello, A., Abbas, M., Allen, A., Ball, S., Bell, S., Bellamy, R., Friel, S., Groce, N., Johnson, A., Kett, M., Lee, M., Levy, C., Maslin, M., McCoy, D., McGuire, B., Montgomery, H., Napier, D., Pagel, C., Patel, J., de Oliveira, J.A., Redclift, N., Rees, H., Rogger, D., Scott, J., Stephenson, J., Twigg, J., Wolff, J., and Patterson, C. (2009) Managing the health effects of climate change. *Lancet* 373: 1693–1733.

Degallier, N., Favier, C., Menkes, C., Lengaigne, M., Ramalho, W.M., Souza, R., Servain, J., and Boulanger, J.-P. (2010) Toward an early warning system for dengue prevention: modeling climate impact on dengue transmission. *Climatic Change* 98: 581–592.

Descola, P. (1994) *In the Society of Nature: A Native Ecology in Amazonia.* Cambridge, UK: Cambridge University Press.

Dye, C. (2008) Health and urban living. *Science* 319: 766–769.

Eisenberg, J.N.S., Cevallos, W., Ponce, K., Levy, K., Bates, S.J., Scott, J.C., Hubbard, A., Vieira, N., Endara, P., Espinel, M., Trueba, G., Riley, L.W., and Trostle, J. (2006) Environmental change and infectious disease: how new roads affect the transmission of diarrheal pathogens in rural Ecuador. *Proceedings of the National Academy of Science* 103: 19,460–19,465.

Ezenwa, V.O., Godsey, M.S., King, R.J., and Guptill, S.C. (2006) Avian diversity and west Nile virus: testing associations between biodiversity and infectious disease risk. *Proceedings of the Royal Society B* 273: 109–117.

Feeley, K. J., Davies, S. J., Ashton, P. S., Bunyavejchewin, S., Nur Supardi, M. N., Rahman Kassim, A., Tan, S., and Chave J. (2007) The role of gap phase processes in the biomass dynamics of tropical forests. *Proceedings of the Royal Society B* 274: 2857–2864.

Fox, J., Fujita, Y., Ngidang, D., Peluso, N., Potter, L., Sakuntaladewi, N., Sturgeon, J., and Thomas, D. (2009) Policies, political-economy, and swidden in southeast Asia. *Human Ecology* 37: 305–322.

Gage, K.L., Burkot, T.R., Eisen, R.J., and Hayes, E.B. (2008) Climate and vector-borne diseases. *American Journal of Preventative Medicine* 35: 436–450.

González, C., Wang, O., Strutz, S.E., González-Salazar, C., Sánchez-Cordero, V., and Sarkar, S. (2010) Climate change and risk of leishmaniasis in North America: predictions from ecological niche models of vector and reservoir species. *PLOS Neglected Tropical Diseases* 4(1): e585 (doi:10.1371/journal.pntd.0000585).

Graham, S.P., Hassan, H.K., Burkett-Cadena, N.D., Guyer, C., and Unnasch, T.R. (2009) Nestedness of ectoparasite–vertebrate host networks. *PLOS ONE* 4(11): e7873 (doi:10.1371/journal.pone.0007873).

Grimm, N.B., Faeth, S.H., Golubiewski, N.E., Redman, C.L., Wu, J., Bai, X., and Briggs, J.M. (2008) Global change and the ecology of cities. *Science* 319: 756–760.

Guedes, G., Costa, S., and Brondizio, E. (2009) Revisiting the hierarchy of urban areas in the Brazilian Amazon: a multilevel approach. *Population and Environment* 30: 159–192.

Harper, K. and Armelagos, G. (2010) The changing disease-scape in the third epidemiological transition. *International Journal of Environmental Research and Public Health* 7: 675–697.

Hemme, R.R., Thomas, C.L., Chadee, D.D, and Severson, D.W. (2010) Influence of urban landscapes on population dynamics in a short-distance migrant mosquito: evidence for the dengue vector *Aedes aegypti*. *PLOS Neglected Tropical Diseases* 4(3): e634 (doi:10.1371/journal.pntd.0000634).

Holdo, R.M., Sinclair, A.R., Dobson, A.P., Metzger, K.L., Bolker, B.M., Ritchie, M.E., and Holt, R.D. (2009) A disease-mediated trophic cascade in the Serengeti creates a major carbon sink. *PLOS Biology* 7: e210 (doi:10.1371/journal.pbio.1000210).

Hotez, P. J. and Kamath, A. (2009) Neglected tropical diseases in sub-Saharan Africa: review of their prevalence, distribution, and disease burden. *PLOS Neglected Tropical Diseases* 3(8): e412 (doi:10.1371/journal.pntd.0000412).

Houlton, B.Z., Sigman, D.M., Schuur, E.A.G., and Hedin, L.O. (2007) A climate-driven switch in plant nitrogen acquisition within tropical forest communities. *Proceedings of the National Academy of Science* 104: 8902–8906.

Irwin, E.G., Jayaprakash, C., and Munroe, D.K. (2009) Towards a comprehensive framework for modeling urban spatial dynamics. *Landscape Ecology* 24: 1223–1236.

Kellman, M. C. and R. Tackaberry. (1997) *Tropical Environments: The Functioning and Management of Tropical Ecosystems*. New York, NY: Routledge.

King, C.H. and Bertino, A.-M. (2008) Asymmetries of poverty: why global burden of disease valuations underestimate the burden of neglected tropical diseases. *PLOS Neglected Tropical Diseases* 2(3): e209 (doi:10.1371/journal.pntd.0000209).

Körner, C. (2009) Responses of humid tropical trees to rising CO_2. *Annual Review of Ecology, Evolution, and Systematics* 40: 61–79.

Levine, R.S., Peterson, A.T., and Benedict, M.Q. (2004) Geographic and ecologic distributions of the *Anopheles gambiae* complex predicted using a genetic algorithm. *American Journal of Tropical Medicine and Hygiene* 70: 105–109.

Lewis, S.L., Lloyd, J., Sitch, S., Mitchard, E.T.A., and Laurance, W.F. (2009) Changing ecology of tropical forests: evidence and drivers. *Annual Review of Ecology, Evolution, and Systematics* 40: 529–549.

Liu, J., Dietz, T., Carpenter, S.R., Alberti, M., Folke, C., Moran, E., Pell, A.N., Deadman, P., Kratz, T., Lubchenco, J., Ostrom, E., Ouyang, Z., Provencher, W., Redman, C.L., Schneider, S.H., and Taylor, W.W. (2007) Complexity of coupled human and natural systems. *Science* 317: 1513–1516.

López, S. and R. Sierra. (2010) Agricultural change in the Pastaza River Basin: a spatially explicit model of native Amazonian cultivation. *Applied Geography* 30: 355–369.

McKey, D., Cavagnaro, T.R., Cliff, J., and Gleadow, R. (2010) Chemical ecology in coupled human and natural systems: people, manioc, multitrophic interactions and global change. *Chemoecology* 20: 109–133.

McNaughton, S.J. (1983) Serengeti grassland ecology: the role of composite environmental factors and contingency in community organization. *Ecological Monographs* 53: 291–320.

Maheu-Giroux, M., Casapía, M., Soto-Calle, V.E., Ford, L.B., Buckeridge, D.L., Coomes, O.T., and Gyorkos, T.W. (2010) Risk of malaria transmission from fish ponds in the Peruvian Amazon. *Acta Tropica* 115: 112–118.

Mark, B.G., Bury, J., McKenzie, J.M., French, A., and Baraer, M. (2010) Climate change and tropical Andean glacier recession: evaluating hydrologic changes and livelihood vulnerability in the Cordillera Blanca, Peru. *Annals of the Association of American Geographers* 100: 794–805.

Millennium Ecosystem Assessment. (2005) *Ecosystems and Human Well-Being: Health Synthesis*. Washington, DC: Island Press.

Montgomery, M.R. (2008) The urban transformation of the developing world. *Science* 319: 761–764.

Nepstad, D.C., Stickler, C.M., Soares-Filho, B., and Merry, F. (2008) Interactions among Amazon land use, forests and climate: prospects for a near-term forest tipping point. *Philosophical Transactions of the Royal Society B* 363: 1737–1746.

Nielsen, J.Ø. and Reenberg, A. (2010) Temporality and the problem of singling out climate as a current driver of change in a small west African village. *Journal of Arid Environments* 74: 464–474.

Ogutu, J.O., Piepho, H.-P., Reid, R.S., Rainy, M.E., Kruska, R.L., Worden, J.S., Nyabenge, M., and Hobbs, N.T. (2010) Large herbivore responses to water and settlement in savannas. *Ecological Monographs* 80: 241–266.

Patz, J.A., Graczyk, T.K., Geller, N., and Vittor, A.Y. (2000) Effects of environmental change on emerging parasitic diseases. *International Journal for Parasitology* 30: 1395–1405.

Patz, J.A., Campbell-Lendrum, D., Holloway, T., and Foley, J.A. (2005) Impact of regional climate change on human health. *Nature* 438: 310–317.

Pickett, S.T.A., Cadenasso, M.L., and Grove, J.M. (2005) Biocomplexity in coupled natural–human systems: a multidimensional framework. *Ecosystems* 8: 225–232.

Polk, M.H., Young, K.R., and Crews-Meyer, K.A. (2005) Biodiversity conservation implications of landscape change in an urbanizing desert of southwestern Peru. *Urban Ecosystems* 8: 313–334.

Poveda, G., Rojas, W., Quiñones, M.L., Vélez, I.D., Mantilla, R.I., Ruiz, D., Zuluaga, J.S., and Rua, G.L. (2001) Coupling between annual and ENSO timescales in the malaria-climate association in Colombia. *Environmental Health Perspectives* 109: 489–493.

Pringle, R.M., Doak, D.F., Brody, A.K., Jocqué, R., and Palmer, T.M. (2010) Spatial pattern enhances ecosystem functioning in an African savanna. *PLOS Biology* 8(5): e1000377 (doi:10.1371/journal.pbio.1000377).

Pueyo, S., Lima de Alencastro Graca, P.M., Imbrozio Barbosa, R., Cots, R., Cardona, E., and Fearnside, P.M. (2010) Testing for criticality in ecosystem dynamics: the case of Amazonian rainforest and savanna fire. *Ecology Letters* 13: 793–802.

Roshanravan, B., Kari, E., Gilman, R.H., Cabrera, L., Lee, E., Metcalfe, J., Calderon, M., Lescano, A.G., Montenegro, S.H., Calampa, C., and Vinetz, J.M. (2003) Endemic malaria in the Peruvian Amazon region of Iquitos. *American Journal of Tropical Medicine and Hygiene* 69: 45–52.

Ruiz, D., Moreno, H.A., Gutiérrez, M.E., and Zapata, P.A. (2008) Changing climate and endangered high mountain ecosystems in Colombia. *Science of the Total Environment* 398: 122–132.

Sayer, E.J., Powers, J.S., and Tanner, E.V.J. (2007) Increased litterfall in tropical forests boosts the transfer of soil CO_2 to the atmosphere. *PLOS One* 2(12): e1299 (doi:10.1371/journal.pone.0001299).

Sitch, S., Huntingford, C., Gedney, N., Levy, P.E., Lomas, M., Piao, S.L., Betts, R., Ciais, P., Cox, P., Friedlingstein, P., Jones, C.D., Prentice, I.C., and Woodward, F.I. (2008) Evaluation of the terrestrial carbon cycle, future plant geography and climate-carbon cycle feedbacks using five Dynamic Global Vegetation Models (DGVMs). *Global Change Biology* 14: 2015–2039.

Thompson, L.G., Mosley-Thompson, E., Brecher, H., Davis, M., León, B., Les, D., Ping-Nan Lin, Mashiotta, T., and Mountain, K. (2006) Abrupt tropical climate change: past and present. *Proceedings of the National Academy of Science* 103: 10536–10543.

Tong, S., Mather, P., Fitzgerald, G., McRae, D., Varrall, K., and Walker, D. (2010) Assessing the vulnerability of eco-environmental health to climate change. *International Journal of Environmental Research and Public Health* 7: 546–564.

Vittor, A.Y., Gilman, R.H., Tielsch, J., Glass, G., Shields, T., Sánchez Lozano, W., Pinedo-Cancino, V., and Patz, J.A. (2006) The effect of deforestation on the human-biting rate of *Anopheles darlingi*, the primary vector of *Falciparum* malaria in the Peruvian Amazon. *American Journal of Tropical Medicine and Hygiene* 74: 3–11.

Wandiga, S.O., Opondo, M., Olago, D., Githeko, A., Githui, F., Marshall, M., Downs, T., Opere, A., Yanda, P.Z., Kangalawe, R., Kabumbuli, R., Kirumira, E., Kathuri, J., Apindi, E., Olaka, L., and Kirumira, E.K. (2010) Vulnerability to epidemic malaria in the highlands of Lake Victoria basin: the role of climate change/variability, hydrology and socio-economic factors. *Climatic Change* 99: 473–497.

Winters, A.M., Staples, J.E., Ogen-Odoi, A., Mead, P.S., Griffith, K., Owor, N., Babi, N., Enscore, R.E., Eisen, L., Gage, K.L., and Eisen, R.J. (2009) Spatial risk models for human plague in the west Nile region of Uganda. *American Journal of Tropical Medicine and Hygiene* 89: 1014–1022.

Young, K.R. (2007) Causality of current environmental change in tropical landscapes. *Geography Compass* 1: 1299–1314.

Young, K.R. (2008) Stasis and flux in long-inhabited locales: Change in rural Andean landscapes. In *Land-Change Science in the Tropics: Changing Agricultural Landscapes*, A. Millington and W. Jepson (eds). New York, NY: Springer, 11–32.

Young, K.R. (2009) Andean land use and biodiversity: humanized landscapes in a time of change. *Annals of the Missouri Botanical Garden* 96: 492–507.

Young, K.R. and Lipton, J.K. (2006) Adaptive governance and climate change in the tropical highlands of western South America. *Climatic Change* 78: 63–102.

Part II

Empirical approaches to injury and infectious disease

5 Buruli ulcer disease

The unknown environmental and social ecology of a bacterial pathogen

Jiaguo Qi, Lindsay P. Campbell, Jenni van Ravensway, Andrew O. Finley, Richard W. Merritt, and M. Eric Benbow

Introduction

Mycobacterium ulcerans *and Buruli ulcer disease*

The scientific community recognizes that anthropogenic impacts on the natural environment—including land use alteration, human migration, habitat encroachment and wildlife translocation, rapid transport, and climate change—play an important role in the distribution of emerging and re-emerging infectious diseases (Confalonieri 2005; Foley *et al.* 2005; Patz *et al.* 2000; Patz and Confalonieri 2005; Wilcox and Gubler 2005). The growing field of landscape epidemiology addresses these problems using ecological and epidemiological data, creating opportunities to mitigate disease incidence in humans based on an understanding of disease vector and reservoir habitats in the environment (Galuzo 1975; Pavlovski 1966; Reisen 2010). A major component of the landscape epidemiological approach is the employment of geospatial data and methodologies. While basic geospatial concepts have been utilized in epidemiology for centuries (e.g., mapping of plague outbreaks in the 1600s; Koch 2005), advances in geographic information systems (GIS) and remote sensing technologies enable studies to take place across vast geographic regions over extended time periods. These technologies, along with research approaches that incorporate both ecological and human epidemiological data, promise continued advancements within the discipline.

A growing body of literature documents relationships between anthropogenic impacts on the natural environment and vector-borne and zoonotic disease emergence (Patz *et al.* 2000; Wilcox and Ellis 2006), but fewer studies investigate human interactions with the ecological niches of environmental bacterial pathogens and resulting disease incidence. Identifying and/or analyzing bacterial pathogens in the environment and linking their presence to human disease incidence pose unique challenges, particularly when the ecological niche is unknown.

An example of such a pathogen is *Mycobacterium ulcerans*, the causal agent of Buruli ulcer (BU) disease. BU, which has been reported to be within the top three most frequent human mycobacterial diseases (Merritt *et al.* 2010), is a rapidly emerging yet neglected tropical disease. Symptoms range from a painless, mobile nodule underneath the skin to, if left untreated, large skin ulcerations and sometimes osteomyelitis, a severe bone infection (World Health Organization 2007). Treatment with antibiotics can be successful in the pre-ulcerative stage of the disease; however, extensive ulcerations require invasive surgery, may cause deformities, and in severe cases require amputation of ulcerated limbs (Chauty *et al.* 2007).

BU is a lesser-known emerging infectious disease that is endemic in at least 32 countries globally (Figure 5.1; Merritt *et al.* 2010), with the majority of disease incidence taking place in sub-Saharan Africa. Although clinically recognized since the 1930s (McCallum *et al.* 1948), a substantial increase in BU prevalence over recent decades, particularly in west Africa and Australia, prompted major concern among researchers and those living in endemic areas. Despite extensive BU research efforts over the last decade, the causes of increased disease incidence are still unknown, a reality that is both alarming to the general public and a source of frustration within the BU community. Interestingly, climatic and socio-economic conditions differ substantially between Australia and most west African countries, leading to the assumption that the disease is likely related to interactions among humans and the natural environment. Currently, however, little is known regarding the ecological niche of the pathogen, the transmission mechanism(s), or the socio-behavioral activities that put humans at risk.

While anthropogenic impacts on the natural environment—specifically, landscape disturbance and potential contributions of climate change on disease distribution—have not been linked to BU emergence directly, research efforts have led to the development of several hypotheses suggesting these activities play a role in present and future distributions of disease incidence. Consequently, identification of the social and ecological drivers behind this disease continues to be the primary focus for much of the BU research community, aiming to mitigate the devastating impacts the disease has across the globe.

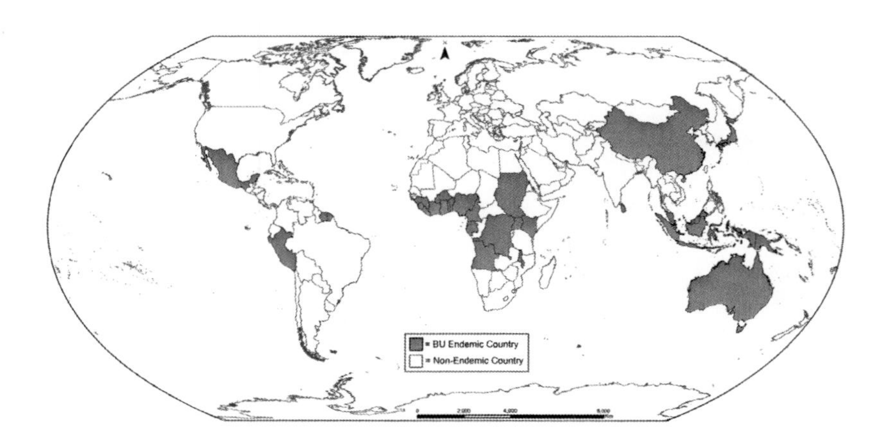

Figure 5.1 Countries with historical or current reports of Buruli ulcer cases.

Source: World Health Organization (2011)

In 1998, the World Health Organization established the Global Buruli Ulcer Initiative with the goals of raising disease awareness, improving treatment access, strengthening surveillance, and providing priority research into disease diagnosis, treatment and prevention (World Health Organization 2007). From a policy standpoint, a focus on prevention using a landscape epidemiological approach that incorporates an understanding of the natural ecology of the pathogen along with human activities that influence its distribution will play a critical role in the mitigation of future BU morbidity.

The focus of this chapter is to discuss the social and ecological components of BU, beginning with a review of social impacts from the disease, challenges in BU research, and pathogen ecology, followed by environmental, socio-behavioral, and coupled human–environmental risk factors. The chapter includes two ongoing case studies, in Victoria (Australia) and Benin (west Africa), that investigate current hypotheses in BU research. In both studies, unknown environmental and socio-behavioral components contribute to BU prevalence, highlighting the need for additional investigations into these factors at broader scales. The chapter closes with a review of the current state of BU research and recommendations for future investigations. Future directions include taking a spatio-temporal analysis approach across broad geographic regions, incorporating climate, landscape, and socio-behavioral components for a more holistic assessment of the drivers behind BU disease. Although the focus of this chapter relates to BU, the challenges encountered and approaches taken to overcome these obstacles may be generalizable to other environmental pathogen investigations.

Social impacts of Buruli ulcer

The rise in BU cases over the last 30 years resulted in considerable socio-economic hardships for endemic communities, especially those within developing countries (World Health Organization 2007). Fortunately, early detection and access to healthcare, along with fewer overall cases, prevented BU from becoming a more serious public health problem in Australia. However, in Victoria, Australia the average cost of treatment in 1997–1998 was nearly seven times the average health expenditure per person, a substantial financial burden, and BU diagnosis frequently resulted in plastic surgical procedures (Drummond and Butler 2004).

Those living in developing countries face greater challenges when seeking treatment, and the majority of cases received at medical facilities are in advanced disease stages (World Health Organization 2000). In west Africa, persons presenting with BU often require extensive surgery and hospital stays, creating economic challenges for families of patients (Aujoulat *et al.* 2003). The World Health Organization (2007) reported that in Ghana the cost of treating a pre-ulcerative nodule in a patient is comparable to 16 percent of the average total family income for the work-year; the cost of treating a patient requiring amputation is equal to 89 percent of the average total family income for the work-year; and the average cost of treating a patient in 1994–1996 exceeded the per capita government spending on healthcare. Compounding the problem is income loss due to incapacitation from the disease or from caring for a family member with BU, and families face additional challenges involving prolonged care for members rendered permanently disabled. Fear of surgery, long distances to treatment facilities, and socio-economic factors contribute to the rural poor being affected most negatively by BU (Raghunathan *et al.* 2005). For an extended discussion of long-term disease impacts on both family and the ability to invest in social and other capitals, see Chapters 3 and 14.

Beyond prohibitive treatment costs and access to medical facilities, a stigma associated with BU exists in many west African countries due to a belief that the infection is a result of witchcraft. In Ghana, Stienstra *et al.* (2002) found that persons without BU often avoided patients suffering from the disease, and that patients may hide symptoms from community members, increasing their risk of serious infection and need for extensive medical care. Some attribute the disease to a lack of patient hygiene, while others believe it is result of a curse and therefore do not consider it to be a "hospital disease" (Stienstra *et al.* 2002). Subsequently, many believe BU should be treated with traditional healing methods, thus preventing patients from seeking necessary medical treatment.

Challenges in Buruli ulcer research

While BU is not transmitted from person to person, the occurrence of an ulcer at the site of previous skin trauma in some patients suggests a direct mode of transmission from the environment (Meyers *et al.* 1974, 1996). Sufficient literature suggests that a range of environmental conditions favor *M. ulcerans* growth, and when humans contact these environments, BU cases occur. However, which environmental attributes—climate, landscape, and socio-behavioral—are specifically responsible for promoting *M. ulcerans* population growth, and in what habitats, have yet to be determined. Unknown lag times between environmental conditions suitable for *M. ulcerans* proliferation and the time when the necessary abundance for inoculation is reached pose challenges in determining when the greatest risk for infection takes place. Further, the human incubation period for BU appears to vary by individual (e.g., from two weeks to seven months; Johnson *et al.* 2007; Quek *et al.* 2007b; Tiong 2005; Veitch *et al.* 1997), as do lag times between symptom presentation and health-seeking behaviors, making time periods of likely transmission difficult to identify.

Additional obstacles exist when investigating BU. One of the most substantial is the inconsistency in the availability of georeferenced, epidemiological data sets across disease-affected regions. Under the Global Buruli Ulcer Initiative framework, a standardized reporting form (referred to as the World Health Organization "BU02 form") served to facilitate a more comprehensive reporting structure across international boundaries in BU endemic regions (World Health Organization 2000). Trained personnel and community volunteers report suspected BU cases to qualified nurses who record confirmed cases using standardized BU02 forms (Sopoh *et al.* 2007). Cases are then aggregated at regional and national levels to monitor the large-scale pattern of disease incidence.

Despite these advances in reporting structure, inconsistencies remain. For example, in some regions, case locations are reported to the town that houses the treatment facility rather than to the town in which the patient resides, introducing uncertainties in the true geographic distribution of the disease. Further, variability in available resources across geographic regions produces disparities in case reporting that impact west African countries disproportionately. In addition to being a developed country with a standardized reporting system, Australia comprises a single continent that is isolated from other BU endemic regions, making the accurate and consistent reporting of BU cases less complicated than in other regions of the world. The transboundary nature of BU in west Africa, in addition to the developing status of each nation that constitutes the region, contributes to a knowledge gap in the contiguous distribution of BU cases in the region.

Challenges also exist in the acquisition of spatially referenced environmental data sets, particularly in developing regions where the majority of disease incidence occurs. For example, continuous, long-term weather station data typically are not available in developing regions due to a lack of infrastructure for reliable data collection and dissemination. In addition, although remotely sensed environmental observations may be available in developing regions, many of these areas are inaccessible to researchers for ground validation due to political or physical barriers. On the other hand, although several environmental data sets are readily accessible in more developed regions, such as Australia, fewer overall cases occurring within smaller geographic regions make it difficult to elucidate the drivers behind the disease.

Despite these challenges, substantial research efforts into the natural, social, and social/environmental ecology of BU have taken place. Major contributions are outlined in the following sections, including two unique ongoing case studies performed by the authors, described in more detail.

Disease ecology

Ecology of Mycobacterium ulcerans

As BU is caused by infection with *M. ulcerans*, research efforts have focused primarily on pathogen reservoirs and transmission mechanisms. *M. ulcerans* is a slow-growing environmental mycobacterium (Merritt *et al.* 2005), considered unusual because of its extracellular activity and mycolactone secretion (George *et al.* 1999). Additional attributes set *M. ulcerans* apart from other related mycobacterial species, particularly that it prefers lower temperatures and has a narrower temperature range for survival than its counterparts (Merritt *et al.* 2010). Further, *M. ulcerans* is intolerant to ultraviolet (UV) light because it lacks pigments found in related mycobacteria, indicating that the pathogen may be protected from sunlight in its natural environment (Stinear *et al.* 2007).

M. ulcerans DNA has been detected in various aquatic environments, insects, fish, and mammals (for a complete review see Merritt *et al.* 2010), although the environmental niche of the pathogen is unknown. Speculation into insects as potential mechanical or biological vectors prompted researchers to investigate numerous aquatic insect species. In west Africa, *M. ulcerans* DNA was detected in certain biting insects (Portaels *et al.* 1999), and the pathogen was able to survive and reproduce in some biting species as well (Marsollier *et al.* 2002; Marion *et al.* 2010). Despite these findings, no evidence surfaced that demonstrated the insects passing the pathogen to humans (Portaels *et al.* 1999). Further research, however, led scientists to examine the potential for trophic transfer of *M. ulcerans* through the environment due to the discovery of *M. ulcerans* colonization on the exoskeleton of certain insects (Mosi *et al.* 2008). The ability for trophic transfer was later confirmed in a laboratory setting, with aquatic insects as the tertiary consumers (Wallace *et al.* 2010). On the other hand, Benbow *et al.* (2008) found no evidence to link the same aquatic insects with *M. ulcerans*. The conflicting findings regarding the role of invertebrates in BU transmission illustrate the need for further insect-related transmission research (Benbow *et al.* 2008; Marion *et al.* 2010; Merritt *et al.* 2010).

Mosquitoes are believed to be a potential vector for *M. ulcerans* transmission in Australia. Speculation began after discovering that areas with BU outbreaks in Victoria, where most Australian cases occur, also had large mosquito populations (Johnson *et al.* 2007). The most common coastal mosquito species in Victoria, *Aedes camptorhynchus* (Dhileepan *et al.* 1997), tested positive for *M. ulcerans* DNA, although it was not determined whether they harbored *M. ulcerans* within their bodies or acquired the pathogen externally from the environment (Johnson *et al.* 2007). However, recent laboratory experiments demonstrated the capacity for mosquito larvae to accumulate *M. ulcerans* in their mouths and midguts over four stages of larval development (Tobias *et al.* 2009). In additional experiments, pathogen DNA was detected in numerous adult mosquitoes that had been infected during the larval stage, although the actual bacteria did not persist (Wallace *et al.* 2010), casting doubt on the hypothesis that mosquitoes may be biological vectors of *M. ulcerans*. On the other hand, Wallace *et al.* (2010) found evidence suggesting the transfer of *M. ulcerans* DNA to external components of adult mosquitoes through feeding processes, offering support of mosquitoes as mechanical rather than biological vectors.

Additionally, researchers discovered that combined notifications of two mosquito-transmitted viruses (Ross River Virus and Barmah Forest Virus) were positively correlated with notifications of BU over a seven-year period (Johnson and Lavender 2009). Although this association may be purely coincidental, no other notifiable infectious disease in Victoria had any correlation with BU (Johnson and Lavender 2009). Further, human BU infections often occur on limbs and have been reported at sites of previous skin trauma, such as an insect bite (Meyers *et al.* 1974, 1996). To that end, a study including residents of Point Lonsdale found those who had mosquito bites on their extremities were at an increased risk for BU infection (compared to other insects bites), while those who wore insect repellent had a significantly lower infection risk (Quek *et al.* 2007a). Although many of these studies appear to support the mosquito hypothesis, no vector competency studies have been published to date, which is a necessary criterion for identifying and describing a possible vector (Hill 1965; Plowright *et al.* 2008; Merritt *et al.* 2010).

Several investigations into potential *M. ulcerans* reservoirs have been conducted. Numerous mammals in Australia, including koalas, alpaca, a domestic cat, potaroos, horses, black rats, and brushtail and ringtail possums all tested positive for *M. ulcerans* (Mitchell *et al.* 1984; Elsner *et al.* 2008; van Zyl *et al.* 2009). A recent study in Australia also detected *M. ulcerans* in ringtail and brushtail possum feces in a BU endemic region, suggesting the possums contribute to maintenance of the bacterium in the environment (Fyfe *et al.* 2010). Conversely, *M. ulcerans* was not identified in any of the hundreds of rodents and shrews collected in Benin during the wet and dry seasons in areas with high and low BU endemicity (Durnez *et al.* 2010). Finally, some researchers believe that migratory birds aid in disseminating *M. ulcerans* between wetlands, but further investigation is needed to determine the potential of this hypothesis (Eddyani *et al.* 2004). Despite continued advances in *M. ulcerans* research, a definitive mode of transmission and the ecological niche of the pathogen remain unknown (Merritt *et al.* 2010).

Environmental risk factors

Environmental characteristics are typically dynamic over time, often exhibiting inter-annual variability or following temporal trends. These ecosystem changes can affect disease host or vector populations, modifying disease outbreak potential (Smith *et al.* 2005). In addition, landscape interactions with climate factors, particularly precipitation and temperature variations, can impact optimal pathogen, reservoir, or vector habitats (World Health Organization 2004). Therefore, the spatio-temporal pattern of climate and land use and land cover (LULC) interactions is a key factor in the emergence of infectious diseases in human and wildlife populations (Farnsworth *et al.* 2005; Smith *et al.* 2005).

The detection of *M. ulcerans* in aquatic environments and the occurrence of BU cases surrounding slow moving or stagnant water bodies or wetlands, prompted speculation into flooding as a potential risk factor for the disease (World Health Organization 2000). Anecdotal evidence from both Australia and West Africa supports this hypothesis. The first presentation of BU infection in Bairnsdale, Australia occurred two to three years after a major flooding event in 1935 (Hayman 1991). In 1978, Bairnsdale had its wettest year in recorded history, and two years later BU cases emerged. Additionally, in 1962 and 1964 severe flooding took place in Uganda, and approximately two to three years later, BU cases surfaced (Dobos *et al.* 1999). The first reported BU cases in Togo occurred in two children living near two separate rivers that experienced seasonal flooding (Meyers *et al.* 1996). In Papua New Guinea, BU cases were reported near the Sepik and Kumusi Rivers, where a spike in incidence occurred following the flooding and widespread environmental destruction from the Mount Lamington volcanic eruption in 1951 (World Health Organization 2000). While observed linkages between flooding events and BU cases exist, quantitative relationships between climate patterns and BU cases have not been explored in west Africa or Australia.

A substantial increase in disease incidence in Victoria, Australia after 1980 raised awareness of BU, initiating several Australian studies over the last two decades. However, previous BU investigations in Australia adopted a localized approach, often focusing on specific outbreaks in Victoria, resulting in a limited understanding of the potential transmission mechanism(s). These studies did not consider the variability in disease incidence over time and also failed to capture landscape-level characteristics, such as land cover, topography and climate conditions, which may differ substantially among BU endemic areas.

Socio-behavioral risk factors

The greatest risk factor for contracting BU is contact with an endemic area (Johnson *et al.* 2007), although certain behavioral patterns may increase risk. Numerous studies examined social and behavioral risk factors related to BU, but results often contradict one another and vary across geographic regions. A matched case-control study in three endemic districts of Ghana confirmed researchers' hypotheses that BU is an environmentally acquired infection associated with rivers and streams (Raghunathan *et al.* 2005). The results also indicated that wading in a river or stream was a risk factor, that using certain soap products and wearing long trousers was protective against BU, and that wearing clothing that covered the upper body while farming was also protective against the disease. In a separate study conducted in Benin, wearing clothing during farming activities was not identified as a protective factor against BU infection, contact with stagnant water increased BU risk, while contact with flowing water and using soap decreased BU risk (Nackers *et al.* 2007). Aiga *et al.* (2004) found that swimming in and water use from rivers were BU risk factors, although water use from piped sources was not a protective factor in Ghana. Conversely, Marston *et al.* (1995) found that swimming in rivers was not a BU risk factor in Côte d'Ivoire. Despite conflicting results, the majority of studies agree that contact with rivers and other water bodies increase the risk of human infection.

Socio-behavioral studies provide invaluable insight into individual risk factors and potential transmission mechanisms, but one limitation is the localized nature of each study. The following section outlines research into human and environmental interactions and their relationship to BU incidence, with a specific focus on activities that impact the landscape across broader scales.

Human–environmental disturbance

BU cases have been associated with both natural and disturbed environments, but human-induced landscape alterations are a recent focus in transmission research. With increasing human expansion into undisturbed landscapes and rapid LULC change, new pathways may be created through which disease pathogens infect susceptible human hosts. Empirical and anecdotal linkages between landscape alterations (e.g., deforestation, agricultural activities, dam creation) and BU emergence exist in west Africa and Australia, promoting speculation about *M. ulcerans* distribution in the environment (Merritt *et al.* 2010).

Hayman (1991) hypothesized that closed rainforests in Victoria contain *M. ulcerans*, and disruption of the rainforest from continued deforestation results in *M. ulcerans*-contaminated runoff. Contaminated runoff could lead to higher concentrations of *M. ulcerans* in water bodies, resulting in a bacteria bloom under favorable environmental conditions (Hayman 1991). In addition, a study in Benin found villages at higher risk for BU were situated within flood-prone landscapes surrounded by an agricultural matrix, where deforestation likely took place (Wagner *et al.* 2008a). Deforestation typically results in soil erosion and increased runoff, creating ideal conditions for nutrient introduction into water bodies, especially where agriculture is present. This process can lead to higher water temperatures, decreased oxygen levels, increased turbidity, and eutrophic conditions, all of which create a habitat that may be suitable to *M. ulcerans* (Merritt *et al.* 2005).

Linkages also exist between BU cases and agricultural environments, with BU risk increasing as percentage agriculture increased within a 20 km buffer surrounding village centers in Benin (Wagner *et al.* 2008b). Cases emerged in Liberia after the replacement of an upland rice field with a swamp rice field (World Health Organization 2000). In addition, irrigated rice fields and banana fields corresponded to high-risk BU zones in Côte d'Ivoire (Brou *et al.* 2008). While sampling for *M. ulcerans* presence did not take place in these studies, cultivated rice fields could provide suitable habitat for the pathogen to thrive, due to their nutrient-rich, semi-aquatic environment.

Several associations were found between water body disturbance, particularly dam creation, and increased BU cases. For example, students on a university campus in Nigeria began contracting BU after construction of a dam on a small river running through campus in order to create an artificial lake (Oluwasanmi 1976). In Liberia, BU cases emerged in areas where the creation of dams expanded wetland areas (World Health Organization 2000). Further, a recent study in Côte d'Ivoire found a relationship between BU rates and patients' proximities to dams (Brou *et al.* 2008). Similar findings have been noted in Australia. Between 1957 and 1958, and in 1962, BU cases occurred following the building of large dams for agriculture irrigation north of Brisbane, Australia (Abrahams and Tonge 1964).

Ongoing case studies

Although the specific transmission mechanism(s) and ecological niche of the pathogen remain unknown, clear indications suggest that BU incidence results from a complex relationship between natural and social ecologies. The current section outlines two ongoing case studies and their preliminary results conducted by the authors. The first case study takes place in Victoria, Australia and investigates climate conditions as a proxy for mosquito abundance and their relationship to disease incidence at a regional scale. As outlined previously, mosquitos are a suspected BU vector in Australia; therefore, exploring linkages between environmental conditions suitable to mosquito abundance and disease incidence could shed light on BU risk and help to promote future research directions and prevention strategies.

The second case study takes place in Benin, West Africa investigating potential anthropogenic landscape disturbances and their relationship to BU incidence. This study complements previous work conducted by Wagner *et al.* (2008a, 2008b) that quantified land cover composition related to BU incidence in the region. While previous studies investigated relationships between specific disturbances and BU incidence, this study will be the first to quantify relationships between BU and potential disturbance across a broad region. Exploring linkages between potential anthropogenic disturbances and BU incidence may help to uncover high-risk activities, leading to more targeted prevention strategies and more comprehensive field sampling efforts.

Victoria, Australia

In response to the lack of larger scale spatio-temporal BU studies in Australia, several authors of this chapter are currently performing a Victoria-wide investigation of the relationships among landscape and climate features with BU incidence across multiple years. Identifying the environmental attributes most related to disease emergence can provide insights into the plausibility of different vector and reservoir hypotheses proposed for this region. Moreover, the analysis includes climate conditions up to two years prior to each reported BU case, providing the researchers an opportunity to quantify lag-times between exposure and disease presentation if significant correlations between previous climate features and disease incidence are identified.

The Victoria study is utilizing a network approach (Dezső and Barabási 2002; Liljeros *et al.* 2001; Newman 2003; Watts 2004) as an initial step to analyze the structure of the BU disease network for 1981–2008. Preliminary results reveal a non-random pattern of BU incidence, suggesting the existence of regional factors driving BU emergence, thereby justifying a larger-scale, Victoria-wide analysis.

The authors are in the process of analyzing monthly BU incidence data in Victoria from 1981 to 2008 (Figure 5.2; Hayman 1991; Johnson *et al.* 1996, 2007; Veitch *et al.* 1997) at the local level in conjunction with corresponding climate and landscape features.

Climate observations play an essential role in this study, not only because regional climate features are an unexplored environmental component in BU research, but also because they act as a proxy for mosquito abundance due to a lack of available mosquito census data. Subsequently, the ongoing study utilizes these specific climate conditions to infer larger mosquito populations.

The Victoria analysis uses a multilevel statistical modeling approach (Bates 2005; Bates and Maechler 2009; Raudenbush and Bryk 2002) to account for the nested structure of the data (i.e., multiple BU cases nested within each locality). Preliminary results suggest that overall, climate features are the most significant predictors of BU incidence, specifically increased precipitation 19 months prior to case emergence, followed by warmer minimum temperatures (i.e., 18 months prior). The initial findings also indicate that localities with greater forest cover and a lower mean elevation have a higher BU risk. Interestingly, the climate conditions are consistent with those associated with greater *Ae. camptorhynchus*

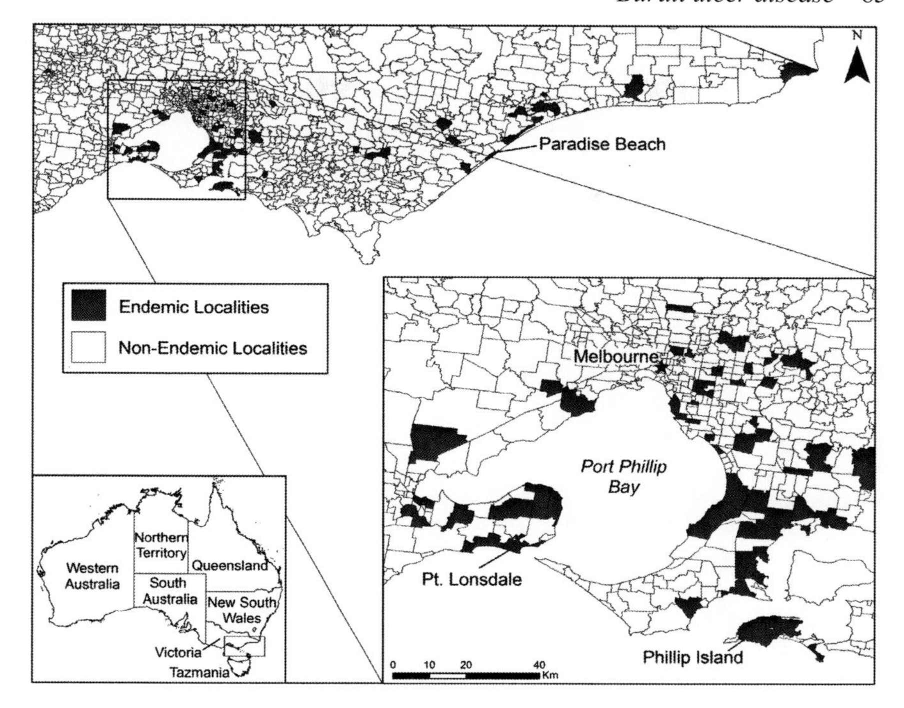

Figure 5.2 The distribution of endemic localities in Victoria, Australia (1981–2008).

populations (Barton *et al.* 2004), however, forests would not provide the ideal habitat for these mosquitoes to thrive. This is primarily because *Ae. camptorhynchus* mate while in flight and therefore need open spaces to successfully reproduce (Bader and Williams 2011) rendering forests as poor habitats for this species. Subsequently, the climate variables may represent something, other than mosquito abundance, interacting with forested land cover, such as flooding.

Previous studies in both Australia and west Africa hypothesized that flooding is a BU risk factor (Hayman 1991; Meyers *et al.* 1996), suggesting that these events could transport the pathogen into regions previously void of *M. ulcerans*. If, in fact, the pathogen flourishes within forested areas, the immediate increase in minimum temperatures following increased rainfall may facilitate *M. ulcerans* growth. Such a process may allow *M. ulcerans* to persist in these previously unoccupied areas, exposing new human populations to the pathogen. Unfortunately, this scenario offers little insight into human incubation periods because documentation does not exist regarding the occurrence of socio-behavioral activities with potentially contaminated environments. However, associations between disease incidence, low elevation, and increased forest cover were also found in west Africa (Pouillot *et al.* 2007; Wagner *et al.* 2008b) suggesting the possibility of similar habitats and transmission mechanisms in two environmentally distinct regions.

Although further analysis is necessary before the authors can offer thorough interpretation of the findings in Victoria, the preliminary results are particularly exciting as they are the first to identify a relationship between climate conditions and BU incidence. This is an area of BU research that has previously been unexplored, but can have serious implications for infectious disease. Over the next century Australia's population is projected to increase by as much as 3.0 percent (Australian Bureau of Statistics 2011), while its climate is expected to become hotter and drier, but with more frequent extreme rainfall events (Solomon *et al.* 2007). The resulting interactions of climate variability and landscape dynamics can provide optimal environmental conditions for disease pathogens and/or vectors (Epstein *et al.* 1998; Epstein 2001; World Health Organization 2004). Consequently, understanding the relationships between environmental factors and BU disease will help to mitigate future risk under changing climate conditions.

The final outcomes of the Victoria study will hopefully offer insights into the ecology of *M. ulcerans* and BU transmission related to climate and landscape features. While previous studies in West Africa examined socio-behavioral BU risk factors and utilized similar methods to analyze and to identify landscape associations with BU, difficulties exist in acquiring climate observations in this region, and therefore, climate factors have been excluded from these studies. The preliminary results from the Victoria study suggest that the inclusion of climate factors in future studies may play a critical role in elucidating social and/ or environmental risk factors in west Africa.

Benin, West Africa

Although previous BU studies investigated proximities of BU cases to anthropogenic landscape disturbances, for example artificial dams, irrigated agricultural plots, or gold mining facilities (Brou *et al.* 2008, Duker *et al.* 2006), these studies did not obtain a broad assessment of potential disturbance in BU endemic regions, prompting the ongoing landscape disturbance study in Benin. The purpose of this study is to quantify land cover patch shapes surrounding BU endemic and non-endemic communities in Benin to determine whether patches indicative of anthropogenic disturbance surround communities with higher BU rates.

Several patterns reflect potential anthropogenic disturbance when measuring land cover patch shapes. For example, more uniformly shaped patches correspond to human activities, such as the construction of roadways or agriculture plots (Krummel *et al.* 1987), while more complexly shaped patches represent undisturbed areas, such as natural wetlands (Iverson 1998). Overall, a higher number of patches suggest potential habitat fragmentation (Narumalani *et al.* 2004). The researchers hypothesize that more fragmented land cover patches with more uniform shapes surround communities with higher BU rates in Benin.

Three land cover categories, forest, wetland, and a mixed forest and agriculture class, are included in this study. Potential linkages between these categories and BU incidence as outlined in the literature provide the basis for class selection.

The second study objective is to determine whether a spatial pattern exists for the variables that contribute to the distribution of BU incidence in Benin. Although the ecological drivers remain unknown, identifying a spatial pattern corresponding to disease incidence would provide several advantages, including the potential to predict BU risk across a landscape from which disease incidence is unknown using spatial statistical methods (Waller and Gotway 2004).

This investigation utilizes epidemiological data provided by the Programme National de Lutte contre la Lèpre et l'Ulcère de Buruli (PNLLUB). The data set consists of a subset of 2004 and 2005 BU positive and BU negative villages in southern Benin (Figure 5.3). Forest, wetland and mixed forest and agricultural land cover patches surrounding villages within the study area were quantified using a land cover map derived from 30 m resolution Landsat satellite imagery, and a suite of landscape metrics measured land cover patch shape characteristics and numbers of patches (Gergel and Turner 2002). The study utilizes a Bayesian hierarchical modeling approach with random spatial effects (Waller and Gotway 2004) to analyze landscape metric values at several distances from village centers.

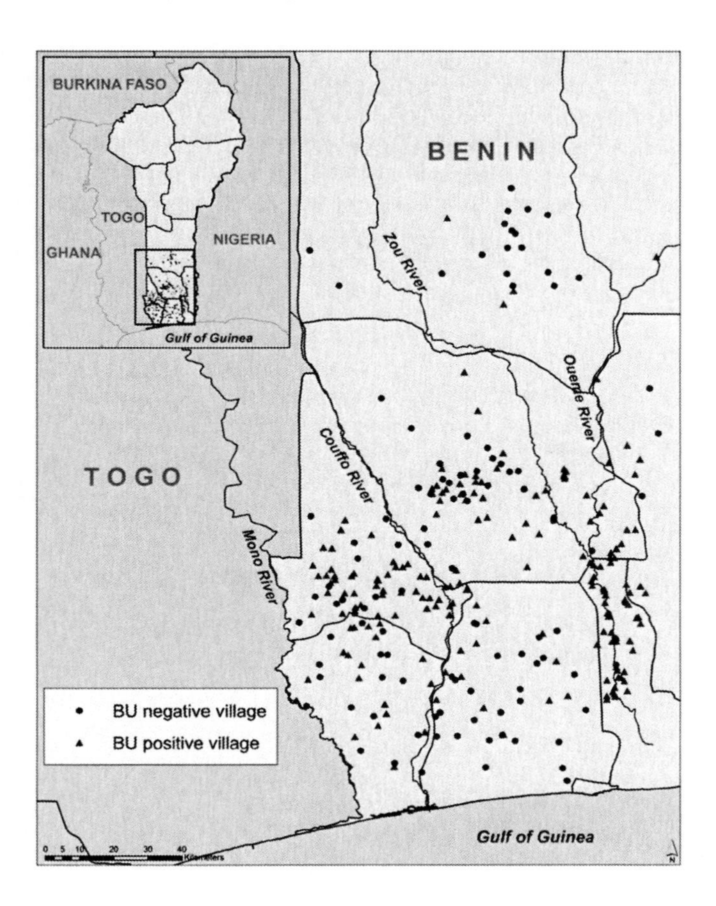

Figure 5.3 Distribution of BU-positive and BU-negative villages in Benin.

Preliminary results do not support the study hypotheses that land cover patches that indicate the presence of human disturbance surround communities with higher BU rates, but suggest that more complexly shaped and aggregated patches, representative of more natural landscapes, surround these communities. The preliminary best fitting model, identified using a root mean square error calculation (Table 5.1), indicates that more aggregated and natural wetland areas within a maximum distance of 1.2 km from village centers surround communities with higher BU rates. These results are consistent with previous research associating BU risk with wetland systems, but provide additional information suggesting that contiguous, undisturbed wetlands may be an important factor in BU risk.

Although the preliminary results are interesting, several factors may have impacted the model outcomes and need further consideration. *M. ulcerans* abundance is likely a dynamic phenomenon, and specific disturbance stages could play an important role in the distribution of the pathogen in the environment. For example, pathogen abundance may increase substantially immediately following a disturbance, such as the initial intrusion into a previously undisturbed forested area for agricultural purposes. Initial disturbances may present more subtle attributes that would not necessarily indicate anthropogenic disturbance under the patch shape analysis framework currently used in this study. Alternatively, secondary succession may play an important role in MU abundance. In this case, the potential exists for the measurements to indicate more natural patch shapes, even though previous disturbance could have an impact on within patch ecology. Future, real-time, quantitative polymerase chain reaction (PCR) genetic-typing applications have the potential to reveal relationships between pathogen abundance and environmental disturbance stages (Merritt *et al.* 2010).

Table 5.1 The root mean square error (RMSE) values between training and verification data sets used in the Bayesian hierarchical modeling approach for Benin (rates per 100,000 people)

Category	*RMSE*
800 m agriculture/forest	507.8119383
800 m forest	632.0489476
800 m wetland	272.2508152
1.2 km agriculture/forest	505.6643283
1.2 km forest	218.4935089
1.2 km wetland	167.053699
1.6 km agriculture/forest	341.9403388
1.6 km forest	491.8787969
1.6 wetland	245.8153601
2 km agriculture/forest	228.7093229
2 km forest	248.449902
2 km wetland	207.4421244

Further, the current study may need to include additional land cover types. The categories selected thus far represent an initial step toward quantifying potential anthropogenic landscape disturbance in the region. The inclusion of additional land cover types could provide a more holistic view of activities taking place and possible relationships to BU disease.

Another important factor may relate to the scale at which the current study is taking place. For example, higher-resolution satellite imagery revealed distinct, uniformly shaped squares, indicative of rice paddy cultivation, within an area identified as natural wetland in the present study. Employment of higher resolution satellite imagery to quantify patch shapes may provide important details corresponding to anthropogenic disturbances that were too subtle to detect using 30 m resolution Landsat imagery. Alternatively, the processes driving BU incidence may be occurring across a broader scale, and patterns representing these processes could be lost when taking a finer-resolution approach. Important questions remain regarding the scale at which to measure potential anthropogenic landscape disturbances and their relationships to BU incidence. Therefore, further investigations at multiple scales are a natural extension of the ongoing study.

Preliminary results support the second hypothesis that a spatial structure exists for drivers behind the pattern of BU incidence in Benin. Advantages of the spatial statistical method used in the current study include improved inference, insight into missing covariates, and increased accuracy and precision of predicted BU risk at unsampled locations (Finley *et al.*2008).

The parameters from the best-fitting landscape disturbance model (Table 5.1), identified using a root mean square error calculation, and the identified spatial structure of missing variables provides the foundation to construct a preliminary BU risk surface using a Bayesian kriging approach (Moyeed and Papritz 2002). The preliminary risk surface spans southern Benin and into the southern region of Togo where reliable BU incidence data is unavailable. One challenge is that reliable census data do not exist in the study region during 2004 and 2005. Therefore, the study area is divided into a grid with points placed at 5 km intervals, each with an assumed population of 100,000, in order to provide location data to predict BU risk.

The result is a preliminary risk surface (Figure 5.4) that identifies several regions with the potential for BU cases to emerge, with an assumption that people encounter these environments and that transmission takes place. Preliminary results demonstrate consistency with known endemic regions within Benin near the town of Tandji and between the Zou and Ouémé Rivers. Additional high-risk areas are predicted along the Couffo River in the west and the Ouémé River in the east. Floodplains exist in both of these regions with an extensive floodplain residing where the Ouémé and Zou Rivers join before draining into the Gulf of Guinea. Wetlands are present in the southern portion of the identified high-risk area where the river flows into Lake Nokoué before emptying into the Bight of Benin and also in the west near the Couffo River and the town of Tandji.

Initial model results predict lower rates along the coast, demonstrating consistency with low disease occurrence in this region (Sopoh *et al.* 2011). One

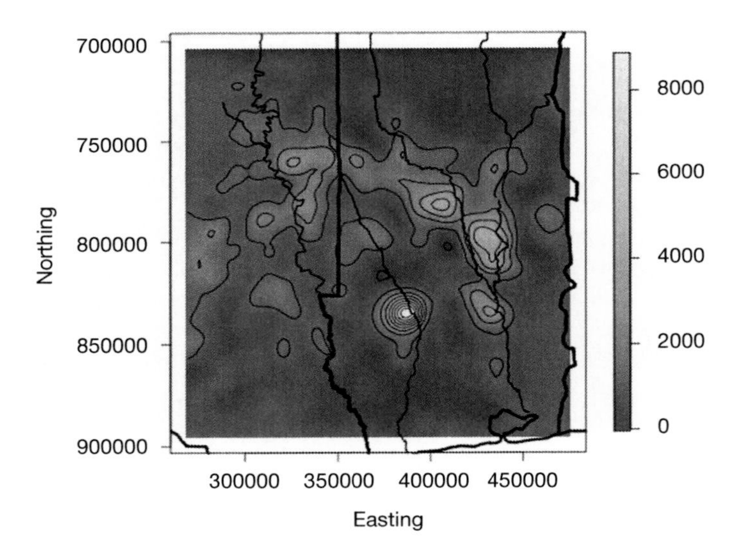

Figure 5.4 Preliminary BU risk surface for southern Benin and Togo (rates per 100,000 people).

hypothesis is that brackish waters may impact environmental suitability for *M. ulcerans* growth due to higher salinity values (Merritt *et al.* 2010). Additionally, model results suggest lower predicted rates in higher elevation regions falling between the Couffo and the Ouémé and Zou Rivers where BU cases do not occur. These results are consistent with a recent study conducted by Sopoh *et al.* (2011) that determined that BU risk increased as elevation decreased.

Preliminary model predictions within the boundary of Togo suggest moderate risk along the Mono River, near the location of Nangbeto Dam. While wetland systems exist in this region, few BU cases have been reported since the construction of the dam 1987 (R. Christian Johnson, personal communication, March 24, 2009). One hypothesis is that controlled water fluctuations, particularly those related to seasonal flooding, may contribute to altered environmental conditions that reduce habitat suitability for *M. ulcerans* (Merritt *et al.* 2010), but further investigation is needed to determine why cases no longer occur in this region.

Generally, the preliminary results suggest more moderate BU risk within the boundary of Togo compared to Benin, and these rates exhibit less variability. This phenomenon may be due to the distance of the predicted locations from the known prevalence locations. As distance increases from known data points, values tend to move toward a mean predictive value (Cressie 1993). Although this phenomenon may have impacted preliminary results within Togo, identification of regions at moderate risk for BU occurrence is a first step in bridging knowledge gaps stemming from data disparities in the region.

The purpose of the preliminary risk surface is not to predict actual transmission rates, but to identify regions at risk for transmission if persons encounter these areas based on environmental variables and the spatial structure of processes driving BU identified in the modeling process. Additionally, the natural environment is not a static phenomenon, nor is BU incidence. The risk surface interpolates values across the study region based on two years of BU case data and one LULC classification, although these components will continue to shift and to evolve over time. Further, reliable climate data at resolutions appropriate for the study are not available, limiting the identification of potentially important environmental variables contributing to BU incidence in the region, and reliable census data to which more precise risk rates could be predicted are not available at this time.

Despite these limitations, observing where BU transmission could likely take place if persons encountered similar environments provides a first step toward surveillance and prevention, while creating a foundation from which to target future environmental sampling and research efforts. Further, the ability to predict potential risk across international borders helps to break down barriers created by data disparities between regions due to lack of resources or infrastructure. These disparities affect the rural poor disproportionately, a large proportion of which are populations most vulnerable to disease.

Conclusions

Considerable progress within the last decade resulted in increased global awareness of BU through collaborations across scientific disciplines and international borders and through the development of innovative methods to examine all aspects of the disease from potential pathogen habitats to improved diagnostic tools in medical facilities. Despite this growing international effort, significant knowledge gaps remain in *M. ulcerans* ecology and BU transmission research.

M. ulcerans DNA detection in numerous environmental samples have yet to provide the information necessary to identify the pathogen's ecological niche. Without additional knowledge, the identification of specific environmental risk factors will remain challenging to discern. Examinations of behavioral risk factors often reveal contrasting results, confounding potential insights into disease transmission mechanisms. Further, limitations in epidemiological and environmental data sources, particularly in developing regions, limit advanced modeling approaches that might lead to additional insight into BU ecology.

The two ongoing research projects outlined in this chapter represent novel investigations in BU research. Initial results from the Australian study indicate the regional disease network in Victoria contains a non-random structure, suggesting that external BU drivers exist at this scale. In addition, preliminary findings indicate that specific climate conditions may be significant predictors of disease emergence. While initial results from the landscape disturbance do not indicate that land cover representative of anthropogenic disturbance surround communities with higher BU rates, additional investigations at multiple scales could reveal important patterns. Importantly, preliminary findings suggest that a spatial pattern exists for drivers of BU incidence in Benin. Therefore, the application of spatial statistical methods may be used to predict risk into remote regions and into territories for which incidence data are not available. While both studies have encountered obstacles in their analyses, the application of temporal and of spatial statistical methods that incorporated data across broader regions promise to result in new contributions to BU research.

A logical next step in BU research is the combination of a spatio-temporal modeling approach that incorporates both environmental and socio-behavioral data. Continued efforts toward a central, standardized reporting policy and cooperation throughout the international community will be a critical component to this research. Such a framework would provide the tools necessary to better understand the social and ecological drivers behind BU incidence, affording an opportunity to learn more about the individual disease system. In addition, this approach has the potential to better predict BU risk in remote regions with limited resources, where the most vulnerable populations reside.

In the face of a changing climate and subsequent ecosystem responses occurring at a global scale, it is imperative that broad-scale approaches in disease ecology research be maintained and continually improved. Equally important, is the exploration and consideration of indirect consequences resulting from human impacts on the environment as potential catalysts for disease emergence. Advances in remote sensing technology, along with the increased quantity and accessibility of remotely sensed imagery, provide a unique and exciting opportunity to explore these hypotheses from global to local scales. Employment of this approach will provide the foundation necessary for comprehensive, geospatial analyses that incorporate climate, landscape, and socio-behavioral variables for a holistic approach to disease ecology research.

References

Abrahams, E.W. and Tonge, J.I. (1964) *Mycobacterium ulcerans* infection in Queensland. *Medical Journal of Australia* 1: 334–335.

Aiga, H., Amano, T., Cairncross, S., Domako, J.A., Nanas, O., and Colemen, S. (2004) Assessing water-related risk factors for Buruli ulcer: a case-control study in Ghana. *American Journal of Tropical Medicine and Hygiene* 71(4): 387–392.

Aujoulat, I., Johnson, C., Zinsou, C., Guedenon, A., and Portaels, F. (2003) Psychosocial aspects of health seeking behaviours of patients with Buruli ulcer in southern Benin. *Tropical Medicine and International Health* 8: 750–759.

Australian Bureau of Statistics. (2011) Population projections, Australia, 2006 to 2011. See www.abs.gov.au/Ausstats/abs@.nsf/mf/3222.0 (accessed 7 October 2011).

Bader, C.A. and Williams, C.R. (2011) Eggs of the Australian saltmarsh mosquito, *Aedes camptorhynchus*, survive for long periods and hatch in instalments: implications for biosecurity in New Zealand. *Medical and Veterinary Entomology* 25(1): 70–76 (doi:10.1111/j.1365–2915.2010.00908.x).

Barton, P.S., Aberton, J.G., and Kay, B.H. (2004) Spatial and temporal definition of *Ochlerotatus camptorhynchus* (Thomson) (Diptera: Culicidae) in the Gippsland Lakes system of eastern Victoria. *Australian Journal of Entomology* 43: 16–22.

Bates, D. (2005) Fitting linear mixed models in R. *R News* 5(1): 27–30. See http://cran.r-project.org/doc/Rnews/Rnews_2005-1.pdf (accessed 17 May 2011).

Bates, D. and Maechler, M. (2009) lme4: linear mixed-effects models using S4 classes. R package version 0.999375–32. See http://CRAN.R-project.org/package=lme4 (accessed 17 May 2011).

Benbow, M.E., Williamson, H., Kimbirauskas, R., McIntosh, M.D., Kolar, R., Quaye, C., Akpabey, F., Boakye, D., Small, P., and Merritt, R.W. (2008) Aquatic invertebrates as unlikely vectors of Buruli ulcer disease. *Emerging Infectious Diseases* 14(8): 1247–1254. See www.cdc.gov/eid/content/14/8/1247.htm (accessed 17 May 2011).

Brou, T., Broutin, H., Elguero, E., Asse, H., and Guegan, J. (2008) Landscape diversity related to Buruli ulcer disease in Côte d'Ivoire. *PLOS Neglected Tropical Diseases* 2(7): e271.

Chauty, A., Ardant, M.F., Adeye, A., Euverte, H., Guedenon, A., Johnson, C., Aubry, J., Nuermberger, E., and Grosset, J. (2007) Promising clinical efficacy of streptomycin-rifampin combination for treatment of Buruli ulcer (*Mycobacterium ulcerans* disease). *Antimicrobial Agents and Chemotherapy* 51(11): 4029–4035.

Confalonieri, U.E.C. (2005) The Millennium Assessment: tropical ecosystems and infectious diseases. *EcoHealth* 2: 231–233.

Cressie, N. (1993) *Statistics for Spatial Data*, revised edition. New York, NY: John Wiley.

Dezső, Z. and Barabási, A.L. (2002) Halting viruses in scale-free networks. *Physical Review E* 65: 055103(R). See www.barabasilab.com/pubs/CCNR-ALB_Publications/200205-21_PhysRevE-HaltingViruses/200205-21_PhysRevE-HaltingViruses.pdf (accessed 15 June 2012).

Dhileepan, K., Peters, C., and Porter, A. (1997) Prevalence of *Aedes camptorhynchus* (Thomson) (Diptera: Culicidae) and other mosquitoes in the eastern coast of Victoria. *Australian Journal of Entomology* 36: 183–190.

Dobos, K.M,, Quinn, F.D., Ashford, D.A., Horsburgh, C.R., and King, C.H. (1999) Emergence of a unique group of necrotizing mycobacterial diseases. *Emerging Infectious Diseases* 5(3): 367–378.

Drummond, C. and Butler, J.R. (2004) *Mycobacterium ulcerans* treatment costs, Australia. *Emerging Infectious Diseases* 10: 1038.

Duker, A.A., Stein, A., and Hale, M. (2006) A statistical model for spatial patterns of Buruli ulcer in the Amansie West district, Ghana. *International Journal of Applied Earth Observation and Geoinformation* 8: 126–136.

Durnez, L., Suykerbuyk, P., Nicolas, V., Barrière, P.,Verheyen, E., Johnson, C.R., Leirs, H., and Portaels, F. (2010) Terrestrial small mammals as reservoirs of *Mycobacterium ulcerans* in Benin. *Applied and Environmental Microbiology* 76(13): 4574–4577.

Eddyani, M., Ofori-Adjei, D., Teugels, G., De Weirdt, D., Boakye, D., Meyers, W.M., and Portaels, F. (2004) Potential role for fish in transmission of *Mycobacterium ulcerans* disease (Buruli ulcer): an environmental study. *Applied Environmental Microbiology* 70(9): 5679–5681.

Elsner, L., Wayne, J., O'Brien, C.R., McCowan, C., Malik, R., Hayman, J.A., Globan, M., Lavender, C.J., and Fyfe, J.A. (2008) Localized *Mycobacterium ulcerans* infection in a cat in Australia. *Journal of Feline Medicine and Surgery* 10: 407–412.

Epstein, P.R. (2001) Climate change and emerging infectious diseases. *Microbes and Infection* 3: 747–754.

Epstein, P.R., Diaz, H.F., Elias, S., Grabherr, G., Graham, N.E., Martens, W.J.M., Mosley-Thompson, E. and Susskind, J. (1998) Biological and physical signs of climate change: focus on mosquito-borne diseases. *Bulletin of the American Meteorological Society* 7(3): 409–417.

Farnsworth, M.L., Wolfe, L.L., Hobbs, T., Burnham, K.P., Williams, E.S., Theobald, D.M., Conner, M.M., and Miller, M.W. (2005) Human land use influences chronic wasting disease prevalence in mule deer. *Ecological Applications* 15: 119–126.

Finley, A.O., Banerjee, S., and McRoberts, R.E. (2008) A Bayesian approach to quantifying uncertainty in multi-source forest area estimates. *Environmental and Ecological Statistics* 15: 241–258.

Foley, J.A., DeFries, R., Asner, G.P., Barfoard, C., Bonan, G., Carpenter, S.R., Chapin, F.S., Coe, M.T., Daily, G.C., Gibbs, H.K., Helkowski, J.H., Holloway, T., Howard, E.A., Kucharik, C.J., Monfreda, C., Patz, J.A., Prentice, I.C., Ramankutty, N., and Snyder, P.K. (2005) Review: global consequences of land use. *Science* 309: 570–574.

Fyfe, J.A.M., Lavender, C.J., Handasyde, K.A., Legione, A.R., O'Brien, C.R., Stinear, T.P., Pidot, S.J., Seemann, T., Benbow, M.E., Wallace, J.R., McCowan, C., and Johnson, P.D.R. (2010) A major role for mammals in the ecology of *Mycobacterium ulcerans*. *PLoS Neglected Tropical Diseases* 4(8): e791.

Galuzo, I.G. (1975) Landscape epidemiology (epizootiology). *Advances in Veterinary Science and Comparative Medicine* 19: 73–96.

George, K.M., Chatterjee, D., and Gunawardana, G. (1999) Mycolactone: a polyketide toxin from *Mycobacterium ulcerans* required for virulence. *Science* 283: 854–857.

Gergel, S. E. and Turner, M.G. (eds) (2002) *Learning Landscape Ecology: A Practical Guide to Concepts and Techniques*. New York, NY: Springer.

Hayman, J. (1991) Postulated epidemiology of *Mycobacterium ulcerans* infection. *International Journal of Epidemiology* 20(4): 1093–1098.

Hill, AB. (1965) The environment and disease association or causation? *Proceedings of the Royal Society of Medicine* 58: 295–300.

Iverson, L.R. (1998) Land-use changes in Illinois, USA: the influence of landscape attributes on current and historic land use. *Landscape Ecology* 2(1): 45–61.

Johnson, P.D.R. and Lavender, C.J. (2009) Correlation between Buruli ulcer and vector-borne notifiable diseases, Victoria, Australia. *Emerging Infectious Diseases* 15: 614–615.

Johnson, P.D.R., Veitch, M.G.K., Leslie, D.E., Flood, P.E., and Hayman, J.A. (1996) The emergence of *Mycobacterium ulcerans* infection near Melbourne. *Medical Journal of Australia* 164: 76–78.

Johnson, P.D.R., Azoulas, J., Lavender, C.J., Wishart, E., Stinear, T.P., Hayman, J.A., Brown, L., Jenkin, G.A., and Fyfe, J.A.M. (2007) *Mycobacterium ulcerans* in mosquitoes captured during outbreak of Buruli ulcer, Southeastern Australia. *Emerging Infectious Diseases* 13(11): 1653–1660.

Koch, T. (2005) *Cartographies of Disease*. Redlands, CA: ESRI Press.

Krummel, J.R., Gardner, R.H., Sugihara, G., O'Neill, R.V., and Coleman, P.R. (1987) Landscape patterns in a disturbed environment. *Oikos* 48(3): 321–324.

Liljeros, F., Edling, C., Amaral, L.N., Stanley, H.E., and Aberg, Y. (2001) The web of human sexual contacts. *Nature* 411: 907–908.

McCallum, P., Tolhurst, J.C., Buckle, G., and Sissons, H.A. (1948) A new mycobacterial infection in man. *Journal of Pathology and Bacteriology* 60: 93–122.

Marion, E., Eyangoh, S., Yeramian, E., Doannio, J., Landier, J., Aubry, J., Fontanet, A., Rogier, C., Cassisa, V., Cottin, J., Marot, A., Eveillard, M., Kamdem, Y., Legras, P., Deshayes, C., Saint-André, J., and Marsollier, L. (2010) Seasonal and regional dynamics of *M. ulcerans* transmission in environmental context: deciphering the role of water bugs as hosts and vectors. *PLoS Neglected Tropical Diseases* 4(7): e731.

Marsollier, L., Robert, R., Aubry, J., Saint André, J.P., Kouakou, H., Legras, P., Manceau, A.L., Mahaza, C., and Carbonelle, B. (2002) Aquatic insects as a vector for *Mycobacterium ulcerans*. *Applied and Environmental Microbiology* 68: 4623–4628.

Marston, B.J., Diallo, M.O., Horsburgh, C.R. Jr., Diomande, I., Saki, M.Z., Kanga, J.M., Patrice, G., Lipman, H.B., Ostroff, S.M., and Good, R.C. (1995) Emergence of Buruli ulcer disease in the Daloa region of Côte d'Ivoire. *American Journal of Tropical Medicine and Hygiene* 52(3): 219–224.

Merritt, R.W., Benbow, M.E., and Small, P.L.C. (2005) Unraveling an emerging disease associated with disturbed aquatic environments: the case of Buruli ulcer. *Frontiers in Ecology and the Environment* 3(6): 323–331.

Merritt, R.W., Walker, E.D., Small, P.L., Wallace, J.R., Johnson, P.D., Benbow, M.E., and Boakye, D.A. (2010) Buruli ulcer disease: a systematic review of ecology and transmission. *PLoS Neglected Tropical Diseases* 4(12): e911.

Meyers, W.M., Shelly, W.M., Connor, D.H., and Meyers, E.K. (1974) Human *Mycobacterium ulcerans* infections developing at sites of trauma to skin. *American Journal of Tropical Medicine and Hygiene* 23: 919–923.

Meyers, W.M., Tignokpa, N., Priuli, G.B., and Portaels, F. (1996) *Mycobacterium ulcerans* infection (Buruli ulcer): first reported patients in Togo. *British Journal of Dermatology* 134: 1116–1121.

Mitchell, P.J., Jerrett, I.V. and Slee, K.J. (1984) Skin ulcers caused by *Mycobacterium ulcerans* in koalas near Bairnsdale, Australia. *Pathology* 16: 256–260.

Mosi, L., Williamson, H., Wallace, J.R., Merritt, R.W., and Small, P.L.C. (2008) Persistent association of *Mycobacterium ulcerans* with west African predaceous insects of the family Belostomatidae. *Applied Environmental Microbiology* 74(22): 7036–7042.

Moyeed, R.A. and Papritz, A. (2002) An empirical comparison of kriging methods for nonlinear spatial point prediction. *Mathematical Geology* 34(4): 365–386.

Nackers, F., Johnson, R.C., Glynn, J.R., Zinsou, C., Tonglet, R., and Portaels, F. (2007) Environmental and health-related risk factors for *Mycobacterium ulcerans* disease (Buruli ulcer) in Benin. *American Journal of Tropical Medicine and Hygiene* 77(5): 834–836.

Narumalani, S., Mishra, D.R., and Rothwell, R.G. (2004) Change detection and landscape metrics for inferring anthropogenic processes in the greater EFMO area. *Remote Sensing of the Environment* 91(3–4): 478–489.

Newman, M.E.J. (2003) The structure and function of complex networks. *SIAM Reviews* 45: 167–256.

Oluwasanmi, J.O., Solanke, T.F., Olurin, E.O., Itayemi, S.O., Alabi, G.O., and Lucas, A.O. (1976) *Mycobacterium ulcerans* (Buruli) skin ulceration in Nigeria. *American Journal of Tropical Medicine and Hygiene* 25(1): 122–128.

Patz, J.A. and Confalonieri, U.E.C. (2005) Human health: ecosystem regulation of infectious diseases. In *Ecosystems and Human Well-being: Current State and Trends: Findings of the Condition and Trends Working Group*, R.M. Hassan, R. Scholes, and N. Ash (eds). Washington, DC: Island Press, 393–415.

Patz, J.A., Graczyk, T.K., Geller, N., and Vittor, A.Y. (2000) Effects of environmental change on emerging parasitic diseases. *International Journal for Parasitology*, 30: 1395–1405.

Pavlovsky, E.N. (1966) *Natural Nidality of Transmissible Diseases with Special Reference to the Landscape Ecology of Zooanthroponses*. Urbana, IL: University of Illinois Press.

Plowright, R.K., Sokolow, S.H., Gorman, M.E., Daszak, P., and Foley, J.E. (2008) Causal inference in disease ecology: investigating ecological drivers of disease emergence. *Frontiers in Ecology and the Environment* 6(8): 420–429 (doi:10.1890/070086).

Portaels, F., Elsen, P., Guimaraes-Peres, A., Fonteyne, P.A., and Meyers, W.M. (1999) Insects in the transmission of *Mycobacterium ulcerans* infection. *Lancet* 353: 986.

Pouillot, R., Matias, G., Wondje, C.M., Portaels, F., Valin, N., Ngos, F., Njikap, A., Marsollier, L., Fontanet, A., and Eyangoh, S. (2007) Risk factors for Buruli ulcer: a case control study in Cameroon. *PLoS Neglected Tropical Diseases* 1(3): e101.

Quek, T.Y.J., Athan, E., Henry, M.J., Pasco, J.A., Redden-Hoare, J., Hughes, A., and Johnson, P.D.R. (2007a) Risk factors for *Mycobacterium ulcerans* infection, south-eastern Australia. *Emerging Infectious Diseases* 13: 1661–1666.

Quek, T.Y.J., Henry, M.J., Pasco, J.A., O'Brien, D.P., Johnson, P.D.R., Hughes, A., Cheng, A.C., Redden-Hoare, J., and Athan, E. (2007b) *Mycobacterium ulcerans* infection: factors influencing diagnostic delay. *Medical Journal of Australia* 187(10): 561–563.

Raghunathan, P.L., Whitney, E.A.S., Asamoa, K., Stienstra, Y., Taylor, T.H. Jr., Amofah, G.K., Ofori-Adjei, D., Dobos, K., Guarner, J., Martin, S., Pathak, S., Klutse, E., Etuaful, S., van der Graaf, W.T.A., van der Werf, T.S., King, C.H., Tappero, J.W., and Ashford, D.A. (2005) Risk factors for Buruli ulcer disease (*Mycobacterium ulcerans* infection): results from a case-control study in Ghana. *Clinical Infectious Diseases* 40: 1445–53.

Raudenbush, S.W. and Bryk, A.S. (2002) *Hierarchical linear models: applications and data analysis methods*, 2nd edition. Thousand Oaks, CA: Sage Publications.

Reisen, W.K. (2010) Landscape epidemiology of vector-borne diseases. *Annual Review of Entomology* 55: 461–483.

Smith, K.F., Dobson, A.P., McKenzie, F.E., Real, L.A., Smith, D.L., and Wilson, M.L. (2005) Ecological theory to enhance infectious disease control and public health policy. *Frontiers in Ecology and the Environment* 3: 29–37.

Solomon, S., Qin, D., Manning, M., Chen, Z., Marquis, M., Averyt, K.B., Tignor, M., and Miller, H.L. (eds) (2007) *Contribution of Working Group I to the Fourth Assessment Report of the Intergovernmental Panel on Climate Change*. Cambridge, UK: Cambridge University Press.

Sopoh, G.E., Johnson, R.C., Chauty, A., Dossou, A.D., Aguiar, J., Salmon, O., Portaels, F., and Asiedu, K. (2007) Buruli ulcer surveillance, Benin, 2003–2005. *Emerging Infectious Diseases* 13: 1374–1376.

Sopoh, G.E., Johnson, R.C., Anagonou, S.Y., Barogui, Y.T., Dossou, A.D., Houézo, J.G., Phanzu, D.M., Tente, B.H., Meyers, W.M., and Portaels, F. (2011) Buruli ulcer prevalence and altitude, Benin. *Emerging Infectious Diseases* 17(1): 153–154 (doi:10.3201/eid1701.100644).

Stienstra, Y., van der Graaf, W.T.A., Asamoa, K., and van der Werf, T.S. (2002) Beliefs and attitudes toward Buruli ulcer in Ghana. *American Journal of Tropical Medicine and Hygiene* 67(2): 207–213.

Stinear, T.P., Seemann, T., Pidot, S., Wafa, F., Yeysset, G., Garnier, T., Meurice, G., Simon, D., Bouchier, C., Ma, L., Tichit, M., Porter, J.L., Ryan, J., Johnson, P.D.R., Davies, J.K., Jenkin, G.A., Small, P.L.C., Jones, L.M., Tekaia, F., Laval, F., Daffé, M., Parkhill, J., and Cole, S.T. (2007) Reductive evolution and niche adaptation inferred from the genome of *Mycobacterium ulcerans*, the causative agent of Buruli ulcer. *Genome Research* 17: 192–200.

Tiong, A. (2005) The epidemiology of a cluster of *Mycobacterium ulcerans* infections in Point Lonsdale. *Victorian Infectious Disease Bulletin* 8(1): 2–4.

Tobias, N.J., Seemann, T., Pidot, S., Porter, J.L., Marsollier, L., Marion, E., Letournel, F., Zakir, T., Azuolas, J., Wallace, J.R., Hong, H., Davies, J.K., Howden, B.P., Johnson, P.D.R., Jenkin, G.A., and Stinear, T.P. (2009) Mycolactone gene expression is controlled by strong siga-like promoters with utility in studies of *Mycobacterium ulcerans* and Buruli ulcer. *PLoS Neglected Tropical Diseases* 3(11): e553.

Veitch, M.G.K., Johnson, P.D.R., Flood, P.E., Leslie, D.E., Street, A.C., and Hayman, J.A. (1997) A large localized outbreak of *Mycobacterium ulcerans* infection on a temperate southern Australian island. *Epidemiology and Infection* 119: 313–318.

Wagner, T., Benbow, M.E., Burns, M., Johnson, R.C., Merritt, R.W., Qi, J., and Small, P.L.C. (2008a) A landscape-based model for prediction *Mycobacterium ulcerans* infection (Buruli ulcer disease) presence in Benin, west Africa. *EcoHealth* 5: 69–79.

Wagner, T., Benbow, M.E., Brenden, T.O., Qi, J., and Johnson, R.C. (2008b) Buruli ulcer disease prevalence in Benin, west Africa: associations with land use/cover and the identification of disease clusters. *International Journal of Health Geographics* 7: 25 (doi:10.1186/1476-072X-7-25). See www.ij-healthgeographics.com/content/7/1/25 (accessed 15 June 2011).

Wallace, J.R., Gordon, M.C., Hartsell, L., Mosi, L., Benbow, M.E., Merritt, R.W., and Small, P.L.C. (2010) Interaction of *Mycobacterium ulcerans* with mosquito species: implications for transmission and trophic relationships. *Applied and Environmental Microbiology* 76(18): 6215–6222.

Waller, L.A. and Gotway, C.A. (2004) *Applied Spatial Statistics for Public Health Data*. Hoboken, NJ: John Wiley & Sons, Inc.

Watts, D.J. (2004) The "new" science of networks. *Annual Review of Sociology* 30: 243–270.

Wilcox, B.A. and Ellis, B. (2006) Forests and emerging infectious diseases of humans. *Unasylva* 224(57): 11–18.

Wilcox, B.A. and Gubler, D.J. (2005) Disease ecology and the global emergence of zoonotic pathogens. *Environmental Health and Preventive Medicine* 10(5): 263–272.

World Health Organization. (2000) *Buruli ulcer:* Mycobacterium ulcerans *infection*. Geneva, Switzerland: World Health Organization.

World Health Organization. (2004) *Using climate to predict infectious disease outbreaks: a review*. Geneva, Switzerland: World Health Organization.

World Health Organization. (2007) Fact sheet: Buruli ulcer disease (*Mycobacterium ulcerans* infection). See www.who.int/mediacentre/factsheets/fs199/en (accessed 17 May 2011).

World Health Organization. (2011) Buruli ulcer: Buruli ulcer endemic countries. See www.who.int/buruli/country/en (accessed 21 December 2011).

van Zyl, A., Daniel, J., Wayne, J., McCowan, C., Malik, R., Jelfs, P., Lavender, C.J., and Fyfe, J.A. (2009) *Mycobacterium ulcerans* infections in two horses in south-eastern Australia. *Australian Veterinary Journal* 88(3): 101–106.

6 The ecology of injuries in Matlab, Bangladesh

Elisabeth D. Root and Michael E. Emch

Introduction

Injuries are responsible for more than five million deaths each year, roughly equal to the number of deaths from HIV/AIDS, malaria, and tuberculosis combined (Gosselin *et al.* 2009). Low- and middle-income countries account for 91 percent of deaths due to unintentional injuries. When compared with high-income countries, the death rate is nearly double in low and middle income countries (65 versus 35 per 100,000), and the rate for disability-adjusted life-years (DALYs) is more than triple (2398 versus 774 per 100,000; Chandran *et al.* 2010). These low- and middle-income countries have higher injury rates, suffer more non-fatal, negative health outcomes, and die more often due to these injuries. Morbidity and mortality rates vary significantly by region. Southeast Asia consistently experiences the highest injury rates, due in large part to the sizeable number of low-income countries in that region and stronger health surveillance systems which can more accurately report injury data. Low- and middle-income countries in southeast Asia contribute over 34 percent of the global unintentional injury deaths (Chandran *et al.* 2010). While deaths due to injury are more often reported, and therefore studied, non-fatal injuries provide a more realistic picture of the overall burden of injuries because many of the most common injuries, such as burns and falls, do not lead to death. While two-thirds of the world's injuries occur in developing countries there is very little research on the prevalence and causes of injury morbidity in these regions (Smith and Barss 1991). This chapter explores the burden of accidental injuries among adults in Matlab, Bangladesh. The population of Matlab has participated in a comprehensive health and demographic surveillance system since the 1960's so that information on unintentional injuries is available. The study not only investigates the prevalence of injuries but also population characteristics associated with accidental injuries, in particular the relationship with socio-economic status. Socio-economic status is measured for all people living in the study area through a comprehensive household-level survey of assets. The chapter provides a review of the literature on injuries in Bangladesh and other developing countries and describes the results of a survey in which non-fatal accidental injuries were systematically identified in an area with an ongoing health and demographic surveillance system.

Theoretical context

This book is concerned with ecologies and politics of health, subjects upon which many specialists of medical and health geography and related fields such as epidemiology have theorized (May 1950, 1958; Hunter 1974). Political and social contexts lay the foundation of many diseases. There are few diseases that people get simply due to chance and poor or politically powerless people get diseases more often (Navarro *et al.* 2006; Laverack 2006; Wagstaff 2002). Injuries are no different than chronic disease such as cancer or heart disease because there are disparities in who is injured and injured individuals live within a socio-political context that affects their chances of injury. This paper focuses on how socio-economic status is fundamentally linked to injuries. Link and Phelan (1995, 2000) theorized that social conditions are usually the fundamental causes of diseases. They argued that social conditions "embody access to important resources, affect multiple disease outcomes through multiple mechanisms, and consequently maintain an association with disease even when intervening mechanisms change" (Link and Phelan 1995: 81). Proximate causes of disease have been studied much more than distal causes such as social status or political context. Mayer (1996) argued that the field of medical geography should incorporate political context into health studies using a political ecology framework. Political ecology is concerned with understanding higher-level or "upstream" causes, such as social or political structures, rather than the "downstream" causes of disease, such as individual-level behaviors or environmental exposures. The political ecology framework is a similar conceptualization of the proximate and distal causes that were described by Link and Phelan (1995; for related discussions on political ecology and health, see Chapters 9 and 14 of the present volume). At the same time, political structures and policy decisions are not solely responsible for disparities in health outcomes. Numerous factors within an individual's environment affect health behaviors and outcomes. Features of the physical environment such as rivers and roads, contaminants such as arsenic or pesticides, social structures such as the community cohesion and the size and strength of a person's social network, all contribute to an individual's health. The strength of the medical geographic approach is that it seeks to understand the multifaceted and multilevel ecology of a disease by placing a person within his or her context. This context varies over time and across space and explains the disparities that arise among different populations. This multilevel framework is one of the most important emerging directions in the field of public health and has been advanced over the past decade by the integration of a geographic perspective of health.

Socio-economic status has been identified as an important determinant of health across a broad range of countries and health issues (Yen and Syme 1999; Wagstaff 2002; Emch *et al.* 2010; Diez-Roux *et al.* 2000). While socio-economic status rarely has a direct influence on health outcomes, it can impact factors in an individual's life that do directly influence health (Bollen *et al.* 2001). These mediating variables may include physical characteristics such as poor sanitation or inadequate and overcrowded housing, which increase exposure to infectious diseases, or behaviors such as condom use, which determine exposure to unprotected sex. Socio-economic status is not just about material deprivation and is thus not sufficient to explain differential disease burden. As Marmot (1986) showed in the Whitehall Study using a cohort of British civil servants, inequality leads to more disease in poorer people. Socio-economic status is a multifaceted concept that may include education, occupation, income, wealth, and residence. However, in less-developed countries conceptualization and measurement of socio-economic status is challenging because adequate data are often not available and because ideas of wealth or income differ by culture or country. Prior research suggests that socio-economic status in developing countries should be measured in terms of assets or wealth rather than income or consumption (Gwatkin *et al.* 2002). In large, informal, predominantly agricultural-based economies, such as Bangladesh, income and consumption expenditure is difficult to measure. Many individuals work on family farms and have no formal income. In addition, price indices for consumable goods fluctuate unpredictably in unstable economies. Both income and consumption are collected using household surveys, so answers are often confounded by recall bias (McKenzie 2004). Household characteristics and asset ownership are typically not subject to these reporting problems and are widely used an indicators of wealth. In this study, we use a comprehensive survey of household assets to construct a measure of socio-economic status for an area of rural Bangladesh. Using this measure of socio-economic status in conjunction with other demographic and environmental risk factors, we examine the risk for accidental injury among adults.

Injuries

Injuries fall into one of two categories. An accidental injury lacks intent in that the injury occurs as an unforeseen or chance outcome of a voluntary action. These types of injuries are also often referred to as unintentional. Examples of accidental injuries include falls, burns or cuts. In Bangladesh, many unintentional injuries are occupational in nature, occurring, for example, when a woman burns herself while cooking, or when a man cuts himself while working in the fields. When an injury is intentional, it occurs as a direct result of an action that was meant to cause harm by that person or another. The simplest example of an intentional injury is suicide, but assault is also often considered intentional. Despite the fact that the injured individual did not intend to get injured, the action that resulted in injury was intentional in nature. Most studies of injuries examine unintentional/accidental and intentional injuries separately, though there are no guidelines for exactly which types of accidents are intentional or accidental. Classification is typically left to the discretion of the researcher.

Severity of injury is also important when understanding the burden of injury-related morbidity and mortality: a minor cut is not the same as a broken bone. It is possible to gauge severity using instruments such as the Abbreviated Injury Scale (Gennarelli and Wodzin 2008), an anatomically based scoring system to determine the severity of injuries based on the survivability of the injury. However, the use of such instruments requires detailed medical record or injury report data that are then reviewed by highly trained health professionals. This type of record review is often not possible in developing countries such as Bangladesh because of both the lack of trained health professionals and detailed injury data. In these cases such, injuries are only studied if they are severe enough to warrant medical intervention, which is used as a proxy for severity. While several recent studies have examined injury-related morbidity and mortality among children in Bangladesh, very little is known about the extent and causes of adult injury.

The few studies that have investigated accidental injury among adults have focused on specific populations, such as women or hospital inpatients. A hospital-based study conducted in 2001 found that 33 percent of the beds in primary and secondary level hospitals in Bangladesh were occupied by injury-related patients, and more than 19 percent of the injuries were related to road traffic accidents. The vast majority of these patients were between the ages of 18 and 45 (Mashreky *et al.* 2010b). Yusuf *et al.* (2000) investigated the major causes of injury-related deaths among women and girls aged 10–50 years in Bangladesh in 1996 and 1997. They found that that almost a quarter of all deaths were injury-related, and that the risk of injury-related death decreased with a woman's age. Mashreky *et al.* (2010a) also found that women had higher rates of accidental death due to burns.

Urban or rural residence also appears to play a role in injuries. A community-based study published in 2008 by Mashreky *et al.* concluded that the rates of non-fatal childhood burns in Bangladesh were significantly different in rural and urban areas, with incidence rates of 435 and 102 per 100,000 children, respectively. In a later study, Mashreky *et al.* (2010a) also found that the injury rate was higher among rural females and many were related to burns. About 90 percent of the burn incidences occurred while cooking and were due to cooking/heating fire and fire from kerosene lamps. Yusef *et al.* (2000) found that 90 percent of all injury-related deaths and 93 percent of unintentional injuries in women came from villages as opposed to cities or towns in Bangladesh.

Studies from other developing countries also provide some insight into differences in the causes of injuries. Overall, road traffic injuries make up the largest proportion of unintentional injury deaths, followed by falls and drowning (Chandran *et al.* 2010). Mock *et al.* (1999) found that in Ghana the causes of injury-related mortality were markedly different in urban and rural areas. They note that urban areas were more burdened by motor vehicle- and transport-related injuries, while rural areas were primarily faced with agriculture related injuries. Sathiyasekaran (1996) conducted a population-based cohort study of the incidence of injury among a low-socio-economic-status population in Madras, India and found that men had a higher rate than women (137 and 118 per 1000, respectively). In addition, men were nearly three times more likely to experience traffic injuries and women were three times more likely to have household-related injuries. A recent study of unintentional home-related injuries in Iran found that burns and lacerations/cuts caused by contact with sharp instruments were the most frequently reported injuries and more common in rural areas and among men. Injury rates were highest among children aged 0–4 years and lowest among the elderly (60 years or over; Mohammadi *et al.* 2005).

Unintentional injuries may be affected by socio-economic status through both the physical environment and behaviors. For example, open cooking and heating fires, more often found in lower-socio-economic status households, may increase the possibility of injury due to burns. Higher levels of education may increase knowledge of the risks associated with water environments or the use of farm equipment, which may in turn reduce risky or unsafe behaviors. While a wide range of studies have examined the relationship between socio-economic status and injuries in the developed world (Alexandrescu *et al.* 2009; Cubbin *et al.* 2000a, b), very few have examined this relationship in developing countries (Nordberg 2000; Murray and Lopez 1997), especially among adults. Only a few studies have investigated the relationship between socio-economic status and injury in Bangladesh, and results were limited to children. Results indicated that poor children were 2.8 times more likely to suffer from injury related mortality than wealthy children, taking into account all the other factors (Giashuddin *et al.* 2009). Rahman *et al.* (2004) found that low literacy and low income were major risk factors for childhood drowning.

Study area, data, and methods

Study area

The research site for the International Centre for Diarrhoeal Disease Research, Bangladesh (icddr,b) and for this project is called Matlab due to the proximity of the centre's hospital to Matlab Town. Matlab is in south-central Bangladesh, approximately 50 kilometers southeast of Dhaka, adjacent to where the Ganges River meets the Meghna River, forming the Lower Meghna River. The majority of the population is engaged in agricultural production and the educational infrastructure is poorly developed. Per capita gross national income is roughly US$590 per year, and approximately 50 percent of the population lives below the poverty level (World Bank 2010). The predominant occupation for rural males is agriculture, with labor force participation rates remaining very high even for older males. Women are largely restricted by convention to activities within the home with relatively little opportunity to venture outside the homestead.

A demographic surveillance system (DSS) has recorded all vital events of the Matlab population since 1966; the study area population has been approximately 200,000 since that time. The database is the most comprehensive longitudinal demographic database of a large population in the developing world (icddr,b 2000). Each individual in Matlab is given a unique ID at birth or entry into the study area, as well as a unique household ID and extended household (*bari*) ID. Matlab DSS data have been used extensively in the demographic literature, and these data are considered to represent one of the few high-quality (i.e. complete, accurate and up-to-date) demographic data sources in the developing world (Fauveau 1994). In 1995, a geographic information system (GIS) database of the Matlab research area was created (Emch 1999; Ali *et al.* 2001). The GIS database includes *bari* locations for all individuals living in the Matlab study area.

Baris often include several households which are clustered around a small central courtyard. The location of each *bari*, defined as the middle point of that courtyard, was collected using handheld GPS units. *Baris* are given the DSS census number within the GIS database so that demographic data and incidence data can be linked to specific *bari* locations. The Matlab field research center has in- and out-patient services, a medical laboratory, and research facilities. One-hundred twenty community health workers (CHWs) visit each household area every month to collect demographic, morbidity, and other data.

Data

A large-scale survey of adult health and socio-economic conditions called the Matlab Health and Socioeconomic Survey (MHSS) was implemented in 1996 (Rahman *et al.* 2001). The MHSS addresses several broad areas of concern to the rural adults. These include the effect of socio-economic and behavioral factors on adult health status and healthcare utilization, the linkages between adult well-being, social and kin network characteristics and resource flows, and the impact of community services and infrastructure on adult health and other human capital acquisition. In addition to detailed data on social networks, life histories and economic activity, the survey collected self-reports on overall health status, activities of daily living (ADLs), and chronic and acute morbidity.

The MHSS used a multistage, multisample household survey that collected information from 11,150 individuals aged 15 and over in 4538 households. At the time of the sample selection, the Matlab surveillance area consisted of 8640 *baris*, of which roughly one-third (31.1 percent or 2687) were randomly sampled. Since *baris* are clusters of patrilineally related groups of households, sampling *baris* rather than households provides a better representation of family networks, a major focus of the MHSS survey. Within each *bari*, up to two households were selected for detailed interviews. Within each selected household, all individuals aged 50 years and over and a sample of individuals below the age of 50 were interviewed. There were 11,150 individuals aged 15 years and over in the MHSS *bari* sample.

The adult questionnaire included a variety of questions pertaining to health, education, employment and marital history. A set of three questions were used to identify the prevalence of injuries for this analysis. The first asked respondents if they "Had any health problems due to accidents during the past month?" The other two questions were asked for individuals who had visited a health facility (as an outpatient or inpatient) in the past year. Respondents were asked if an accident was the "purpose of your visit to the health facility/provider." If a respondent answered yes to any of these questions, he or she was coded as having an accidental injury. Separate questions on the survey asked about injuries that may have been intentional in nature, such as assault or suicide, and participants were asked to only report injuries that occurred as a result of an accident. Prompts included broken bones, serious cuts and burns. Although we were unable to directly measure severity of the injury due to the nature of the questions asked on the survey, we only captured injuries that were severe enough to require medical intervention.

The MHSS also asked questions related to individual characteristics such as age, sex, religion, marital status and occupation and household characteristics such as ownership of land, livestock and other assets. Of particular relevance to this study, the MHSS asked all heads of household a detailed list of questions about housing construction and household assets. These questions asked if any member of the household owned items such as a clock or watch, hurricane lamp, furniture, quilt, and so on. The MHSS recorded the *bari* number to which each household belongs, enabling researchers to link survey data to outside data sources using geographic location. For the purposes of this study, we focus on 10,414 respondents aged 15 years and older for whom we have complete information. 737 individuals from 35 households were dropped from the analysis due to missing household characteristic data. This analysis was cross-sectional, meaning that data were only collected at one point in time (May–August 1996).

Methods

Socio-economic status measurement

A categorical socio-economic status variable was developed using principle components analysis (PCA) in SAS version 9.2 software. PCA is a data compression method that facilitates the reduction of many variables by outputting new variables ("components") in order of greatest explanatory power and that are orthogonal to each other in that data space. Each component effectively captures the "essence" of the original variables. In this case, we created a single household-level measure of socio-economic status from multiple MHSS variables that record household assets. The individual measures of household assets by themselves do not give us a clear picture of the overall socio-economic state of the household. But the composite variable created using PCA captures the essence of household socio-economic status.

The socio-economic status measure created from the MHSS data reflects a composite of six dummy variables of ownership of household assets and one ordinal variable of household wall material (Table 6.1). Roof material and ownership of agricultural land were also collected but both were excluded because most residents have a tin roof and own farmland. socio-economic status scores were first created for each household in the study sample. The household-level socio-economic status scores were then collapsed by *bari*, and the mean score represents *bari*-level socio-economic status. Both household- and *bari*-level socio-economic status scores were sorted from lowest to highest and divided into equal quintiles. Higher quintiles reflect higher socio-economic status. We chose to classify the socio-economic status variable rather than use the raw scores because the actual numeric values of the socio-economic status variable do not have any inherent meaning aside from larger scores denoting "higher" socio-economic status and smaller scores denoting "lower" socio-economic status. Thus, some type of an ordinal scale is most appropriate for representing this variable.

There are many methods for classifying indicators such as socio-economic status including quantiles and standard deviation. Quantiles divide the sample into a specified number of groups, usually four (quartiles) or five (quintiles), with an even number of observations in each group. Standard deviation is typically used when a variable is normally distributed while quantiles are used when a variable is skewed or has serious outliers. It is well known that different classification methods may affect results (Monmonnier 1996; Huff 1993; Openshaw and Taylor 1979). However, when a variable is normally distributed, standard deviations and quantiles typically result in nearly identical classifications. In public health research, it is common to use quintiles when using PCA to create a new variable due to the way the new variable is created. The new variable is standardized in relation to a standard normal distribution, meaning that the variable has a mean of zero and a standard deviation of one. Due to the fact that the socio-economic status variable was normally distributed and given the standard classification method in public health, we chose to use quintiles to classify socio-economic status.

We hypothesize that increasing household and *bari* socio-economic status will lead to a decreased risk of an injury. We modeled the effect of both household- and *bari*-level socio-economic status scores in order to examine whether the socio-economic condition of the larger *bari* was equally important as the socio-economic status of the household in which the individual lives.

Environmental conditions

The MHSS also recorded the DSS *bari* number for each household, which allows us to link MHSS data to the spatial database. Thus, survey data are linked to specific *bari* locations and indicators of proximity to environmental characteristics can be developed. Using the GIS, we calculated the linear distance in meters between each *bari* included in the MHSS and the nearest river and road. Essentially, a straight line was drawn between each *bari* and the closest river and road. These distance calculations were then linked back to individual MHSS respondents using the *bari* number. This method of measuring distance is appropriate for rural Bangladesh as very few roads exist and people typically walk through fields in the straightest possible line toward their destination. Drawing from prior literature that suggests road traffic accidents and drowning are two of the most prevalent injury types, we

Table 6.1 Variables included in principal components analysis to create socio-economic status index

Household assets (1 = yes, 0 = no)	*Wall material*
Had cows/buffaloes in the past year?	Main wall type of this household?
Does any member of the household own furniture?	1 = Pucca/cement
Does any member of the household own a quilt?	2 = Wood
Does any member of the household own a bike?	3 = Bamboo
Does any member of the household own a clock/ watch?	4 = Dirt, mud, straw, and leaves
Does any member of the household own a radio?	

hypothesize that individuals living in households and *baris* closer to a river or road will have a greater risk of an injury because they are more likely to cross or use these geographic features than individuals living farther from a river or road. Since our sample only includes adults aged 15 and older, we do not expect to see a strong effect of distance to river since drowning is most prevalent among children under the age of 5 (Iqbal *et al.* 2007). However, during the monsoon season, flooding does cause drowning deaths even among the adult population. Road traffic accidents are most prevalent among adolescents and adults age 15 and older (Yusuf *et al.* 2000; Rahman *et al.* 2004). Since we did not have information on the specific cause of the accident, we cannot explore the relationship between these environmental variables and injury type, a limitation we discuss further below.

Statistical methods

Univariate analyses were used to examine the relationship between each explanatory variable and the outcome. Variables showing a p-value of 0.2 or less derived from chi-squared or t-tests were retained for the multivariate analysis. Relationships among injuries, demographic characteristics, environmental variables, and socio-economic status were measured using generalized estimating equations (GEE) with a logit link function to account for the *bari*-level correlation and cluster sampling scheme of the survey. These models were built using independent and exchangeable within-*bari* correlation matrices to control for the correlation. The dependent variable was occurrence of an injury (yes or no). The demographic, environmental and socio-economic status variables were included as explanatory variables. Coefficients of independent variables in the models were exponentiated to estimate the odds ratio (OR) of injury associated with different variables. Standard errors for coefficients were used to estimate p-values and associated 95 percent confidence intervals (CIs) for the ORs. All statistical tests were interpreted using a two-tailed distribution. Two separate models were built to differentiate between the effect of household-level socio-economic status and *bari*-level socio-economic status. Univariate and multivariate analyses were carried out in SAS 9.2.

Results

A total of 142 individuals aged 15 and older reported an accidental injury during the year prior to the survey, for an overall rate of 12.7 injuries per 1000 people. Figure 6.1 shows the spatial distribution of the population sampled for the MHSS survey by village and the location of accidental injury events. Villages shaded in dark grey have a greater number of individuals sampled during the survey and a larger population. The black dots show the *bari* of individuals that reported an injury during the past year; larger dots show that two injuries were reported for that *bari*. Figure 6.2 shows the rate of injuries by socio-economic quintile. In general, injury rates appear to decrease as household socio-economic status increases, ranging from a high of 18.2 per 1000 persons to a low of 9.9 per 1000 persons. While this trajectory is not smooth, it does show an overall trend in injuries by socio-economic status.

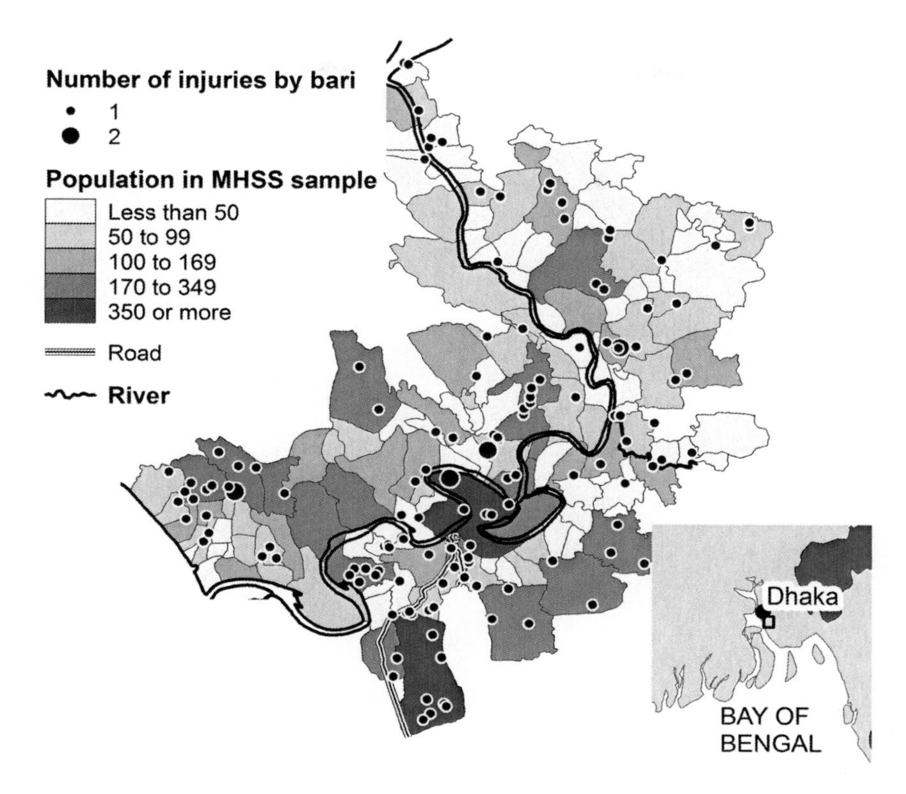

Figure 6.1 Spatial distribution of Matlab Health and Socioeconomic Survey population and accidental injury events (1996).

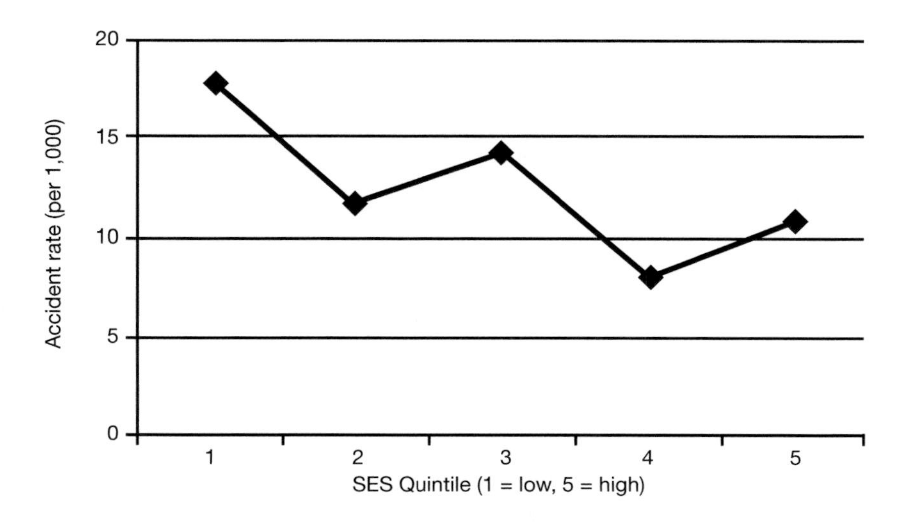

Figure 6.2 Injury rate per 1000 by household socio-economic status quintile.

Table 6.2 shows the characteristics of the population and results from the univariate analysis. In general, individuals experiencing injuries were younger, male, unmarried, and more likely to be able to read. Neither environmental variable showed a statistically significant univariate association with injuries. In addition, neither the household-level socio-economic status or *bari*-level socio-economic status indicators showed a significant univariate association with injuries at the $p<0.05$ level.

Table 6.2 Sample characteristics and results from univariate analyses

	Case	*Control*	*p**
Age (mean years)	34.3	40.3	<0.0001
Sex			
Male	80	5003	
Female	62	6006	0.0096
Religion			
Muslim	131	9809	
Other	11	1200	0.2301
Marital status			
Married	88	7993	
Never married, divorced, or separated	54	3016	0.0048
Literacy			
Can read easily or with some difficulty	79	4830	
Cannot read	63	6176	0.0050
Occupation			
Farming	31	2178	
No farming	111	8831	0.5431
Distance to river (mean meters)	215	193.4	0.1418
Distance to road (mean meters)	261.1	288.5	0.2838
HH-level SES			
1—lowest	26	1430	
2	25	2117	
3	38	2643	
4	11	1109	
5—highest	42	3675	0.2009
Bari-level SES			
1—lowest	33	1920	
2	24	2117	
3	28	2368	
4	26	2032	
5—highest	31	2565	0.4892

Note: **p*-value derived from chi-squared tests (categorical variables) or *t*-tests (continuous variables).

Table 6.3 presents the results of the multivariate models estimating the risk of an accidental injury. Household socio-economic status is significantly associated with the occurrence of an injury (OR=0.88, 95 percent CI: 0.77–1.00): the higher the socio-economic status score, the lower the occurrence of an injury. The odds ratio shows that for every quintile increase in socio-economic status, the odds of an injury drop approximately 12 percent. Thus, there is nearly a 50 percent decrease in the odds of an injury between the lowest and highest quintiles of socio-economic status. There was no significant relationship, however, between *bari*-level socio-economic status and injury occurrence. Age was the strongest predictor of an injury (OR=0.98, 95 percent CI=0.97–0.99). Older individuals were at a decreased risk for accidental injury than younger individuals; the odds of an injury drop by approximately 2 percent for each year of life. Sex and literacy show marginally significant associations, with men and individuals who can read experiencing greater odds of an injury.

Discussion

Findings from this study confirm and augment results from the few studies that have been done on socio-economic status and injuries in Bangladesh. Prior studies show an increased risk of injury among children in low-socio-economic-status families (Giashuddin *et al.* 2009; Rahman *et al.* 1998) but no studies to date have examined this relationship among adults. This study suggests that a similar relationship exists between low socio-economic status and risk for injury among adults. Socio-economic status was measured using an index of household assets, which has successfully been used in other health-related studies in Bangladesh (Emch *et al.* 2010; Giashuddin *et al.* 2009; Mobarak *et al.* 2008). The collection and use of household assets, rather than *bari*-level assets, may explain the significant relationship between household-level socio-economic status and injury risk, but not between *bari*-level socio-economic status and injury risk.

Table 6.3 Predictors of injury risk

Variables	*Model 1: household-level SES*			*Model 2:* bari-*level SES*		
	OR*	95% CI	*p*-value	OR*	95% CI	*p*-value
Age	0.98	0.97–0.99	0.016	0.98	0.97–0.99	0.015
Sex	1.39	0.97–1.99	0.067	1.40	0.97–2.00	0.065
Literacy	1.51	1.00–2.26	0.046	1.45	0.96–2.19	0.075
Marital status	0.86	0.58–1.26	0.449	0.86	0.58–1.27	0.451
River distance	1.00	0.99–1.00	0.088	1.00	0.99–1.00	0.088
Household SES	0.88	0.77–1.00	0.051	–	–	–
Bari SES	–	–	–	0.91	0.79–1.04	0.185

Note: *Multivariate odds ratio for the cited variable, adjusted for all other variables in the table, in a model using GEE with the logit link function.

Construction of a different variable, which more accurately captures *bari*-level socio-economic status and may not be related to household assets at all, could lead to completely different results. More research should be conducted to examine what exactly constitutes high or low *bari*-level (or even neighborhood- or village-level) socio-economic status. The construction of household asset indices is a hotly debated topic in development studies and there is very little agreement about exactly which assets should be included, what methods should be used to combine assets into an index or even with which health outcomes asset indices should correlate. While the asset index we chose to employ has been used extensively, there is need for more research on how to construct and categorize socio-economic status. The quintile method of socio-economic status construction used here should also be explored. While this is one standard of classification in public health, including fixed effects of household metrics or continuous variables with and without polynomials, could also reveal interesting results.

The association between injury and literacy was statistically significant in both the univariate and multivariate analysis, though not in the expected direction. Individuals who reported the ability to read "well" or "with some difficulty" had higher odds of injury, even after controlling for age, sex, literacy, marital status, and socio-economic status. Literacy was originally included in the PCA used to develop the socio-economic status index, but was removed because it did not load with the other factors (e.g., household assets). In light of these results, it is possible literacy captures a completely different relationship between socio-economic status and injury than does the index of household assets. More puzzling is the direction of the relationship. Most studies of socio-economic status and health suggest that lower levels of literacy or education are inversely related to health. Less education leads to higher morbidity and mortality rates and poorer health related behaviors. It may be that poor health leads to lower levels of schooling, since individuals in poor health typically miss more school due to illness. Additionally, more highly educated individuals may be exposed to, or are more likely to adopt, new behaviors and attitudes that improve overall health. Less educated individuals are more likely to engage in activities that lead to adverse health outcomes, presumable because they do not fully understand the risks involved in those behaviors. The unexpected relationship between literacy and injuries in this study may be due to the way in which literacy and socio-economic status are measured and interact in this population. While literacy was significantly associated with injuries in the univariate analysis, controlling for other demographic and socio-economic variables in the multivariate analysis removed the statistical significance. It is likely that the relationship between injuries and literacy is mediated by additional variables, and that literacy and socio-economic status capture similar causal mechanisms, such as improved access to injury prevention programs or ownership of household assets that may increase risk for injury. In Matlab, households involved in a microcredit loan program often purchase items such as boats, motor vehicles or heavy farm equipment. Some researchers suggest that uneducated or socio-economically disadvantaged individuals may have difficulty engaging in microcredit activities because they cannot meet the participation and eligibility requirements. Lack of education or low literacy may also compromise the ability of potential microcredit clients to understand the benefits of credit or successfully utilize credit (Evans *et al.* 1999; Rahman *et al.* 1992). To date, no study has specifically examined the relationship between microcredit programs and accidental injury, though such an undertaking would be interesting given the proliferation of microcredit programs in the developing world.

Multivariate results also indicate that younger individuals have greater odds of accidental injury, as do men and unmarried individuals. These results support prior studies, which have also shown that younger individuals and men are more prone to accidental injury (Mashreky *et al.* 2010b; Rahman *et al.* 1998; Yusuf *et al.* 2000). Given the differential in gender roles in Matlab, the increased risk of injuries found among men in our analysis is not surprising. Bangladeshi men engage in heavy farm labor or fishing and are highly mobile—leaving the *bari* often to engage in labor and to procure resources for the household. All of these activities place men at a greater risk for accidental injury. Women, on the other hand, have very limited mobility and stay in the *bari* conducting housework and childcare, raising livestock, and providing after-harvest agricultural support such as sorting ad drying of rice and lentils. Most of these activities pose no great risk for accidental injury. The relationship between young age and accidents may be attributable to the type of work or risk behaviors in which young people engage. In Matlab, younger men conduct more of the heavy labor which may increase their risk for injury. At the same time, older individuals may have acquired higher levels of skill at their respective occupations, decreasing the risk for injury. Risk-taking behavior is also a normal part of young adulthood and injuries are a likely outcome of risk taking (Cohen and Potter 1999).

Several environmental variables were examined, including distance to major roads and rivers. These measures were included because prior literature indicates that road or traffic accidents make up a large percentage of accidental injuries in developing countries (Chandran *et al.* 2010; Mashreky *et al.* 2010b). We did not find a significant association between injuries and proximity to roads, which may be due to the rural nature of the study area. Matlab has only one road that can support motor vehicle traffic, a narrow paved road connecting Matlab Town to a highway. It is therefore unusual for a traffic accident to occur, though such events do occasionally occur in and around Matlab Town. In addition, the condition of the road is poor and drivers cannot typically drive fast. We did find a marginally significant association between proximity to rivers and injuries, which may reflect the high rates of drowning and other water-related injuries reported by other studies in the region (Chowdhury *et al.* 2009; Ahmed *et al.* 1999). The results for both distance to river and road may also be influenced by the type of injury that occurs in these locations and how these injuries are reported in the MHSS survey. Drowning and road traffic accidents are often fatal and would not have been picked up in the MHSS since only living individuals were surveyed. A more complete picture of accidental injury would require mortality data, which might well alleviate any systematic underreporting of more severe injuries.

Several limitations to this study should be mentioned, especially given the marginally significant relationship between socio-economic status and accidental injury. First, data on injuries were collected by self-report during the course of the MHSS survey. While interviewers prompted survey respondents to only report accidental injuries and injuries that were severe enough to restrict daily activities or warrant medical intervention, it is certainly possible that some injuries were incorrectly reported. Second, the current analysis would be stronger if some measure of severity of injury was collected. It may be that socio-economic status is more strongly predictive of severe injuries or certain types of injuries. In addition, the environmental variables (distance to river and road) could potentially be strong predictors of certain types of injuries, near-drowning and traffic accidents in particular which are often more severe. It is entirely possible that the addition of a severity index would reveal stronger relationships between ecological factors and injury risk. However, due to the fact that we did not collect data on the type of injury we cannot link information on environmental context to specific injury type, which may explain the poor relationship observed between river and road distance and odds of an injury. The second round of this survey (MHSS2, scheduled for 2012) will collect more detailed injury information, including severity and injury type. An alternative enhancement would be to include information on injury mortality that occurred in the region along with the survey data on morbidity. Addition of mortality data would certainly capture a certain level of severity and may lead to additional information about the role of social and environmental factors in injury risk. For example, it is possible that our data on occupation showed no relationship to injury because agricultural injuries (such as those involving heavy machinery) are more likely to lead to death. Alternatively, individuals engaged in non-farm occupations in Matlab often work in nearby factories and the risk factors related to occupational injuries may not adequately be captured by variables in the MHSS survey. This suggests that measurement and construction of variables in a survey of this nature should be carefully considered so that situations and environments that lead to injuries are correctly measured.

Conclusion

In their early and important study, Link and Phelan (1995) suggested that in order to understand the fundamental cause of disease public health scientists should focus on social conditions, rather than individually based risk factors. While fully acknowledging the importance of individual risk factors, they argued that these factors need to be considered within a broader social context. People are exposed to situations that influence individual behavior through interactions with their social and physical environment. In addition, some social conditions affect access to results that help individuals avoid diseases or modify unhealthy behavior. Link and Phelan based their argument on clear evidence that lower socio-economic status is associated with lower life expectancy, higher overall mortality and higher rates of infant and perinatal mortality (Link and Phelan 1995, 2000). These findings continue to obtain support as increasingly more studies examine social, economic and racial disparities in health.

Health and medical geographers have contributed a tremendous amount to understanding social and environmental context and health. Through the use of techniques and methodologies such as GIS, spatial cluster analysis, and spatial and multilevel regression, geographers have linked social and physical environment to individual behaviors and outcomes in order to study the impact of these environments on a variety of health outcomes. The study discussed in this chapter used several of these techniques to integrate social and economic data with individual-level information on injury. But health and medical geographers contribute more than just techniques. Fundamental to the field of geography is an understanding of the importance of human–environment interactions, and how social and environmental contexts change with geographic scale. The world exists at multiple scales, and some health-influencing processes, such as health or development policies, exist at a much higher scale (e.g., national or regional) than other processes, such as poverty or water quality (e.g., local or individual). By integrating social and physical context and a variety of geographic scales, we can increasingly understand how and at what scale certain aspects of context influence health outcomes. Medical and health geographers are uniquely positioned to place people in their context. This is, possibly, one of the more important contributions of geography to public health—the capacity to think at multiple scales and integrate political and ecological factors into a cohesive and holistic understanding of human health.

The study presented in this chapter, while limited in scope, exemplifies how we can integrate ecological and individual data using geographic methods. Within the literature on injuries, it is one of the first which attempts to integrate environmental, household, and individual-level factors. Since injuries are caused by a complex set of circumstances which exist within and outside the individual, this multilevel framework is an important way in which to frame and study this aspect of health and can be used as a model for future research. In this study, the distal (upstream) causes of injuries (socio-economic status) were found to be as important as the proximate (downstream) causes (age and sex). Thus, in Link and Phelan's words, socio-economic status may be a fundamental cause of injuries in rural Bangladesh. This finding is not surprising given that there are disparities in just about any health-related outcome one might study including a wide range of outcomes such as cardiovascular disease, obesity, depression, psychiatric disorders, the common cold, self-rated health and cholera (Diez-Roux *et al.* 2000; Pollack *et al.* 2007; Vegso *et al.* 2007; Morenoff *et al.* 2007; Cohen *et al.* 2008; Fernald *et al.* 2008; Williams *et al.* 2008; Clougherty *et al.* 2009; Subramanyam *et al.* 2009; Thurston and Matthews 2009; Emch *et al.* 2010). It is, however, surprising that there was such a clear socio-economic gradient in rural Bangladesh considering that almost everyone is "poor" by Western standards. In other words, there are even disparities in the poorest tail of the socio-economic distribution for injuries. Studies such as this should be used to direct additional research and inform injury prevention policy. Policies targeting lower-socio-economic-status households may succeed in reducing injury in this population. At the same time, more research is needed on the mechanisms by which socio-economic status affects risk. Do women in low- socio-economic-status households use unsafe cooking equipment which places them at greater risk for burns? Do men of greater socio-economic status typically not engage in heavy farm work, thereby decreasing their risk of injury? Once these mechanisms are understood more target injury prevention policies and interventions can be planned.

References

Ahmed, M.K., Rahman, M., and van Ginneken, J. (1999) Epidemiology of child deaths due to drowning in Matlab, Bangladesh. *International Journal of Epidemiology* 28: 306–311.

Alexandrescu, R., O'Brien, S.J., and Lecky, F.E. (2009) A review of injury epidemiology in the UK and Europe: some methodological considerations in constructing rates. *BMC Public Health* 9: 226.

Ali, M., Emch, M., Ashley, C., and Streatfield, P.K. (2001) Implementation of a medical geographic information system: concepts and uses. *Journal of Health, Population, and Nutrition* 19: 100–110.

Bollen, K., Glanville, J.E., and Stecklov, G. (2001) Socioeconomic status and class in studies of fertility and health in developing countries. *Annual Review or Sociology* 27: 153–185.

Chandran, A., Hyder, A.A., and Peek-Asa, C. (2010) The global burden of unintentional injuries and an agenda for progress. *Epidemiological Reviews* 32: 110–120.

Chowdhury, S.M., Rahman, A., Mashreky, S.R., Giashuddin, S.M., Svanström, L., Hörte, L.G., and Rahman, F. (2009) The horizon of unintentional injuries among children in low-income setting: an overview from Bangladesh health and injury survey. *Journal of Environmental Public Health* 2009: Article ID435403.

Clougherty, J.E., Eisen, E.A., Slade, M.D., Kawachi, I., and Cullen, M.R. (2009) Workplace status and risk of hypertension among hourly and salaried aluminum manufacturing employees. *Social Science and Medicine* 68: 304–313.

Cohen, L.R. and Potter, L.B. (1999) Injuries and violence: risk factors and opportunities for prevention during adolescence. *Adolescent Medicine* 10: 125–135.

Cohen, S., Alper, C.M., Doyle, W.J., Adler, N.E., Treanor, J.J., and Turner, R.B. (2008) Objective and subjective socio-economic status and susceptibility to the common cold. *Health Psychology* 27: 268–274.

Cubbin, C., LeClere, F.B., and Smith, G.S. (2000a) Socioeconomic status and the occurrence of fatal and nonfatal injury in the United States. *American Journal of Public Health* 90: 70–77.

Cubbin, C., LeClere, F.B, and Smith G.S. (2000b) Socioeconomic status and injury mortality: individual and neighbourhood determinants. *Journal of Epidemiology and Community Health* 54: 517–524.

Diez-Roux, A., Link, B.G., and Northridge, M. (2000) A multilevel analysis of income inequality and cardiovascular disease risk factors. *Social Science and Medicine* 50: 673–687.

Emch, M.E. (1999) Diarrheal disease risk in Matlab, Bangladesh. *Social Science and Medicine* 49: 519–530.

Emch, M., Yunus, M., Escamilla, V., Feldacker, C., and Ali, M. (2010) Local population and regional environmental drivers of cholera in Bangladesh. *Environmental Health* 9: 2.

Evans, T.G., Adams, A.M., Mohammed, R., and Norris, A.H. (1999) Demystifying nonparticipation in microcredit: a population-based analysis. *World Development* 27: 419–430.

Fauveau, V. (1994) *Matlab: Women, Children and Health*. Dhaka, Bangladesh: International Centre for Diarrhoeal Disease Research.

Fernald, L., Burke, H., and Gunnar, M. (2008) Salivary cortisol levels in children of low-income women with high depressive symptomatology. *Development and Psychopathology* 20: 423–436.

Gennarelli, T.A. and Wodzin, E. (2008) *The Abbreviated Injury Scale 2005: Update 2008*. Des Plaines, IL: American Association for Automotive Medicine (AAAM).

Giashuddin, S., Rahman, A., Rhaman, R., Mashreky, S.R., Chowdury, S.M., Linnan, M., and Shafinaz, S. (2009) Socioeconomic inequality in child injury in Bangladesh: implication for developing countries. *International Journal for Equity in Health* 8: 7.

Gosselin, R., Spiegel, D., Coughlin, R., and Zirkle, L. (2009) Injuries: the neglected burden in developing countries. *Bulletin of the World Health Organization* 87: 246.

Gwatkin, D.R., Ruston, S., Johnson, K., Paned, R.P., and Wagstaff, A. (2000) *Socioeconomic Differences in Health, Nutrition and Population in Bangladesh*. Washington, DC: The World Bank.

Huff, D. (1993) *How to Lie with Statistics*. New York, NY: W.W. Norton & Company.

Hunter, J.M. (1974) The Challenge of Medical Geography. In *The Geography of Health and Disease*, J.M. Hunter (ed.). Chapel Hill, NC: Department of Geography, 1–31.

icddr,b (2000) Health and demographic surveillance system, Matlab. *Registration of Health and Demographic Events* 33.

Iqbal, A., Shirin, T., Ahmed, T., Ahmed, S., Islam, N., Sobhan, A., and Siddique, A.K. (2007) Childhood mortality due to drowning in rural Matlab of Bangladesh: magnitude of the problem and proposed solutions. *Journal of Health, Population, and Nutrition* 25: 370–376.

Laverack, G. (2006) Improving health outcomes through community empowerment: a review of the literature. *Journal of Health, Population, and Nutrition* 24: 113–120.

Link, B.G. and Phelan, J. (1995) Social conditions as fundamental causes of disease. *Journal of Health and Social Behavior* 35: 80–94.

Link, B.G. and Phelan, J. (2000) The fundamental cause concept as an explanation for social disparities in disease and death. In *The Handbook of Medical Sociology*, C. Bird, P. Conrad, and A. Fremont (eds). Upper Saddle River, NJ: Prentice Hall, 3–17.

McKenzie, D.J. (2004) *Measuring Inequality with Asset Indicators*. BREAD working paper no. 42. Cambridge, MA: Bureau for Research and Economic Analysis of Development.

Marmot, M.G. (1986) Social inequalities in mortality: the social environment. In *Class and Health: Research and Longitudinal Data*, R.G. Wilkinson (ed.). London, UK: Tavistock, 21–33.

Mashreky, S.R., Rahman, A., Chowdury, S.M., Giashuddin, S., Svanstrom, L., Linnan, M., Shafinaz, S., Uhaa, I.J., and Rahman, F. (2008) Epidemiology of childhood burn: yield of largest community based injury survey in Bangladesh. *Journal of the International Society for Burn Injuries* 34(6): 856–862.

Mashreky, S.R., Rahman, A., Svanström, L., Khan, T.F., and Rahman, F. (2010a) Burn mortality in Bangladesh: findings of national health and injury survey. *Injury* 41: 792–795.

Mashreky, S.R., Rahman, A., Khan, T.F., Faruque, M., Svanström, L., and Rahman, F. (2010b) Hospital burden of road traffic injury: major concern in primary and secondary level hospitals in Bangladesh. *Public Health* 124: 185–189.

May, J.M. (1950) Medical geography: Its methods and objectives. *Geographical Review* 40: 9–41.

May, J.M. (1958) *The Ecology of Human Disease*. New York, NY: MD Publications.

Mayer, J.D. (1996) The political ecology of disease as one new focus for medical geography. *Progress in Human Geography* 20: 441–456.

Mobarak, A.M., Kuhn, R., and Peters, C. (2008) *Marriage Market Effects of a Wealth Shock in Bangladesh*. See www.du.edu/korbel/health/pdf/work/marriage_market.pdf (accessed 13 October 2011).

Mock, C.N., Abantanga, G., Cummings, P., and Koepsell, T.D. (1999) Incidence and outcome of injury in Ghana: a community-based survey. *Bulletin of the World Health Organization* 77: 955–964.

Mohammadi, R., Ekman, R., Svanström, L., and Gooya, M.M. (2005) Unintentional home-related injuries in the Islamic Republic of Iran: findings from the first year of a national programme. *Public Health* 119: 919–924.

Monmonnier, M. (1996) *How to Lie with Maps*. Chicago, IL: University of Chicago Press.

Morenoff, J.D., House, J.S., Hansen, B.B., Williams, D.R., Kaplan, G.A., and Hunte, H.E. (2007) Understanding social disparities in hypertension prevalence, awareness, treatment, and control: the role of neighborhood context. *Social Science and Medicine* 65: 1853–1866.

Murray, C.J. and Lopez, A.D. (1997) Mortality by cause for eight regions of the world: Global Burden of Disease Study. *Lancet* 349: 1269–1276.

Navarro, V., Muntaner, C., Borrell, C., Benach, J., Quiroga, A., Rodríguez-Sanz, M., Vergés, N., and Pasarín, M.I. (2006) Politics and health outcomes. *Lancet* 368: 1033–1037.

Nordberg, E. (2000) Injuries as a public health problem in sub-Saharan Africa: epidemiology and prospects for control. *East African Medical Journal* 77: S1–S43.

Openshaw, S. and Taylor, P. (1979) A million or so correlated coefficients. In *Statistical Applications in the Spatial Sciences*, N. Wrigley (ed.). London, UK: Pion, 127–144.

Pollack, K.S., Sorock, G.S., Slade, M.D., Cantley, L., Sircar, K., Taiwo, O., and Cullen, M.R. (2007) The association of body mass index and acute traumatic workplace injury in hourly manufacturing workers. *American Journal of Epidemiology* 166: 204–211.

Rahman, F., Andersson, R., and Svanström, L. (1998) Medical help seeking behaviour of injury patients in a community in Bangladesh. *Public Health* 112: 31–35.

Rahman, F., Rahman, A., Linnan, M., Giersing, M., and Shafinaz, S. (2004) The magnitude of child injuries in Bangladesh: a major child health problem. *Injury Control and Safety Promotion* 11: 153–57.

Rahman, H.S., Hossain, M., Chowdhury, O.H., Sen, B., and Hamid, S. (1992) *Rethinking Rural Poverty: The Case of Bangladesh*. Dhaka, Bangladesh: Bangladesh Institute of Development Studies.

Rahman, O., Menken, J., Foster, A., and Gertler, P. (2001) *Matlab Health and Socioeconomic Survey (MHSS)*, 1996. Computer file. Santa Monica, CA: RAND [producer]; Ann Arbor, MI: Interuniversity Consortium for Political and Social Research [distributor].

Sathiyasekaran, BW. (1996) Population-based cohort study of injuries. *Injury* 27: 695–698.

Smith, G.S. and Barss, P.G. (1991) Unintentional injuries in developing countries: the epidemiology of a neglected problem. *Epidemiology Reviews* 13: 228–266.

Subramanyam, M., Kawachi, I., Berkman, L., and Subramanian, S.V. (2009) Relative deprivation in income and self-rated health in the United States. *Social Science and Medicine* 69: 327–334.

Thurston, R.C. and Matthews, K.A. (2009) Racial and socio-economic disparities in arterial stiffness and intima media thickness among adolescents. *Social Science and Medicine* 68: 807–813.

Vegso, S., Cantley, L., Slade, M., Taiwo, O., Sircar, K., Rabinowitz, P., Fiellin, M., Russi, M.D., and Cullen, M.R. (2007) Extended work hours and risk for acute occupational injury: a case-crossover study of manufacturing workers. *American Journal of Industrial Medicine* 50: 597–603.

Wagstaff, A. (2002) Poverty and health sector inequality. *Bulletin of the World Health Organization* 80: 97–105.

Williams, D.R., Herman, A., Stein, D.J., Heeringa, S.G., Jackson, P.B., Moomal, H., and Kessler, R.C. (2008) Twelve-month mental disorders in South Africa: prevalence, service use and demographic correlates in the population-based South African Stress and Health Study. *Psychological Medicine* 38: 211–220.

World Bank (2010) World development indicators database. See http://data.worldbank.org/data-catalog/world-development-indicators (accessed 13 October 2011).

Yen, I.H. and Syme, S.L. (1999) The social environment and health: a discussion of the epidemiologic literature. *Annual Reviews of Public Health* 20: 287–308.

Yusuf, H.R., Akhter, H.H., Rahman, M.H., Chowdhury, M.K., and Rochat, R.W. (2000) Injury-related deaths among women aged 10–50 years in Bangladesh, 1996–97. *Lancet* 355: 1220–1224.

7 Human settlement, environmental change, and frontier malaria in the Brazilian Amazon

Marcia C. Castro and Burton H. Singer

Introduction

Between 1970 and 2007, the Brazilian Amazon grew from 7.8 million people to 23.6 million (IBGE 2009). This increase was mostly a result of in-migration in response to governmental incentives in support of agriculture, mineral extraction, cattle ranching, and wide-ranging human settlement (Benchimol 1985; Moran 1985; Sawyer 1986; Mahar 1989; Schmink and Wood 1992; Browder and Godfrey 1997). At the same time, the Amazon experienced significant environmental changes. In 1978, approximately 169,900 km^2 of the forest had been removed, and by 2003 the cleared area amounted to 648,500 km^2 (16.2 percent of the initially forested area of the Amazon; Fearnside 2005b). Since 2000, an average of about 18,000 km^2 of forest cover was removed annually (INPE 2009). Much of this environmental change was a direct result of the aforementioned governmental incentives, but some was a consequence of illegal activities in the region (e.g., timber extraction). Consequences of these transformations were observed in many aspects of the life of the Amazonian population. In this chapter we focus on health, particularly on malaria.

A series of reports released in the early 1900s revealed the poor sanitary conditions of the Amazon region and portrayed malaria as very severe, particularly proximate to rivers, swamps, and rubber extraction areas (Chagas 1903, 1913; Cruz 1913; Peixoto 1917). In 1945, indoor residual spraying with dichlorodiphenyltrichloroethane (DDT) was first utilized in the region and progressively expanded over time (Deane *et al.* 1948). In 1970, the region recorded almost 32,000 cases of malaria (61 percent of the national total). Since then, the disease has continued to become a burden in the Amazon, causing severe morbidity and accounting for the virtual majority of cases in Brazil.

Malaria cases are heterogeneously distributed in the Amazon and are mostly concentrated in agricultural settlement areas, mining camps, and areas of land invasion. While significant (human) migration of a largely malaria-naïve population (Sawyer and Sawyer 1987; Singer and Castro 2001) and increased deforestation are critical factors associated with increases of malaria transmission (Tauil *et al.* 1985; Sawyer 1992a; Patz *et al.* 2000; Olson *et al.* 2010), each factor cannot solely account for the disease burden (Sawyer 1992b). It is neither the deforestation process per se nor the migratory movement per se that leads to an increase in the number of malaria cases. Instead, it is the social context (which, broadly defined, also includes the impact of policies on people's lives) in which both occur and interact that determines the levels and patterns of malaria transmission. In that regard, there are examples of development projects coordinated by private and public firms that were implemented in the Amazon and that were able to prevent malaria outbreaks (Chagas *et al.* 1982; Couto *et al.* 2001). These projects involved both migration and deforestation. However, tight surveillance and strict rules regarding individual prevention, treatment, and vector control were critical for minimizing malaria risk.

Understanding this context, the interactions that it fosters, and the effects for the health of those involved requires a multidisciplinary analytical approach that combines aspects related to the environment, the political scenario, demographics, individual knowledge and behavior, culture, and economic conditions. In addition, since the majority of these aspects are dynamic and specific to each location, the analytical approach also needs to incorporate temporal and spatial dimensions. Consequently, a new concept known as frontier malaria was introduced to characterize malaria transmission in settlement areas in the Amazon (Sawyer 1988; Castro *et al.* 2006a). Frontier malaria highlights the importance of the social sciences for the study of malaria (Sawyer and Sawyer 1992), and the fact that the determinants of transmission are not static, but rather vary in importance over time and across different geographical scales (Castro 2002; Castro *et al.* 2006b; da Silva *et al.* 2010).

The importance of a multidisciplinary approach is at the core of the Malaria Eradication Research Agenda (malERA) initiative, which assembled a team of scientists to develop a research and development agenda for eradicating malaria (Alonso *et al.* 2011). It is also expressed in the need for multisectoral initiatives aimed at improving local health conditions (e.g., joint and coordinated efforts between governmental sectors that focus on public health, urban planning, agriculture, and sanitation; Singer and Castro 2007). Further, the need for a multifaceted approach has been increasingly recognized in different research arenas; for example, bringing together multiple disciplines has been linked to environmental conservation and management in the Amazon (Fearnside 2010).

In this chapter we provide a comprehensive assessment of the context in which frontier malaria emerges in settlement areas. Our results indicate that malaria transmission in the early years of occupation is mostly driven by environment-related conditions. By comparison, after approximately ten years, social, economic, and behavioral aspects become more relevant. This overall pattern, however, shows local variability, which has important implications for the selection and implementation of control interventions.

The remainder of this chapter is organized in four sections. We start with a brief description of the environmental transformations that have occurred in the Amazon in the recent past, with a special focus on potential consequences of those changes on the patterns of malaria transmission. The following section provides a comprehensive discussion of a systemic view of malaria, introduces the concept of frontier malaria, and characterizes the framework required for the study of malaria risk in the Amazonian context. The next section presents a thorough description of malaria risk profiles in one settlement area of the Amazon, utilizing an analytical framework that allows the use of multidisciplinary information as well as the incorporation of spatial and temporal dimensions. Finally, we conclude with some remarks on policy implications of our findings with regard to malaria and to the continued opening of new settlements.

The Brazilian Amazon: an evolving ecosystem

As of 2007, almost 2500 agricultural settlement projects had been opened in the Brazilian Amazon (MDA/INCRA/DTI 2007) and varying in terms of available resources, average size of land, and soil quality. More recent projects contained smaller plots (30–40 ha on average, compared to the more than 100 ha plots of earlier projects) so that a larger number of people could be settled (Becker 1990). However, newer projects often had poorer soil quality, with most of the area demanding the use of fertilizers in order to become suitable for agricultural. Unfortunately, the vast majority of settlers could not afford fertilizers, machinery, or extra labor (Fearnside 1986), compromising the chances of successful crop production.

The negative impacts of these projects were equally varied, including land turnover, disease outbreaks, deforestation, and land concentration (Brondízio *et al.* 2002; Browder 2002; Wood and Porro 2002). Malaria outbreaks were one of the factors that led to turnover of settlers and contributed to land concentration (Sawyer and Sawyer 1987). Martine (1990) highlighted the fact that, in the early 1970s, some settlers would exchange their plots for medical treatment, a practice that made some of the first physicians that came to Rondônia state large landowners. Moreover, some migrants who did not succeed in obtaining land returned to their place of origin or kept moving to alternative locations in the Amazon, often serving as carriers of the malaria parasite and contributing to the spread of the disease.

Initially, the government did not consider pasture as an option for land use in settlement projects, since the forest provided varied ways for sustainable economic use (Wittern and Conceição 1982). Nevertheless, the 2006 Agrarian Census revealed that 19 percent of the area used for pasture in the country was located in the Amazon (IBGE 2009). Most importantly, however, is the fact that in 1970 this percentage was approximately 5 percent. The pasture area in three states of the Amazon region, namely Rondônia, Acre, and Pará, increased by 3804, 1540, and 326 percent, respectively, between 1970 and 2006. Rondônia, in particular, witnessed intense government sponsored settlement efforts during the 1980s and 1990s. However, land turnover for varied reasons (e.g., high malaria transmission, poor soil quality) resulted in land concentration (INCRA/CRUB/UNB 1997), which was often connected with a change of land use to pasture (Sawyer and Sawyer 1987; Amaral 2007; Grego *et al.* 2007).

Regarding urbanization, the number of municipalities in the region more than doubled between 1970 and 2000, and the percentage of people living in urban areas shifted from 37 percent in 1970 to 69 percent in 2000 (IBGE 2009). Part of this growth was driven by unplanned occupations of peripheral areas, which resulted in deforestation and alteration of watercourses (among other environmental modifications). This pattern of urban growth has been associated with increases in malaria incidence in Amazonian cities (Gonçalves and Alecrim 2004).

Much transformation was also brought about by infrastructure projects that resulted in population displacement, deforestation, illegal occupation, land speculation, and significant health burdens (Moran 1981; Martine 1982; Fearnside 1989; Schmink and Wood 1992; Fearnside 1999; Nepstad *et al.* 2000; Fearnside and Laurance 2002; Fearnside 2005a). Despite changes initiated in 1981 in the Brazilian legislation, which culminated in the requirement that any infrastructure project should have an environmental impact assessment (EIA) and a plan documenting mitigation actions and compensation strategies (Machado 1995), much controversy still takes place with newly approved and planned projects (Agra Filho 1993; Ninio *et al.* 2008). The most recent examples include the approval of three dams in the Amazon (Jirau and Santo Antônio in Rondônia state, and Belo Monte in the state of Pará) that are expected to, among other impacts, result in severe environmental degradation, create heavy social disruptions, and increase the transmission of malaria (Fearnside 2006; Rosa 2006; Katsuragawa *et al.* 2010).

Meanwhile, novel initiatives of sustainable forest management are being promoted locally through varied programs supported by the government or by community groups and non-governmental organizations. Of note is an innovative sustainable land development strategy implemented in the state of Acre, which is based on a program of socio-environmentalism (Viana 2007).

Environmental change and malaria

Tropical rainforests provide good conditions for a diversity of insects, given the high temperature, humidity, and rainfall throughout most of the year. In the case of the main malaria vector in the Amazon, *Anopheles darlingi*, natural breeding places are often observed in the forest margins and near rivers and small streams, becoming more abundant at the beginning and end of the rainy season, which starts in October and ends in March (Tadei *et al.* 1998). Inside the undisturbed forest, however, the ideal conditions for *An. darlingi* are seldom found, since standing water is acidic and the partial shade favored by this species is absent. This natural equilibrium is often broken by human-made modifications that tend to contribute to an increase in the number of water habitats suitable for *Anopheles* breeding (Coimbra Jr. 1988).

A common agricultural practice in the Amazon is slash-and-burn. A poor clearing and burning process, however, can cause the obstruction of steams (a result of fallen trees that block the flow of water) and can leave the taller trees standing, providing the necessary partial shade for *An. darlingi* breeding. This often creates the forest fringe, a frontier between the forest and the property, where the risk of malaria transmission is very high. In the early years of occupation, settlers tend to live and work in close proximity to the forest fringe, since the initially cleared area is usually small. Houses are of poor quality and offer little, if any, protection against mosquitoes (initially, houses are made of sawmill leftovers, palm thatch, cardboard, and plastic). In addition, the opening of roads often precedes human occupation and creates corridors that facilitate the spread of mosquitoes (Smith 1982; Sawyer 1992b). For a related discussion on development as a contributing factor to infectious disease, see Chapter 9.

Transformations resulting from mining activities also have the potential to increase the risk of malaria transmission. The association between malaria incidence and gold mining can be attributed to five major factors: (i) workers have very high mobility, facilitating the spread of the disease; (ii) lack of infrastructure in gold mines that leaves the workers without basic healthcare and appropriate housing; (iii) high exposure to malaria vectors due to intense work shifts, usually amounting to more than 12 hours a day; (iv) the nomadic characteristic of the activity itself – when the gold in the area is exhausted the area is abandoned, and gold miners move on to other areas, most of the time carrying malaria; and (v) most workers understand malaria as a disease that is inevitable, and do not adopt personal protection (Barbieri and Sawyer 1996).

Another important habitat transformation that can facilitate the spread of malaria in the Amazon is the process of extended urbanization (Monte-Mór 1997, 2004). A few years after initial occupation, colonization areas expand in a process of extended urbanization that moves beyond cities and towns to encompass villages, hamlets, proto-urban sites of various types, and rural areas such as modern farms and cattle ranches, and mining sites. Diverse urban–rural arrangements produce integrated local micro-regions where mobile families and individuals divide time and resources between localities on a regular and seasonal basis. Although allowing for further cooperation within and between communities, this process has often produced additional environmental changes and intense human mobility that may favor the spread of malaria.

Malaria transmission: a dynamic process

Understanding the dynamics of malaria transmission demands a systemic view that describes the relationships among human, vector, and parasite that occur in a specific environmental setting (Singer and Castro 2011). This understanding requires a multidisciplinary analytical approach, combining epidemiology, environmental sciences, demography, anthropology, entomology, economics, politics, geography, hydrology, and molecular biology (Castro *et al.* 2006a).

The human component of the systemic view of malaria includes conditions that characterize individuals (e.g., their susceptibility to infection, their knowledge, behavior, demographics, and ways through which they can transform the local environment) and the context in which individuals interact (e.g., political, social, and economic). The vector component includes characteristics that are crucial for choosing the ideal package of interventions to launch an integrated vector management program tuned to a given ecosystem (Beier *et al.* 2008) and for measuring local exposure to malaria infection (e.g., vector feeding and biting). The parasite component refers to issues that determine the effectiveness of drug-based intervention, and that indicate levels of disease severity. Finally, the environment includes both the natural environment (e.g., climate, elevation, and hydrography) and the built environment (e.g., land use, sanitation, and housing quality). These components are not static but rather interact with each other, producing local unique profiles of the disease.

Frontier malaria

Frontier malaria is defined by a set of characteristics operating over time and at three spatial scales (Sawyer 1988, 1992a; Castro *et al.* 2006a). Initially, and at a micro/individual level, vector densities are high with little seasonal variation as a consequence of ecosystem transformations that promote *An. darlingi* larval habitats (partial shade near the forest fringe and along river edges, and clear standing water of high pH; Charlwood 1980, 1996; Tadei *et al.* 1998). Human exposure is intense, reflecting limited knowledge of disease transmission mechanisms among settlers: *An. darlingi* has a bimodal biting pattern (Klein and Lima 1990), occurring at dawn and dusk just when settlers are going to and returning from their agricultural fields. *Plasmodium falciparum* (the most lethal form) is the primary parasite augmented by limited abundance of *P. vivax* (often associated with high morbidity and low mortality) in the early stages of settlement. This pattern reverses as the settlement becomes more stable (Sawyer 1992b). Acquired immunity is low among new settlers (who mostly come from malaria-free areas). Housing quality is poor, thereby rendering indoor residual spraying ineffective. Curative health services are sparsely available, thus limiting anti-malarial drug distribution.

Second, at a community level, frontier malaria is characterized by weak institutions, minimal community cohesion, political marginality of settlers, and high rates of both in- and out-migration. Third, at a state and national level, frontier malaria is characterized by unplanned development of new settlement areas (e.g., areas with poor soil quality), stimulated by agricultural failures at previous settlement localities and by a desire of people to avoid further malaria episodes. This process, however, only serves to promote further transmission.

Frontier malaria also follows a distinctive time path (Sawyer and Sawyer 1992; Castro 2002). At the opening of a settlement area, malaria transmission rises rapidly. After 6–8 years, the unstable human migration and the highly variable ecological transformations (driven by variation in land clearance practices and local ecology) is replaced by a more organized process of urbanization and development of community cohesion. The final stage of frontier malaria is a gradual transition to more stable low levels of transmission.

The complexity of frontier malaria becomes even more evident if we consider distinct types of settlement that have occurred in the Amazon, as proposed by Sawyer (1988) and summarized in Table 7.1. Although this is not an exhaustive list, it highlights the heterogeneity that has been observed regarding the potential for malaria transmission between different types of human settlement in the Amazon. Also, each type of settlement is associated with substantial internal variation in the risk of malaria transmission, thereby adding to the complexity of frontier malaria. Moreover, the close proximity between different types of settlement can alter the expected pattern of malaria (e.g., indigenous areas may observe outbreaks of malaria due to gold mining activities nearby). Although the broad characteristics of frontier malaria apply to each type of settlement where transmission is observed, detailed aspects of the dynamics of malaria depend on how different elements of the malaria system interact in each location.

Table 7.1 Types of human settlement in the Brazilian Amazon

Type of settlement	*Characteristics*
I Rural	
A. Extractive	
1 Rubber estates	Currently, these areas have low population density, low human mobility, and high levels of acquired immunity among adults.
2 Open mining	Miners come and go freely, without any systematic control of the government. They often lack proper housing, clothing, and overall infrastructure. These areas frequently register very high malaria prevalence, and it is extremely hard to control the disease.
3 Closed mining	Miners are under strict control of the government. Despite the potential for high malaria transmission, disease control is easier.
B. Agricultural	
4 Old colonization	Following the proposed time path of frontier malaria, old colonization areas have lower transmission, and good opportunities for effective control since better infrastructure is in place, housing improved, and deforestation is less intense.
5 New colonization	While the context in which the project is implemented can have an impact on the pattern of malaria transmission, the majority of new colonization areas observed outbreaks of malaria. In this chapter we detail the dynamics in one of such areas.
6 Old ranching	Areas where most forest cover was removed, with very little population density, and therefore very low malaria prevalence.
7 New ranching	Implies heavy forest clearing and large number of temporary workers highly exposed to the risk of malaria infection.
II Urban	
A. New isolated area	
8 Spontaneous town	Following the concept of extended urbanization, these areas are exposed to high human mobility (rural–urban), and can be in close proximity to the forest.
9 Company town	Organized small towns or camps, under strict control of public or private companies.
10 Peri urban area	Sprawling suburbs, often nearby breeding habitats, but with easier access to medical services.
B. Old area	
11 Old town	Often have very little or no malaria.
III Indigenous	
12 Indigenous area	Impacted by activities happening in the vicinity of the area (e.g., farming, ranching, and mining), and therefore under the risk of malaria outbreaks.

Source: Adapted from Sawyer (1988)

In the next section we focus on one of these types of settlement, a colonization project, and detail the variations in malaria risk locally and temporally.

Framework of frontier malaria transmission

The defining characteristics of frontier malaria comprise a framework for understanding transmission, which has three important features: it is spatial, it is temporal, and it is multidisciplinary. The framework recognizes that, in a constantly changing ecosystem such as the Amazon, malaria transmission does not depend solely on any one of environmental changes, personal behavior, migratory patterns, or the biology of mosquitoes and parasites. It depends on a complex interplay of factors, and the context in which each of them are operating.

The framework was empirically assessed using data from the Machadinho settlement project, located in the western Amazon (Singer and Castro 2001; Castro 2002). Started in late 1984, Machadinho had an original and carefully planned plot design in which the shape of the plots was irregular, following the course of rivers and streams, so that every plot would have a natural source of water in its back. The average plot size was approximately 40 ha. Roads were planned in such a way that during the rainy season the increased water volume of rivers and streams would not leave them impassable. Areas with very irregular elevation were assigned as protected forest reserves. With regard to malaria, an outbreak was soon observed, following the arrival of the first settlers in late 1984. In 1985, the Annual Parasite Index (API; the number of positive blood slides per total population per year) reached 3400 positive slides per thousand people, 65.7 percent of the population had malaria at least once, and this number jumped to 90.1 percent in the next year. Also in 1986, 55.9 percent of people had malaria episodes during more than five months of the year (Sawyer 1986; Sydenstricker 1992). By 1995, following the consolidation of the settlement project, however, the API had been reduced, corroborating the proposed temporal pattern of frontier malaria (Sawyer and Sawyer 1992; Castro *et al.* 2006a).

Four field surveys were conducted in Machadinho from the inception of the project and covered a period of ten years (1985, 1986, 1987, and 1995). Issues investigated in these surveys included a variety of topics ranging from demographics, personal behavior, socio-economic characteristics, and local environmental conditions. In addition, all data were georeferenced to the plot level. Therefore, these data facilitated empirical evaluation of frontier malaria, considering its spatial, temporal, and multidisciplinary features.

Malaria risk profiles in the Brazilian Amazon

Based on variables collected during the surveys conducted in Machadinho, distinct malaria risk profiles were constructed by conditions observed in two broad domains: environmental and behavioral/economic. Within the environmental domain, three sets of conditions were defined by survey variables that reflect exposure to *An. darlingi* larval habitats and/or adult mosquito biting preferences:

- housing/plot characteristics (e.g., soil quality, elevation, quality of roof and walls, effectiveness of house sealing, source of water, type of bathroom, distance to the hospital, and road quality);
- proximity to features of importance for exposure (e.g., distance to the nearest forest, distance to the forest fringe, distance to a culvert, distance to the main river and to smaller streams, distance to the nearest neighbor, and distance to bodies of temporary water); and
- land transformation (e.g., crop area planted, pasture area, cleared area, and type of crop—coffee, cocoa, rubber tree, etc.).

Within the behavioral/economic domain, three sets of conditions were defined by survey variables that describe the capacity of individuals to protect themselves, and behaviors that may put them at more or less risk for acquiring malaria:

- demographics (e.g., education of the household head, education of the household head's spouse, occupation, number of people in the house, and status of family members, i.e. living in Machadinho or not);
- migratory history, use of protective measures, and malaria knowledge (e.g., place of origin and time of arrival at the settlement program, time of arrival in the Amazon region and in Rondônia, region of residence in the 12 months before moving to Machadinho, use of insecticides, house spraying with DDT, belief that government should spray houses with DDT, use of repellent, use of herbs to treat malaria, belief that malaria is transmitted by dirty water, knowledge that malaria is transmitted by a mosquito, and membership to a community association); and
- economics (e.g., ownership of a chainsaw and/or a planter, number of household goods owned, settler is the first owner of the plot, household owns chickens and/or pork, and received a bank loan for agriculture, pasture, or equipment purchase).

Knowledge on the vector and the parasite (Giglioli 1956; Tauil *et al.* 1985; Tadei 1991; Charlwood 1996; Consoli and Oliveira 1998; Tadei *et al.* 1998) was used to select conditions for the environmental and human components that directly or indirectly could impact or be impacted by vector and parasite characteristics. Also, the spatial dimension was incorporated through the identification of subareas in Machadinho that were homogenous regarding the spatial distribution of malaria (Castro *et al.* 2006a, b). A total of 15 subareas were considered over four periods of time and using 46–63 categorical variables representing 160–213 conditions included in the two domains described above (the number of variables and conditions varied by time period due to slight changes in the survey questionnaire).

Profiles of low and high malaria risk

Identification and description of high and low malaria risk profiles in a heterogeneous landscape (such as Machadinho), where no combination of conditions occurred at particularly high frequency, were facilitated through the use of grade of membership (GoM) models (Singer 1989; Castro *et al.* 2006a). A total of 120 malaria risk profiles were obtained: 15 subareas, each over four periods of time for two levels of risk (high and low). As an example, Table 7.2 shows the profile description of high and low malaria risk for one subarea in Machadinho in 1986 (Castro 2002).

Interpretation of the profiles was facilitated by ethnographic information collected through personal observation (R.L. Monte-Mór, personal communication, 2000) and by focus groups with worker's unions and community associations (Castro *et al.* 2006a). Based on the results of the GoM analysis, two scenarios were identified. The first described the initial three years of the settlement project. At that time there was less diversity in terms of social and economic aspects. The settlers were basically poor, had no resources to improve the quality of the soil or to carry out mechanized agricultural production, had no immunity against malaria, and lacked basic knowledge regarding malaria risk factors. At the same time, there were large differences in malaria transmission, and traditional control measures were not effective. Under that scenario, profiles of low and high malaria risk in 1985, 1986, and 1987 were mostly determined by conditions related to the natural and human-made environment. The most important conditions at that time were the quality of and characteristics near the house, including distance to the forest and to sources of water, cleared, planted

Table 7.2 Description of high and low malaria risk profile in subarea 4, Machadinho, 1986

High malaria risk	*Low malaria risk*
The settler was not working one year before he or she came to Machadinho, arrived in Machadinho in 1986, and the region of residence one year before he or she came to Machadinho was either south or center-west. The house has poor quality walls and roof, has only one or two rooms, and is located in an area with low elevation. The household owns less than five goods, and lacks a chainsaw. The distance to the nearest neighbor is greater than 1 km. The cleared area in the plot is smaller than 16 ha, no major crops are cultivated, and the settler does not have chickens and pigs in the plot.	The settler was engaged in rural activities one year before he or she came to Machadinho, and the region of residence one year before he or she came to Machadinho was North. The household owns more than five goods, including a planter. The distance from to the house to the health post in the urban area is less than 20 km. The plot is not within a zone of 900 meters from a protected forest reserve, and there is no source of permanent water near the house. The planted area in the plot is between 16 and 45 ha, and different crops are cultivated. The settler has chickens and pigs in the plot, and the house was sprayed with DDT.

Note: Subarea 4 had the highest malaria rate among all subareas defined in 1986: 43.8 percent. Description of malaria risk profiles obtained through GoM modeling. The model is specified by extreme profiles (here defined as low and high malaria risk), and each individual observation is assigned a grade of membership score. These scores are a quantitative measure of the degree of similarity of the individual to each of the profiles. In addition, the model estimates the probability that conditions (categories) of each variable will occur in a certain profile. These estimated probabilities are compared against the marginal distribution of each condition in order to define which conditions are significant in each profile (Singer 1989; Castro *et al.* 2006a).

and pasture area, bathing place, topography, distance to the nearest neighbor, and the existence of alternative sources of blood meals for mosquitoes (chickens and pigs).

The second scenario described the situation at later stages of the settlement project, here represented by the 1995 data. At that time, settlers were more diverse in terms of social and economic aspects. Some improved their education, knowledge about the disease, and financial resources. Others were newcomers to the area and faced the same adverse conditions observed in the early years of occupation. Despite this diversity, malaria was more homogenous in space, and occurred at much lower levels than in the first years. Conditions related to the human-made environment were still contributing to the definitions of low and high-risk profiles, although personal characteristics started to contribute consistently to these definitions.

In summary, while both environmental and behavioral/personal domains were important in describing the mechanisms of frontier malaria transmission in Machadinho, their relative importance depended on time and varied across space. These trends have direct implications for selecting and targeting malaria control interventions. One of the great advantages of GoM models for the purpose of this analysis is that by characterizing the conditions of the groups of people who effectively avoid malaria, but also of those who consistently become infected, it is possible to observe that the conditions that characterize high risk are not the exact opposite of those that describe low risk, and that evidence facilitates policy making that combines a description of situations to avoid and others to adopt, with the aim of minimizing the risk of transmission.

Profile changes over time and space

Results of the GoM analysis highlighted important differences in malaria risk profiles over time and space. First, each one of the 120 profiles of high and low malaria risk was unique. This finding reflects how dynamic and complex the process of frontier malaria transmission is, thereby highlighting the need to devise combinations of control interventions targeted locally and temporally.

Second, some conditions were significant contributing factors for either high or low risk, depending on the area and on the time period in question. For example, in 1985 settlers that did not use medicinal plants to treat malaria were better off than those who did; but in 1995 this same condition contributed to higher malaria risk. It could be the case that available medicinal plants could only be effective under less intense transmission and possibly with changed immunological conditions of settlers. Less intense transmission is, in turn, linked to environmental risk factors, such as the proximity of the house to the forest and the quality of the house. Moreover, in the first year of the settlement it is not expected that settlers could have had the chance to get acquainted with local knowledge, depending much more on the use of drugs for treatment. Another example is the effect of owning a chainsaw. In areas with low malaria rates which have reached a certain level of stability, lacking a chainsaw was a contributing condition to low risk. This finding suggested that settlers did not work in the clearing process outside their plots, an activity that could increase their exposure to the disease. However, lacking a chainsaw in areas with higher malaria rates was a condition defining a high-risk profile. The absence of a chainsaw here frequently implied association with a poor land-clearing process in the plot, which in turn is linked to proliferation of breeding habitats and increased vector density. The distance to the nearest neighbor also showed changing effects over time. In 1985, when most plots had a very low cleared area and faced similar environmental risks, having neighbors only at distances greater than 2 km was a condition defining a low malaria risk. In this case, settlers were protected by staying isolated. In 1986, when malaria was extremely high in Machadinho, very short distances were contributing factors to low malaria risk, possibly reflecting the support provided in the event of an emergency. In the following years, distances between 1 and 2 km were associated with high malaria risk, but some subareas did not have this condition as a contributing factor for the definition of profiles. This trend is likely to be related to the amount of cleared area in the neighboring plot.

Conclusions

Malaria is a complex disease. Proper understanding of its transmission depends on factors driven by its social, political, environmental, behavioral, economic, biological, and epidemiological dimensions. Most importantly, it depends on how each dimension interacts with and modifies the others. Also, the dynamics vary spatially and temporally, as exemplified by the analysis of frontier malaria transmission in human settlements presented in this chapter. As a result, attempts to effectively control malaria in these settings are equally complex. Depending on where the settlement is located, how old it is, and the specific social context, a factor that a priori might be considered as a condition determining a high risk of malaria may in fact be conducive to low risk. Therefore, interventions need to be targeted and developed according to local and temporal vulnerabilities. Implementing targeted interventions in the entire Amazon region is far from trivial. It demands intersectoral collaboration (Singer and Castro 2007), local coordination, capacity-building, multidisciplinary research to generate evidence, and a strong political will that facilitates not only sustainable health interventions but also a development policy that protects, respects, and therefore promotes rational use and growth of the region.

Although much information has been compiled in the recent past, proper knowledge of the conditions that characterize groups more vulnerable to malaria in different types of settlement in the Amazon is not always available. In this chapter we have proposed a framework for the study of frontier malaria, and a methodological approach that allows the description of risk profiles based on survey variables. The methodology has the advantage of indicating the degree to which each individual in the sample belongs to a profile, with the flexibility of allowing individuals to share characteristics of more than one profile.

Our analysis focused on one type of human settlement. It is expected that profiles of frontier malaria risk would present variations between settlement types (Table 7.1), but also within the same type (as exemplified by the Machadinho case). Yet, our results for Machadinho shed light on the processes observed in other old and new colonization projects (Table 7.1, types B4 and B5) in the Brazilian Amazon. It provides a starting point for improved control strategies in existing projects, and offers varied recommendations on how to proper plan new ones. While the concept of frontier malaria cannot be generalized to other regions in the world where malaria is endemic (just as fringe malaria only applies to Africa), the multidisciplinary framework and the methodological approach proposed in this chapter could be applied to any other settlement type, and to other areas where malaria is endemic, upon the availability of data.

Currently, new settlements continue to be opened in the Amazon, many of them as part of planned agrarian reform initiatives (Sparovek 2003). In 2001 new legislation was approved demanding the issuing of licenses for the establishment of new settlement projects in the country, including an evaluation of the susceptibility of the area for malaria transmission if the settlement is located in the Amazon region. But the implementation of the law has been far from ideal, and less than 10 percent of all settlement projects implemented in Brazil (more than 7000) have the required licenses (Araújo 2006).

Our results indicate feasible strategies that could be employed to mitigate malaria transmission in newly opened settlement, such as:

1. match the soil composition and agricultural potential of plots to the capacity of settlers for farming (based on their past experience) at the time plots are allocated;
2. encourage rapid initial clearance of land for agriculture, and the construction of houses that offer protection against mosquitoes;
3. make available an effective agricultural extension service to provide technical support to settlers; and
4. install health clinic facilities in parallel with opening of the settlement area.

Items (1) and (3) contribute to effective farming, while item (2) is a preventive measure that minimizes the exposure time of new settlers to *An. darlingi* habitats upon initial arrival at the settlement site. Finally, item (4) provides a basis for effective diagnosis and treatment of malaria, as well as for promotion of health education, emphasizing preventive measures throughout the settlement area (Castro *et al.* 2006a).

In addition to current challenges in controlling frontier malaria in the Amazon, three issues are likely to add further complexity to the task (Castro and Singer, 2011). First, natives and long-term residents who were continually exposed to malaria are likely to be asymptomatic. They are unlikely to search for healthcare, but they are able to infect mosquitoes (Alves *et al.* 2005). Asymptomatics can act as reservoirs of malaria, facilitating the spread of the disease to non-immune populations (Coura *et al.* 2006), and potentially contributing to sustained levels of transmission (Macauley 2005). The prevalence of asymptomatic infections in the Amazon, and its spatial distribution, is largely unknown. Second, large-scale development and integration plans (e.g., dams, roads, and hydroways) will likely bring about further changes to the social and environmental landscapes of the region (IDB 2006), with important health consequences. Third, the Amazon's fragile ecosystem can be significantly impacted by extreme climatic events, resulting in drought, flooding, excessive heat, climate-driven population displacement, reduced crop production, and threatened water and food security, all of which impact individuals' vulnerability to infections including malaria.

In summary, the social, economic, political, and environmental dynamics of frontier malaria can contribute to increase inequality among settlers. On the one hand, those that have limited financial and technical resources can become worse off, abandoning their land or exchanging it for medical care (as was observed in Rondônia in the early 1970s). On the other hand, those that have better resources are more likely to succeed, improving even more their conditions. The complexity and dynamics of frontier malaria transmission, exemplified in this chapter by the Machadinho case, suggests that a region-wide uniform control strategy may only reduce transmission up to a point, after which combinations of interventions that consider local and temporal idiosyncrasies will be needed. The analytical approach used in this chapter facilitates the identification of these idiosyncrasies, and provide evidence for the design of locally and temporally targeted interventions.

References

Agra Filho, S.S. (1993) *Os estudos de impactos ambientais no Brasil: uma análise de sua efetividade*. Brasília, Brazil: Instituto de Pesquisa Econômica Aplicada (IPEA).

Alonso, P.L., Brown, G., Arevalo-Herrera, M., Binka, F., Chitnis, C., Collins, F., Doumbo, O.K., Greenwood, B., Hall, B.F., Levine, M.M., Mendis, K., Newman, R.D., Plowe, C.V., Rodrıguez, M.H., Sinden, R., Slutsker, L., and Tanner, M. (2011) A research agenda to underpin malaria eradication. *PLoS Medicine* 8(1): e1000406 (doi:10.1371/journal.pmed.1000406).

Alves, F.P., Gil, L.H.S., Marrelli, M.T., Ribolla, P.E.M., Camargo, E.P., and Silva, L.H.P. (2005) Asymptomatic carriers of *Plasmodium* spp. as infection source for malaria vector mosquitoes in the Brazilian Amazon. *Journal of Medical Entomology* 42: 777–779.

Amaral, J.J.O. (2007) *Os Latifúndios do INCRA*. Porto Velho, Brazil: Editora da Universidade Federal de Rondônia (EDUFRO).

Araújo, F.C. (2006) Reforma Agrária e Gestão Ambiental: Encontros e Desencontros. Master's thesis, Universidade de Brasília, Brasília, Brazil.

Barbieri, A.F. and Sawyer, D.O. (1996) Malária nos garimpos do Norte de Mato Grosso: diferenciais na homogeneidade. In *Anais do X Encontro Nacional de Estudos Populacionais, 1996 Caxambú*. Caxambú, Brazil: Associação Brasileira de Estudos Populacionais (ABEP), 2413–2426.

Becker, B.K. (1990) Estratégia do Estado e povoamento espontâneo na expansão da fronteira agrícola em Rondônia: interação e conflito. In *Fronteira Amazônica: questões sobre a gestão do território*, B.K. Becker, M. Miranda, and L.O. Machado (eds). Brasília, Brazil: Editora UnB, 147–164.

Beier, J.C., Keating, J., Githure, J.I., Macdonald, M.B., Impoinvil, D.E., and Novak, R.J. (2008) Integrated vector management for malaria control. *Malaria Journal* 7: S4.

Benchimol, S. (1985) Population changes in the Brazilian Amazon. In *Change in the Amazon Basin*, J. Hemming (ed.). Manchester, UK: Manchester University Press, 37–60.

Brondízio, E.S., McCracken, S.D., Moran, E.F., Siqueira, A.D., Nelson, D.R., and Rodriguez-Pedraza, C. (2002) The colonist footprint: toward a conceptual framework of land use and deforestation trajectories among small farmers in the Amazonian frontier. In *Deforestation and Land Use in the Amazon*, C.H. Wood and R. Porro (eds). Gainesville, FL: University Press of Florida, 133–161.

Browder, J.O. (2002) Reading colonist landscapes: social factors influencing land use decisions by small farmers in the Brazilian Amazon. In *Deforestation and Land Use in the Amazon*, C.H. Wood and R. Porro (eds). Gainesville, FL: University Press of Florida, 218–240.

Browder, J.O. and Godfrey, B.J. (1997) *Rainforest Cities: Urbanization, Development, and Globalization of the Brazilian Amazon*. New York, NY: Columbia University Press.

Castro, M.C. (2002) Spatial configuration of malaria risk on the Amazon frontier: the hidden reality behind global analysis. PhD thesis, Princeton University, Princeton, NJ.

Castro, M.C. and Singer, B.H. (2011) Malaria in the Brazilian Amazon. In *Water and Sanitation Related Diseases and the Environment: Challenges, Interventions and Preventive Measures*, J.M.H. Selendy, (ed.). John Wiley & Sons, Inc, 401–419.

Castro, M.C., Monte-Mór, R.L., Sawyer, D.O., and Singer, B.H. (2006a). Malaria risk on the Amazon frontier. *Proceedings of the National Academy of Sciences* 103: 2452–2457.

Castro, M.C., Sawyer, D.O., and Singer, B.H. (2006b). Spatial patterns of malaria in the Amazon: implications for surveillance and targeted interventions. *Health and Place* 13: 368–380.

Chagas, C. (1903) *Estudos Hematológicos no Impaludismo*. Rio de Janeiro, Brazil: Typographia da Papelaria União.

Chagas, C. (1913) *Notas Sobre a Epidemiologia do Amazonas*. Rio de Janeiro, Brazil: Instituto Oswaldo Cruz.

Chagas, J.A., Barroso, M.A., Amorim, R.D., and Robles, C.R. (1982) Controle da malaria em projeto hidreletrico no estado do Amazonas. *Revista Brasileira de Malariologia e Doenças Tropicais* 34: 68–81.

Charlwood, J.D. (1980) Observations on the bionomics of *Anopheles darlingi* Root (Diptera: Culicidae) from Brazil. *Bulletin of Entomological Research* 70: 685–692.

Charlwood, J.D. (1996) Biological variation in *Anopheles darlingi* Root. *Memórias do Instituto Oswaldo Cruz* 91: 391–398.

Coimbra JR., C.E.A. (1988) Human factors in the epidemiology of malaria in the Brazilian Amazon. *Human Organization* 47: 254–260.

Consoli, R.A.G.B. and Oliveira, R.L. (1998) *Principais mosquitos de importância sanitária no Brasil*. Rio de Janeiro, Brazil: Editora Fiocruz.

Coura, J.R., Suárez-Mutis, M., and Ladeia-Andrade, S. (2006) A new challenge for malaria control in Brazil: asymptomatic *Plasmodium* infection: a review. *Memórias do Instituto Oswaldo Cruz* 101: 229–237.

Couto, Á.A., Calvosa, V.S., Lacerda, R., Castro, F., Rosa, E.S., and Nascimento, J.M. (2001) Controle da transmissão da malária em área de garimpo no Estado do Amapá com participação da iniciativa privada. *Cadernos de Saúde Pública* 17: 897–907.

Cruz, O. (1913) *Relatório sobre as condições médico-sanitárias do Valle do Amazonas*. Rio de Janeiro, Brazil: Typographia do Jornal do Comércio, de Rodrigues & C.

Da Silva, N.S., Silva-Nunes, M.D., Malafronte, R.S., Menezes, M.J., D'Arcadia, R.R., Komatsu, N.T., Scopel, K.K.G., Braga, É.M., Cavasini, C.E., Cordeiro, J.A., and Ferreira, M.U. (2010) Epidemiology and control of frontier malaria in Brazil: lessons from community-based studies in rural Amazonia. *Transactions of the Royal Society of Tropical Medicine and Hygiene* 104: 343–350.

Deane, L.M., Ledo, J.F., Freire, E.P.S., Cotrim, J., Sutter, V.A., and Andrade, G.C. (1948) Contrôle da Malária na Amazônia pela aplicação domiciliar de DDT e sua avaliação pela determinação do índice de transmissão. *Revista do Serviço Especial de Saúde Pública* 2: 545–560.

Fearnside, P.M. (1986) Settlement in Rondônia and the token role of science and technology in Brazil's Amazonian development planning. *Interciencia* 11: 229–236.

Fearnside, P.M. (1989) Brazil's Balbina dam: environment versus the legacy of the pharaohs in Amazonia. *Environmental Management* 13: 401–423.

Fearnside, P.M. (1999) Social impacts of Tucuruí dam. *Environmental Management* 24: 483–495.

Fearnside, P.M. (2005a) Brazil's Samuel Dam: lessons from hydroelectric development policy and the environment in Amazonia. *Environmental Management* 35: 1–19.

Fearnside, P.M. (2005b) Deforestation in Brazilian Amazonia: history, rates, and consequences. *Conservation Biology* 19: 680–688.

Fearnside, P.M. (2006) Dams in the Amazon: Belo Monte and Brazil's hydroelectric development of the Xingu River Basin. *Environmental Management* 38: 16–27.

Fearnside, P.M. (2010) Interdisciplinary research as a strategy for environmental science and management in Brazilian Amazonia: potential and limitations. *Environmental Conservation* 37: 376–379.

Fearnside, P.M. and Laurance, W.F. (2002) O futuro da Amazônia: os impactos do Programa Avança Brasil. *Ciência Hoje* 31: 61–65.

Giglioli, G. (1956) Biological variations in *Anopheles darlingi* and Anopheles gambiae: their effect on practical malaria control in the neotropical region. *Bulletin of the World Health Organization* 15: 461–471.

Gonçalves, M.J.F. and Alecrim, W.D. (2004) Non-planed urbanization as a contributing factor for malaria incidence in Manaus-Amazonas, Brazil. *Revista Panamericana de Salud Pública* 6: 156–166.

Grego, C.R., Miranda, E.E., Valladares, G.S., Custódio, D.O., Franzin, J.P., and Silva, C.F. (2007) *Análise exploratória e dinâmica espacial e temporal dos sistemas de produção em Machadinho d'Oeste (RO), entre 1986 e 2005.* Documentos 64. Campinas, Brazil: Embrapa Monitoramento por Satélite.

IBGE. (2009) SIDRA—Sistema IBGE de Recuperação Automática. See www.sidra.ibge.gov.br (accessed December 2009).

IDB. (2006) *Building a New Continent: A Regional Approach to Strengthening South American Infrastructure. Initiative for the Integration of Regional Infrastructure in South America (IIRSA).* Washington, DC: Inter-American Development Bank.

INCRA/CRUB/UNB. (1997) *I Censo da Reforma Agrária do Brasil.* Brasília, Brazil: DF, Instituto Nacional de Colonização e Reforma Agrária (INCRA).

INPE. (2009) Projeto PRODES – Monitoramento da Floresta Amazônica Brasileira por Satélite. See www.obt.inpe.br/prodes (accessed October 2009).

Katsuragawa, T.H., Gil, L.H.S., Tada, M.S., Silva, A.A., Costa, J.D.A.N., Araújo, M.S., Escobar, A.L., and Silva, L.H.P. (2010) The dynamics of transmission and spatial distribution of malaria in riverside areas of Porto Velho, Rondônia, in the Amazon region of Brazil. *PLoS ONE* 5(2): e9245 (doi:10.1371/journal.pone.0009245).

Klein, T.A. and Lima, J.B.P. (1990) Seasonal distribution and biting patterns of Anopheles mosquitoes in Costa Marques, Rondônia, Brazil. *Journal of the American Mosquito Control Association* 6: 700–707.

Macauley, C. (2005) Aggressive active case detection: a malaria control strategy based on the Brazilian model. *Social Science and Medicine* 60: 563–573.

Machado, P.A.L. (1995) *Direito Ambiental Brasileiro.* São Paulo, Brazil: Malheiros Editores.

Mahar, D.J. (1989) *Government policies and deforestation in Brazil's Amazon Region.* Washington, DC: World Bank.
Martine, G. (1982) Colonization in Rondônia: continuities and perspectives. In *State Policies and Migration: Studies in Latin America and the Caribbean*, P. Peek and G. Standing (eds). London: Croom Helm, 147–172.
Martine, G. (1990) Rondônia and the fate of small producers. In *The Future of Amazonia: destruction or sustainable development?*, D. Goodman and A.L. Hall (eds). Basingstoke, UK: Macmillan, 23–48.
MDA/INCRA/DTI. (2007) Projetos de Reforma Agrária Conforme Fases de Implementação—Período da Criação do Projeto: 01/01/1900 até 05/10/2007 [Online]. Brasília, Brazil: Ministério do Desenvolvimento Agrário (MDA); Instituto Nacional de Colonização e Reforma Agrária (INCRA); Diretoria de Obtenção de Terras e Implantação de Projetos de Assentamento (DT); Coordenação-Geral de Implantação – DTI – Sipra. See http://pfdc.pgr.mpf.gov.br/atuacao-e-conteudos-de-apoio/publicacoes/reforma-agraria/questao-fundiaria/assentamentos_2001_a_2010.pdf (accessed 4 October 2007).
Monte-Mór, R.L. (1997) Urban and rural planning: impact on health and the environment. In *International Perspectives on Environment, Development, and Health: Toward a Sustainable World*, G.S. Shahi, B.S. Levy, A. Binger, T. Kjellstrom, and R. Lawrence (eds). New York, NY: Springer, 554–566.
Monte-Mór, R.L. (2004) Modernities in the jungle: extended urbanization in the Brazilian Amazonia. PhD thesis, PhD, University of California, Los Angeles, CA.
Moran, E.F. (1981) *Developing the Amazon.* Bloomington, IN: Indiana University Press.
Moran, E.F. (1985) An assessment of a decade of colonization in the Amazon Basin. In *Change in the Amazon Basin: The Frontier After a Decade of Colonization*, Hemming, J. (ed.). Manchester, UK: Manchester University Press, 91–102.
Nepstad, D., Capobianco, J.P., Barros, A.C., Carvalho, G., Moutinho, P., Lopes, U., and Lefebvre, P. (2000) Avança Brasil: os custos ambientais para a Amazônia. Belém, PA: Instituto Sócio Ambiental (ISA); Instituto de Pesquisa Ambiental da Amazonia (IPAM).
Ninio, A., Batmanian, G., Bonilla, J.P., Margulis, S., Quintero, J., and Maurer, L. (2008) *Environmental Licensing of Hydroelectric Projects in Brazil: A Contribution to the Debate.* Report no. 40995-BR.Washington, DC: World Bank.
Olson, S.H., Gangnon, R., Silveira, G.A., and Patz, J.A. (2010) Deforestation and malaria in Mâncio Lima county, Brazil. *Emerging Infectious Diseases* 16(7). See wwwnc.cdc.gov/eid/article/16/7/09-1785.htm (accessed 15 June 2012).
Patz, J.A., Graczyk, T.K., Geller, N., and Vittor, A.Y. (2000) Effects of environmental change on emerging parasitic diseases. *International Journal of Parasitology* 30: 1395–1405.
Peixoto, A. (1917) *O Problema Sanitário da Amazônia.* Rio de Janeiro, Brazil: Imprensa Nacional.
Rosa, S.A. (2006) Parecer técnico sobre saúde pública. In *Relatório de Análise do Conteúdo dos Estudos de Impacto Ambiental (EIA) e do Relatório de Impacto Ambiental (RIMA) dos Aproveitamentos Hidrelétricos de Santo Antonio e Jirau, no Rio Madeira, Estado de Rondônia*, MPRO and COBRAPE (eds). October, part B, vol. II: Aspectos Socioeconômicos: Ministério Público do Estado de Rondônia; Cobrape—Cia Brasileira de Projetos e Empreendimentos. See www.mp.ro.gov.br/web/guest/Interesse-Publico/Hidreletrica-Madeira (accessed 28 May 2007).
Sawyer, D.R. (1986) Malaria on the Amazon frontier: economic and social aspects of transmission and control. *Southeast Asian Journal of Tropical Medicine and Public Health* 17: 342–345.

Sawyer, D.R. (1988) *Frontier Malaria in the Amazon Region of Brazil: Types of Malaria Situations and Some Implications for Control.* Brasília, Brazil: PAHO/WHO/TDR.

Sawyer, D.R. (1992a) *Deforestation and Malaria on the Amazon Frontier.* Campinas, Brazil: Seminar on Population and Deforestation in the Humid Tropics, International Union for the Scientific Study of Population.

Sawyer, D.R. (1992b) *Malaria and the Environment.* Brasília, Brazil: Instituto SPN.

Sawyer, D.R. and Sawyer, D.O. (1987) *Malaria on the Amazon Frontier: Economic and Social Aspects of Transmission and Control.* Belo Horizonte, Brazil: CEDEPLAR.

Sawyer, D.R. and Sawyer, D.O. (1992) The malaria transition and the role of social science research. In *Advancing the Health in Developing Countries: The Role of Social Research*, L.C. Chen (ed.). Westport, CT: Auburn House, 105–122.

Schmink, M. and Wood, C.H. (1992) *Contested Frontiers in Amazonia.* New York, NY: Columbia University Press.

Singer, B.H. (1989) Grade of membership representations: concepts and problems. In *Probability, Statistics, and Mathematics: Papers in Honor of Samuel Karlin*, S. Karlin, T.W. Anderson, K.B. Athreya, and D.L. Iglehart (eds). Boston, MA: Academic Press, 317–334.

Singer, B.H. and Castro, M.C. (2001) Agricultural colonization and malaria on the Amazon frontier. In *Population Health and Aging: Strengthening the Dialogue between Epidemiology and Demography*, M. Weinstein, A.I. Hermalin, and M.A. Stoto (eds). New York, NY: Annals of the New York Academy of Sciences, 184–222.

Singer, B.H. and Castro, M.C. (2007) Bridges to sustainable tropical health. *Proceedings of the National Academies of Science* 104: 16,038–16,043.

Singer, B.H. and Castro, M.C. (2011) Reassessing multiple-intervention malaria control programs of the past: lessons for the design of contemporary interventions. In *Water and Sanitation-Related Diseases and the Environment: Challenges, Interventions, and Preventative Measures*, J.M.H. Selendy (ed.). John Wiley & Sons, Inc, 151–165.

Smith, N.J.H. (1982) *Rainforest Corridors: The Transamazon Colonization Scheme.* Berkeley, CA: University of California Press.

Sparovek, G. (2003) *A Qualidade dos Assentamentos da Reforma Agrária Brasileira.* São Paulo, Brazil: Páginas & Letras Editora e Gráfica.

Sydenstricker, J.M. (1992) Parceleiros de Machadinho: história migratória e as interações entre a dinâmica demográfica e o ciclo agrícola em Rondônia. Master degree dissertation, Universidade de Campinas, Campinas, Brazil.

Tadei, W.P. (1991) *Considerações Sobre as Espécies de Anopheles e a Transmissão da Malária na Amazônia.* Manaus, Brazil: Instituto Nacional de Pesquisas da Amazônia (INPA).

Tadei, W.P., Thatcher, B.D., Santos, J.M.M., Scarpassa, V.M., Rodrigues, I.B., and Rafael, M.S. (1998) Ecologic observations on anopheline vectors of malaria in the Brazilian Amazon. *American Journal of Tropical Medicine and Hygiene* 59: 325–335.

Tauil, P., Deane, L., Sabroza, P., and Ribeiro, C. (1985) A malária no Brasil. *Cadernos de Saúde Pública* 1: 71–111.

Viana, J. (2007) Mais que administrar, cuidar! *Revista do Serviço Público* Edição Especial: 49–57.

Wittern, K.P. and Conceição, M. (1982) *Levantamento de Reconhecimento de Média Intensidade dos Solos e Avaliação da Aptidão Agrícola das Terras em 100,000 Hectares da Gleba Machadinho, no Município de Ariquemes, Rondônia.* Boletim de Pesquisa no. 16. Rio de Janeiro, Brazil: EMBRAPA/SNLCS.

Wood, C.H. and Porro, R. (2002) *Deforestation and Land Use in the Amazon.* Gainesville, FL: University Press of Florida.

Part III

Disease histories, the state, and (mis)management

8 Vaccines, fertility, and power

The political ecology of indigenous health and well-being in lowland Latin America

Kendra McSweeney and Zoe Pearson

Introduction

By most measures, the health of indigenous peoples[1] in Latin America's lowlands is terrible. These small ethnic groups of rarely more than 50,000 individuals, and commonly fewer than 5000, live in or near some of the world's last resource frontiers, including the Amazon and Orinoco basins, the Pacific lowlands of the Andes, and the Caribbean littoral. For the past half-century, if not much longer, they have faced a "perfect storm" of epidemiological misery (Coimbra and Santos 2000, 2004). Land-invading colonists bring new diseases (such as dengue and whooping cough), and mining and dam-building create ideal environments for malaria and schistosomiasis vectors (Souza-Santos *et al.* 2008; Gracey and King 2009). Loss of land and resources intensifies diseases associated with crowding, rapid dietary change, and food insecurity, such as tuberculosis, diarrheal infection, pneumonia, and parasitic infection (Hames and Kuzara 2004; Raich 2004; Coimbra and Basta 2007). Moreover, distance, poverty, discrimination, and government apathy contribute to appalling health service delivery (Nawaz *et al.* 2001; Gracey and King 2009).

Women and children suffer particularly (Ribas *et al.* 2001; Coimbra and Garnelo 2003; Arps 2009; Gracey and King 2009). Maternal morbidity and mortality rates for indigenous societies are among the highest in the world, many times more than for non-native women in respective nations (PAHO 2004a; Montenegro and Stephens 2006; Gracey and King 2009). Mortality rates for indigenous infants are routinely higher than for non-indigenous babies (Montenegro and Stephens 2006). Depressingly, these estimates are probably conservative, as very little is known about the health and demography of indigenous populations, or about the effectiveness of existing health programs (Ribas *et al.* 2001; Hurtado *et al.* 2005; Montenegro and Stephens 2006; Goicolea *et al.* 2008; Gracey and King 2009).

Given that this volume is dedicated to socio-political and ecological dimensions of human health, readers will quickly recognize that the health of lowland natives in Latin America is inextricably tied up with cultural and ecological devastation, and the violent and ongoing usurpation of indigenous lands and labor (Salzano and Hurtado 2004; Montenegro and Stephens 2006). Indeed, indigenous peoples of the New World arguably offer one of the most straightforward examples of the ways in which disease ecologies can only be fully understood in the context of social groups' subjugation by colonial and capitalist political economic formations. But is this the only way to think about how health, ecologies, and power come together in Latin America's lowland resource frontiers? Might it be possible to recognize the agency and resilience of native societies in the context of health and ecological change? Could, for example, a closer look at health lead to clearer insights on indigenous peoples' ongoing *resistance* to their social and ecological marginalization?

This chapter argues that, yes, attention to health can contribute important insights into the nature of individual and collective struggles for access to land and resources, and to the reverse. In the process, we contribute to and extend "political ecologies of health" scholarship (King 2010) by considering the relatively under-explored ways in which health can be mobilized as a tool of resistance to shape socio-political and environmental outcomes. Additionally, our case studies contribute to an understanding of health as fundamentally "biosocial" (Mansfield 2008; Mansfield 2011), demonstrating how health is *always simultaneously* social, political, environmental, and biological.

We draw from our own and others' work with native populations in lowland Central and South America to focus on two indigenous health issues in particular: maternal morbidity and childhood vaccination rates. We show how the health of indigenous mothers and children is profoundly tied up with the struggle over indigenous lands and resources, and also in ways that challenge conventional narratives about indigenous epidemiological and ecological vulnerabilities. The case studies also reveal how the links between health, power, and ecology are multidirectional; specifically, that considerations of indigenous health are incomplete without attention to political–ecological struggles, and, in turn, that the nature of those political–ecological struggles is better understood by attention to the health of the individuals and communities that are engaged in them.

Background: indigenous health and political mobilization in lowland Latin America

The epidemiological history of indigenous peoples in the Americas is well known. With little immunity to so-called "Old World" diseases (including smallpox, measles, yellow fever, malaria, pneumonia, influenza, and more), indigenous populations following first contact (whether in the sixteenth or the twentieth century) often experienced mortality rates up to an unimaginable 90 percent (Denevan 1992; Montenegro and Stephens 2006). Today, many of these diseases continue to present acutely, even among indigenous populations with long histories of contact (Escobar *et al.* 2004). Fatality rates typically exceed those of the non-indigenous rural poor (Coimbra and Basta 2007). Moreover, indigenous peoples' relative nutritional reliance on their immediate environment means they can be disproportionately exposed to contaminants such as lead or mercury (e.g., Witzig and Ascencios 1999; Raich 2004).

Indigenous peoples, of course, have not been passive in the face of these attacks on their lives, lands, and livelihoods. Since the 1960s, indigenous peoples across Amazonia and Central America have been organizing politically. The

Shuar Federation in Ecuador was one of the earliest, forming in 1964 in response to *mestizo* colonization (Federación Shuar 1976; Rubenstein 2001). They and other lowland indigenous organizations played a crucial early role in a process of political resurgence that, by the late 1980s, had united upland and lowland organizations and become a pan-ethnic and pan-hemispheric indigenous movement (Perreault 2001;Yashar 2005; Wilson 2010). International conservation groups concerned with the fate of neotropical diversity were early and important allies of the movement (Conklin and Graham 1995; Zimmerman *et al.* 2001).

The achievements of the indigenous movement have been significant, particularly in the last two decades. They include: (i) the homologation of often massive territories to native control, such as the 34 million ha that Colombia set aside for indigenous communities (Velasco 2011); (ii) constitutional reforms that recognize the rights of indigenous peoples (Martí i Puig 2010), particularly over the governance of their lands and educational and health programs; (iii) the 2007 adoption of the United Nations (UN) Declaration on the Rights of Indigenous Peoples, which asserts their right to improve and administer health programs (United Nations 2008); and (iv) multiple electoral successes, including the historic 2008 inauguration of indigenous President Evo Morales in Bolivia (Morales 2011).

Many of these successes, however, have been tempered by ongoing challenges. Seemingly emancipatory constitutional provisions, for example, may actually constrain or distort autonomous projects (Rosengren 2003; Hale 2005). Further, land reforms have been stalled or reversed by slow implementation and ever-changing political–economic configurations around indigenous territories (Kennedy and Perz 2000; Stocks 2005). For example, Brazil's economic ascendance throughout the 2000s has allowed it to self-finance a host of new energy and transportation projects in the Amazon basin, with massive impacts on indigenous lands (Tollefson 2011). The Colombian government, meanwhile, has stalled and backtracked in its indigenous land-titling program (Velasco 2011). Peru, in turn, while acknowledging indigenous land title to 13.5 percent of the Peruvian Amazon, had as of 2009 leased 41.2 percent of the same area for oil and gas development (Orta-Martínez and Finer 2010). Many observers are now describing the current era as a distinctly new phase of intensified resource extraction and land dispossession in Latin America (Morales 2009; Renfrew 2011). As a result, indigenous political and spiritual leaders are increasingly targeted by paramilitaries and landowning elites, often in collusion with drug cartels exploiting the trafficking potential of remote indigenous territories (Chomsky 2005; Fearnside 2008; BBC News 2011; Carroll 2011; Collyns 2011).

At the same time, indigenous organizations' once-vital alliances with the powerful conservation lobby have soured around native claims of neocolonialism (Chapin 2004; Lauer 2005). Conservationists, in turn, complain that erstwhile "forest stewards" are not living up to conservationists' image of them—as, for example, when they wander from "fixed" territories, hunt too much game, or, perhaps most vexingly, have too many babies (see, e.g., Holt 2005; McSweeney 2005; McSweeney and Pearson 2009).

To be sure, indigenous families across lowland Latin America are unusually large, due especially to unusually high fertility rates among indigenous women.

Across ethnic groups, total fertility rates in excess of 6 are the norm, and within specific groups, total fertility rates of 8 are common. Both are significantly higher than neighboring non-indigenous populations (Kennedy and Perz 2000; McSweeney and Arps 2005a, b). In fact, indigenous populations living across Latin America's lowlands typically have more in common demographically with each other than with non-indigenous populations within their respective countries (McSweeney and Arps 2005a). This high fertility combines with declining mortality rates to accelerate the growth rate of indigenous populations. As a result, the majority are characteristically young, with as much as half the population under age 15 (McSweeney and Arps 2005a; Pagliaro *et al.* 2005).

Population expansion has characterized indigenous demographic trends in lowland Latin America since at least the 1980s. But the process was widely overlooked until the results of the 2000 round of national censuses became available (Perz *et al.* 2008). Policy observers have tended to view the rapid growth with alarm, concerned about the pressures that young populations create on already over-stretched social services (United Nations 2009), while conservationists worry about the implications of population growth for biodiversity (for reviews see McSweeney 2005; Oldham 2006). Meanwhile, a minority of observers, particularly Brazilian scholars, have suggested a different view that accounts for the historical and political context of demographic expansion. They point out that the growth of indigenous populations across Latin America's lowlands can be read as a happy reversal of 500 years of catastrophic demographic decline (Gomes 2000; Penna 1984). Viewed in terms of population size and structure, they note that indigenous peoples are more demographically robust than they have been for centuries. Further, they point to ways in which population growth, cultural revitalization, and political resurgence have been mutually dependent (Gomes 2000; Kennedy and Perz 2000; Pagliaro *et al.* 2005).

In what follows, we consider these unorthodox observations in the context of two of the most pressing health issues faced by indigenous populations in lowland Latin America: maternal morbidity and the uneven vaccination and immunization of indigenous children. In the process, we show how population size, family size, maternal and child health, indigenous political mobilization, and ecological change are related in surprising ways that raise new questions for political ecologies of health research.

Health as contested terrain in lowland Latin America

Maternal morbidity and mortality

Gracey and King (2009) review the multiple risks that indigenous mothers face, including under-nutrition during pregnancy, anemia caused by nutrient deficiencies and underlying disease (e.g., malaria), deficiencies in other nutrients, high rates of urinary tract infections, gestational diabetes, and "scant human, clinical and laboratory resources for safe pregnancy, delivery, and postnatal care" (Gracey and King 2009: 67). Compounding these risks are the barriers faced by mothers seeking care. Multiple studies show that even when indigenous women reach a health clinic or hospital, they receive culturally insensitive and degrading treatment, if they are seen at all (see for example Paulson 2000; Azevedo 2003; Arps 2009). All of these factors contribute to the exceptionally high rates of sickness and death reported for indigenous mothers or inferred from demographic evidence of missing women (e.g., Azevedo 2003; UNICEF 2007).

Multiple programs have sought to address the problem, including initiatives within the Pan American Health Organization's "Health of Indigenous Peoples in the America's program" (2003–2007), but achievements have been elusive (Hughes 2004; Goicolea *et al.* 2008). We argue below that the lack of progress may be partially explained by programmatic misunderstandings of the logic behind indigenous women's reproductive health decisions, particularly regarding the size of their families.

High parity as pathology

Many health interventions that target indigenous women focus on fertility reduction. In addition to providing modern contraceptives, efforts also include support for initiatives that keep indigenous teen girls in school longer, which is seen to have the dual effect of delaying the age of first birth and ultimately raising maternal education levels overall (McKinley 2003; Goicolea *et al.* 2008; Gracey and King 2009). These initiatives are grounded in the assumption that most indigenous women desire smaller families but lack the means to achieve them. Key support for this logic comes from large-scale surveys of rural populations, during which indigenous women *themselves* say that they want fewer children than they already have or are likely to have, indicative of an "unmet need" for contraception (CEPAR 2004; Goicolea *et al.* 2008; Bremner *et al.* 2009).

There is no question that pregnancy is particularly risky for young first-time mothers and for high-parity older mothers; lowland indigenous populations hold relatively large numbers of both. But the focus of reproductive health programs on the number of children that indigenous women have reflects more than biomedical science; it also reflects an ideological commitment to seeing birth rates through the dominant modernist lens of demographic transition (and related neo-Malthusian concerns; see McKinley 2003). This is conveyed when high parity is understood in terms of incomplete or failed socio-economic development among indigenous populations (Hughes 2004; Kramer and Greaves 2007). Indigenous women have many children, the orthodoxy goes, because they lack what more developed, low-parity women have, particularly with respect to education and income (see, e.g., Hughes 2004). Further, their ongoing transition from "traditional" to "modern" means indigenous women are caught in a particularly unstable acculturation phase in which standard methods for fertility control (e.g., polygynous marital arrangements, the use of herbal contraceptives and abortifacients) are lost before modern ones can be effectively adopted (e.g., Hern 1992; see McSweeney 2005). As a result, indigenous women may want, but cannot on their own achieve, smaller families.

Ultimately, then, reproductive health programs in indigenous areas tend to approach the care of indigenous mothers based on the premise that repeat motherhood is itself pathological, at least under the conditions that indigenous women achieve it. In other words, women are *primarily* sick because they are having yet another child. Once women cease to be so persistently pregnant, the logic goes, improved health will result (see, e.g., Hughes 2004).

High parity as healthful

But what if we suspend, for a moment, the assumption that high parity is itself bad for indigenous women's health? Clearly, women's fertility and their health are inseparable, but must they necessarily be in opposition? Is there an alternative to the socio-economic lenses through which reproductive health is conventionally understood? Could women's high fertility, for example, actually contribute in some substantive way to *improved* health? As heretical as it might seem, this is a possibility that has been suggested repeatedly, if not universally, by research conducted with multiple different indigenous Amazonian and lowland societies in Latin America. These studies are typically conducted by anthropologists or microdemographers using immersive ethnographic methods rather than large-scale surveys. Thus, indigenous reproductive behaviors are assessed in their particular historic and political contexts; emic explanations for health and reproductive behaviors are emphasized, and women's views are often privileged (Pagliaro *et al.* 2008). Finally, in keeping with indigenous epistemologies of health throughout Amazonia, many studies consider individual health in terms of the family, the community, and population or ethnic group health (Levin and Browner 2005; King *et al.* 2009). As Azevedo (2003: 178) states, "All events that occur during the reproductive life of women are immediately inscribed within the totality of her social–cultural–environmental context" (our translation).

Taken together, these studies set out a series of propositions that offer new ways to interpret indigenous mothers' health. Most significantly, researchers have found that both indigenous men and women often talk about family size and high fertility not in terms of women's health but in terms of *collective* (kin group, village, faction, ethnic group) health (e.g., Browner 2000). Specifically, in two respects high fertility is explicitly understood to be an important mechanism for achieving specific group goals regarding ethno-cultural continuity and population size. First, high fertility is related to the survival and long-term health of the group. Not surprisingly, small ethnic groups with a collective memory of recent and harrowing population losses seem particularly keen to regrow their populations (see Flowers 1994; Hern 1994; Price 1994; Pagliaro and Junquiera 2005). In such cases, collective optimism about the group's future, particularly with respect to available land and resources, seems to be essential to the process. This inspires couples' willingness to bear and raise multiple children, which in turn fuels optimism about group survival in place (see also McSweeney and Arps 2005a; Pagliaro *et al.* 2005).

Second, high fertility is considered in terms of political strategies around defending and securing access to land and resources—i.e., the survival and well-being of the group in the face of external threats, particularly from *mestizos* invading group lands (Browner 2000; Azevedo 2003; Radcliffe and

Pequeño 2010; see also McSweeney and Arps 2005a; Pagliaro *et al.* 2005). Brazilian demographer Túlio Penna (1984: 1576) referred to this demographic strategy for land occupation as a "*Reconquista*" by indigenous Amazonians. That large families are perceived to be one vital strategy for filling up, patrolling, and otherwise defending homelands is understandable given persistent claims by the dominant culture that indigenous peoples want "too much for too few" (Stocks 2005; see also Kennedy and Perz 2000).

Importantly, these reproductive strategies for territorial occupation and defense appear to work, with positive ecological effects. For example, satellite imagery reveals that densely populated indigenous reserves remain forested oases compared with the thinly populated settler-dominated landscapes around them (Nepstad *et al.* 2006; Stocks *et al.* 2007).

A significant body of research, then, is offering compelling historical, political, and ecological reasons for indigenous societies to be strongly pronatalist: *intentional* in their pursuit of large families. This is no post-hoc explanation: studies also shed light on the specific mechanisms by which indigenous men and women consciously adjust their nuptial and reproductive behaviors in deliberate pursuit of collective population goals. These include the cessation of infanticide or a reduction in the use of abortifacients (Meireles 1988; Hern 1994), and the deliberate loosening of customs restricting exogamy and age of marriage (e.g., Pagliaro *et al.* 2005). Significantly too, studies relate ways in which women are creatively combining traditional and modern contraceptives, and do so not necessarily to limit their total births but instead to space their births according to their own understandings of what is best for their own and their children's health[2] (e.g., Hern 1994; UNICEF 2007).

Some studies also point to ways in which indigenous women derive personal satisfaction and enhanced social status from their large families (e.g., Azevedo 2003; Pagliaro *et al.* 2008). If they express concern over childbearing, it is often with reference to concern for their children's futures, not the number of children per se (Radcliffe and Pequeño 2010). These observations resonate with critical demographic–anthropological scholarship on fertility, particularly regarding pregnancy intention and contraceptive use (see, e.g., Greenhalgh 1995; Bledsoe *et al.* 1998). Together, these empirical and theoretical observations call into question large-scale biomedical health survey findings regarding indigenous women's dissatisfaction with their high parity and their assumed desire for fertility-*limiting* contraception. Thus indigenous women's multiple births should not, a priori, be considered the unhappy and highly risky result of their socio-economic, cultural and environmental impoverishment. Rather, it appears that high parity can be read as a vital if insufficient requirement for the improvement of the group's collective physical, political, and ecological health in an era of profound challenge.

New questions about maternal health

This new perspective does not change the fact that indigenous women are unacceptably sick and require vastly improved medical attention. But it does lay bare the ways in which the state of a woman's body (pregnant, anemic, lactating) is inseparable from her social and ecological health (her own sense of contentment, the current and future well-being of the collective of which she forms a part, and the environmental spaces that her group uses and defends). This is not to reify women as embodied "Nature." Rather, it is to assert that in Latin America's lowlands, indigenous women's health is deliberately and directly shaped by the ongoing contest among social groups over land and resources with vital ecological consequences.

This view raises questions about health providers' emphasis on fertility reduction as a first step towards improving indigenous women's health. Once health-care providers recognize that women might consider birth-limiting contraceptive practice antithetical to a collective pronatalist goal or worse, an echo of past genocides (see, e.g., Federación Shuar 1976), they can adjust their focus from fertility reduction to practices that ensure safe motherhood for women no matter what their reproductive plans. Another lesson is to recognize that territorial security, and with it, food and resource security, cultural revitalization, and the elimination of the need for biological reproduction to secure it, is perhaps the single most critical precondition for improving indigenous women's health.

These new perspectives also raise urgent new questions. Foremost among them is the degree to which indigenous women have power over the decisions regarding their reproduction. Clearly, they can and do manipulate their fertility in response to collective goals. But research to date suggests that they may do so against their will, in response to strong pressures from male partners, leaders, or even other women (e.g., Browner 1986; Hern 1994; Browner 2000; Radcliffe and Pequeño 2010). But there is also evidence that women subvert and negotiate these pressures; power over reproductive decision-making appears diffuse and complex (Greenhalgh 1995). A related question is the degree to which members of pronatalist societies consider an explicit tradeoff between the risks to women's health and the collective pursuit of social, political, and ecological health.

Childhood vaccination and immunization

These urgent questions echo in a distinct but related health issue for the growing numbers of indigenous children across Latin America's lowlands: vaccination against disease. We have argued above that compared with standard narratives, new perspectives on maternal morbidity point to more hopeful interpretations about women's health and its contribution to group well-being. In what follows, we offer a contrasting case by reviewing how indigenous organizations are taking greater control over efforts to improve the vaccination of their children. But this process is complicated and potentially undermined when gains to children's health come at the cost of social and ecological well-being. Those familiar with the role of vaccination in the colonial enterprise will perhaps not be surprised by our conclusions (see, for example, Lee and Fulford 2000). Nevertheless, the specific dynamics of the case are important for illustrating how antigens, land, and power come together in the context of indigenous political resurgence in contemporary lowland Latin America.

The vaccination of indigenous populations across the region, especially since the 1960s, is widely credited with lowering mortality rates for all ages, but especially children (de la Hoz *et al.* 2005; Gracey and King 2009). At the same time, however, vaccination coverage is spatially and temporally uneven, haphazardly implemented, and routinely under-monitored (de la Hoz *et al.* 2005). Consider, for example, that within most lowland environments, the basic recommendation is that children be vaccinated against tuberculosis, measles, mumps, rubella, polio, hepatitis B, diphtheria, pertussis, tetanus, and yellow fever. For these vaccinations to result in immunization, first doses must be administered within specific age ranges and boosters within defined intervals (de la Hoz *et al.* 2005). Based on the scant data available, these schedules remain a remote ideal. The majority of indigenous children are under-vaccinated while some are dangerously over-vaccinated and others are missed completely (see UNICEF 2007, for example). Equally troubling is the dismal number of follow-up studies on the effectiveness of vaccines in lowland environments (Escobar *et al.* 2004; de la Hoz *et al.* 2005; Hurtado *et al.* 2005; Gracey and King 2009). As a result, the degree to which incomplete immunization contributes to childhood illness and death remains an open question (Cardoso *et al.* 2005; Montenegro and Stephens 2006).

Vaccination rates for indigenous children in lowland societies are also routinely lower than for proximate non-indigenous rural populations (de la Hoz *et al.* 2005). This has been explained in terms of the isolation of indigenous societies relative to *mestizo* populations (de la Hoz *et al.* 2005), and due to poor coordination and political apathy among responsible national and international agencies (Hurtado *et al.* 2005). But even well-coordinated and well-funded vaccination efforts repeatedly report a key stumbling block: the "invisibility" of a large share of indigenous children, meaning that their parents have not registered their births (e.g. see Teixeira 2005; UNICEF 2007). Low registration rates are attributed to the distance of most registration offices from indigenous settlements, and to the fear and ignorance of indigenous parents regarding state registration (UNICEF 2003). A contradictory explanation is that many indigenous societies find state registration insulting and neocolonial and so avoid it (e.g. see Federación Shuar 1976).

Whatever the reason, one result of non-registration is that many indigenous children in lowland societies are effectively non-existent in the eyes of the state and the entities with which the state collaborates on vaccination brigades (e.g., UNICEF, PAHO). Nor are there typically any reliable proxy sources for such data, such as health records or national census data (PAHO 2004b).

This becomes a problem for indigenous organizations themselves. As ethnic federations assume greater control over health resources, they find health planning to be hamstrung by the lack of basic data on their own populations (Montenegro and Stephens 2006). In response, a growing number of indigenous organizations have begun to develop and administer their own demographic and health surveys. An early example is the 1992 survey of almost 17,000 individuals conducted by the Federation of Indigenous Groups of the Upper Rio Negro in Brazil (Azevedo 2003). Subsequent surveys include those conducted by Sateré-Mawé (Teixeira 2005; Teixeira and Brasil 2005), the Shuar Federation (Jokisch and McSweeney 2011) and others (Coimbra and Santos 2004). All were run by or in close collaboration with indigenous federations, with native enumerators administering surveys in native languages. All were also understood as a means for native organizations to access information for both internal planning purposes *and* in order to better leverage health (and other) resources from the state and international organizations (Jokisch and McSweeney 2011).

Because the surveys were expensive to implement and process, all relied on funding from a mix of domestic and foreign governments and UN agencies, including UNFPA and UNICEF, with academics and health practitioners often providing technical support (Azevedo 2003; Jokisch and McSweeney 2011). Clearly, then, the surveys cannot be considered fully "autonomous" (cf. Azevedo 2003: 76). But for indigenous political organizations, the very act of counting and assessing their self-defined constituents in their own language is to perform a function that characterizes autonomous nations (Anderson 1983). In the process, indigenous federations push back against neocolonial state registries and problematic censal instruments while asserting their rights to nationhood (see also Kertzer and Arel 2002).

In this way, the auto-generation of data on their own health status not only becomes a tactic by which indigenous organizations secure funding for specific health initiatives, but also a means to serve a larger and explicitly ethno-territorial goal as well (see Cabral in Teixeira 2005). This is because native-run surveys signal federations' authority to *find and render visible* indigenous bodies within contested spaces, thus asserting indigenous authority over both. The Shuar's Director of Health made the territorial and ecological connections clear in proposals for the Shuar Federation's survey, noting that data needs were urgent in the context of "the struggle against oil companies' efforts at penetration" and in light of "the growth and presence of mining companies in different zones of Shuar settlement" (Tiwi 2005; our translation).

In the process, indigenous children's vaccination status becomes enrolled in a series of political tactics that seek to assert indigenous control over indigenous individuals and the spaces they occupy. At first glance, this might appear to be a straightforward example of an emancipatory political ecology of health; one in which disadvantaged actors mobilize information about their (children's) health to demand their rights to state resources *while* signaling their vigilance over lands and resources coveted by external elites. But closer scrutiny of the dynamics at work, particularly in light of recent work within political ecology, suggests an alternative and less hopeful interpretation: one in which children's health plays no less a strategic role, but one that potentially undermines autonomous control over indigenous well-being, lands, and ecologies. We refer here to recent scholarship on indigenous mapping initiatives in Latin America (Hale 2005; Wainwright and Bryan 2009). Like the act of self-censusing, self-mapping constitutes a critical element of national imagining (Anderson 1983), and appears to be an important expression of growing autonomous control over indigenous bodies and lands. But these putatively autonomous projects can ultimately serve to enroll and confine indigenous subjects within state structures more efficiently and completely than the state could have done itself (Hale 2005). But how, exactly? And how is this related to the vaccination of indigenous children?

To answer these questions, it is instructive to recall that the Shuar Federation's and the Sateré-Mawé surveys were partially funded through the UN's Children's Fund (UNICEF). UNICEF's interest in supporting the demographic/health surveys was not based on a direct interest in children's health per se, but rather in indigenous children's unclaimed Ecuadorian/Brazilian *citizenship*. Indeed, UNICEF considers community-run surveys as "cost-effective short cut[s]" for identifying unregistered or otherwise unaccounted-for indigenous children (UNICEF 2003: 20). UNICEF's support for the indigenous surveys thus formed part of its hemispheric effort to support the rights of *all* children under national laws. These rights are only guaranteed when children become visible to the state through the creation of birth certificates and citizenship cards (see Teixeira 2005). As UNICEF-Ecuador (2007) puts it:

> Young Shuar children and adolescents have been denied the right to a name and the Ecuadorian nationality; the majority of births are not documented, or are documented late ... "Put your name to Ecuador" is the motto of the National Registration Programme that ... ensure[s] children's rights to a name and nationality ...

Thus, when Shuar or Sateré-Mawé use UNICEF monies to survey their own populations, several things happen. Indigenous federations access critical information about health needs, including vaccination schedules. In the process, however, data on formerly "invisible" children are made available to UNICEF and other international organizations. These data are essential to "grant" those children Ecuadorian or Brazilian (not Shuar or Sateré-Mawé) citizenship, including an identity card attesting to their legal status, and a vaccination card re-affirming their right to health (see also de la Hoz *et al.* 2005). At the same time, data on these children enter the national databases that seek to monitor all indigenous individuals (Teixeira 2005).

We do not mean to suggest that the vaccination of children becomes some sinister *pretext* for rendering remote and defiant indigenous peoples as national citizens. Our point here is to emphasize how the seemingly emancipatory effects of native demographic/health surveys are to some degree undermined through the very act of their execution. Shuar and other indigenous federations are seasoned and savvy negotiators of their rights to be sure (Rubenstein 2001), but, through a self-census that holds the promise of improving health of their children, they effectively offer the state more and better data on themselves and their whereabouts that the state could have ever generated itself. This contributes to the state's ability to spatially and temporally track otherwise remote populations within contested landscapes rich in oil and gold. At the same time, Shuar children are inculcated into Ecuadorian, not Shuar, nationhood. Both processes fundamentally undermine the authority and possible autonomy of the Shuar Federation, which becomes ever more deeply enmeshed within, and legible to, state bureaucracies. By weakening Shuar power, the state and allied capital interests move one step closer to unlocking the resources held in native-controlled lands.

At this point, a health practitioner might reasonably ask: "Why does this matter? Why the cynical take on such an unproblematic good as vaccination?" We would argue that our insights matter because they insist that those committed to extending the lives of native children understand what their particular actions mean for the collective well-being of the group, and the future well-being of the child. To the extent that vaccination teams have always acted as an advance guard in the assault on native lands (see Lee and Fulford 2000), there is an implicit and profound tradeoff between the biophysical health of individual children and the survival of their people as *healthy*, autonomous populations able to live within a functioning and intact ecosystem.

This does *not* mean children should not be vaccinated. It does imply, however, that those involved in the seemingly altruistic one-off act of vaccination also bear responsibility for thinking through potentially troubling longer-term political–ecological impacts. Contemplation of the power-laden context of vaccination would also help practitioners to better understand how it would be possible for a parent, a village, or other collective to refuse vaccination (see for example de la Hoz *et al.* 2005). This response is often put down to sheer ignorance, or, worse yet, to the repellent intrusion of "politics" into a life-or-death decision. A political–ecological perspective, in contrast, would demand considering the possibility that the refusal of a health service might be an agonizing decision that weighs the cost of individual illness against a collective future in place.

If the relationship between ecology and power helps us to better understand key dimensions of vaccination, so too does attention to health issues like vaccination help us better understand indigenous political–ecological struggles. This case study is a reminder that no terrain is sacrosanct in the battle over lands and resources. Access to indigenous lands is, after all, an extremely high-stakes game in which even unvaccinated newborns necessarily play a part. For all of the above reasons, the bloodstream of a baby is just the sort of terrain where political–ecological struggles are increasingly likely to play out (see Peet *et al.* 2011).

Discussion

In this chapter, we focus on indigenous mothers and children in lowland Latin America to show how their health is profoundly enmeshed in the dynamics of power and control over the ecosystems in which they live. One case appears hopeful, the other less so. Combined, they yield four "take home" points of broad interest to research and praxis related to political ecologies of health.

Health as political–ecological struggle

First, the case studies highlight the importance of considering health itself as the material embodiment of the efforts of individuals and groups to assert their power in the access, occupation, and ownership of lands and resources. To date, much of the scholarship on political ecologies of health has focused on pathologies resulting from ecological usurpation, degradation, or dispossession, or on the ways in which illness itself impacts resource use and access (see Mayer 1996;

DeWalt 1998; Collins 2002; Hanchette 2008; King 2010). If we parse the field of political ecology according to the dominant narratives of the field suggested by Robbins (2004), it might be said, then, that the bulk of health research in political ecology falls within two of Robbins's theses: degradation and marginalization, and conservation and control.[3]

We suggest here that there is productive analytical purchase in extending complementary scrutiny to health in the context of Robbins's other two theses: environmental conflict and identity, and social movements. For example, by conceptualizing health as the embodiment of environmental conflict, the two-way play of power can be foregrounded. To be sure, health can reveal much about social and ecological subjugation. But this does not foreclose the possibility that subjugated peoples can mobilize their individual or collective health to fight back. As we have shown here, the health of indigenous mothers has long been read through the lens of their victimization by poverty, ignorance, or cultural and ecological devastation. We show, in contrast, ways in which indigenous women's health, through their high parity, can be read as a critical (if insufficient) condition for indigenous political resurgence. This interpretation in no way diminishes the severity of indigenous women's suffering, nor their acute need for medical attention. But it does demand that we need to do much greater justice to women's aspirations, motivations, and constraints as they negotiate their health in the context of social and ecological crises.

Individual and collective health

Second, our explorations of health as struggle expose important questions about the relationships between individual and collective health. We do not imply a simple relationship between the self and some ontologically given collective Self, but recognize the always fluid and negotiated character of this relationship (see, e.g., Rosengren 2003). This being so, the majority of ethnic groups in Amazonia might be best described as thriving: birth rates are up, mortality rates are down, and they seem finally poised, after 500 years, to regain numbers comparable to pre-Conquest populations. Consider further that this demographic "turnaround" has enabled profound cultural revitalization and greatly enhanced political stature (Gomes 2000), and it would seem fair to assert that they are experiencing a historically high degree of population-level health.

Obviously, this assessment contrasts sharply with the aggregate statistics on indigenous morbidity, particularly for women and children. Our point here is *not* that the conclusions drawn about native health depend on the scale of inquiry (slice the data one way to get one result; a different way for another result). Quite the opposite. We want to instead draw attention to the ways in which a broad notion of health recognizes health as simultaneously and iteratively experienced by the individual and the collective, and therefore should not be "sliced" separately. This is particularly true in non-Western contexts in which individual aspirations cannot be considered in isolation from group goals (e.g. see Rosengren 2003).

Health as well-being

Third, the entangling of individual and collective health raises the question of just what is meant by "health." Clearly, health demands a broader definition than "absence of disease." As we have shown, it is possible that the processes that allow indigenous children to be vaccinated against disease can contribute to undesirable cultural assimilation and the devastation of their homelands: collective ill-health, if you will. Anthropologists inspired by political–economic approaches to health are exploring this issue through the concept of well-being (e.g., Izquierdo 2005; Levin and Browner 2005). They have found that biomedical models of health are highly inadequate for capturing the way that people, particularly in non-Western cultures, think about themselves, their relationships to one another, and their relationships to the environment. Within political–*ecological* scholarship too, Richmond *et al.* (2005) found that Namgis First Nations respondents defined health and well-being with respect to economic, political and social/cultural "health" dimensions of the community, including access to natural resources (see also Middleton 2010).

In our own experience, we have encountered indigenous women who, while suffering from multiple ailments, are profoundly satisfied with their families and the social, cultural, and economic futures of their children and grandchildren in their homeland. For practitioners, the challenge is to respect that contentment—that deep sense of well-being—while attending to the physical factors that may prematurely end a woman's life. For political ecologists more generally, it suggests the potential for fruitful inquiry around *understandings* of health and well-being in contexts of socio-political and ecological change.

Health as biosocial

Our final remarks concern the ways that health can be viewed as fundamentally biosocial (Mansfield 2008). This view contributes to a broader move within human geography and nature–society theories to refuse rigid distinctions between environmental, political, economic, and social aspects of life (Panelli 2010). The case studies presented here exemplify the fluidity and simultaneity of these aspects by showing how the health of individuals can be deliberately modified in ways that are *at once* biological (the joining of gamete cells, the introduction of antigens) and social (part of a conscious policy of population expansion, the result of demands for state health support). Moreover, we also point to ways that environments are themselves biosocial, not merely places or settings with which people interact to produce health outcomes (Guthman and Mansfield 2008). As we have shown in this chapter, to the extent that indigenous lands and resources, and the fight to maintain them, are inseparable from the well-being of indigenous groups, those lands and resources are themselves biosocial elements of indigenous health.

Acknowledgments

We are most grateful for the ongoing advice and support of Marta Azevedo, Heloisa Pagliaro, and Ricardo Ventura Santos. Portions of this chapter were presented by K. McSweeney and Brad Jokisch at the XXVIII International Congress of the Latin American Studies Association in Rio de Janeiro, Brazil, 11 June 2009. Joel Wainwright's comments on that presentation are appreciated. Thanks too to the editors of this volume for their suggestions on ways to improve the chapter.

Notes

1 We use this term in the knowledge of its imperfection but for lack of an equally parsimonious way to refer to the individuals and collectives that self-identify as indigenous.

2 That Shuar women use contraceptives to concentrate parity in a relatively short span contrasts with Bledsoe *et al.*'s (1998) finding that women use contraceptives to space their births more widely. Nevertheless, both outcomes support Bledsoe *et al.*'s primary argument: "in contexts with high levels of reproductive morbidity and mortality, a health model, not a demographic one, dominates people's thinking about contraception, superseding by far any specific worries about family size" (Bledsoe *et al.* 1998: 17).

3 We recognize a parallel and important body of work within biological anthropology (e.g., Leatherman and Thomas 2001; Salzano and Hurtado 2004), which space constraints have not allowed us to elaborate here.

References

Anderson, B. (1983) *Imagined Communities: Reflections on the Origin and Spread of Nationalism*. London, UK: Verso.

Arps, S. (2009) Threats to safe motherhood in Honduran Miskito communities: local perceptions of factors that contribute to maternal mortality. *Social Science and Medicine* 69(4): 579–586.

Azevedo, M.M. (2003) *Demografia dos Povos Indígenas do Alto Rio Negro/AM: Um Estudo de Caso de Nupcialidade e Reprodução*. Campinas, Brazil: Programa de Pós Graduação em Demografia, Brazil, Universidade Estadual de Campinas (UNICAMP).

BBC News. (2011) Brazil indigenous Guarani leader Nisio Gomes killed. *BBC News Online* (18 November). See www.bbc.co.uk/news/world-latin-america-15799712 (accessed 18 November 2011).

Bledsoe, C., Banja, F., and Hill, A.G. (1998) Reproductive mishaps and western contraception: an African challenge to fertility theory. *Population and Development Review* 24(1): 15–57.

Bremner, J.L., Bilsborrow, R.E., Feldacker, C., and Holt, F.L. (2009) Fertility beyond the frontier: indigenous women, fertility, and reproductive practices in the Ecuadorian Amazon. *Population and Environment* 30(3): 93–113.

Browner, C.H. (1986) The politics of reproduction in a Mexican village. *Signs: Journal of Women in Culture and Society* 11(4): 710–724.

Browner, C.H. (2000) Situating women's reproductive activities. *American Anthropologist* 102(4): 773–788.

Cardoso, A.M., Santos, R.V., and Coimbra Jr., C.E.A. (2005) Mortalidade infantil segun do raça/cor no Brasil: o que dizem os sistemas nacionais de informação? *Cadernos de SaúdePública* 21(5): 1602–1608.

Carroll, R. (2011) Drug barons accused of destroying Guatemala's rainforest. *Guardian Online* (13 June). See www.guardian.co.uk/world/2011/jun/13/guatemala-rainforest-destroyed-drug-traffickers (accessed 22 June 2011).
CEPAR. (2004) *Situación de Salud de los Pueblos Indígenas en el Ecuador.* Encuesta Demografica y de Salud Materna e Infantil (ENDEMAIN). Quito, Ecuador: El Centro de Estudios de Población y Desarrollo Social (CEPAR).
Chapin, M. (2004) A challenge to conservationists. *World Watch* (November/December): 17–31.
Chomsky, A. (2005) It seems impossible to believe: a survivor describes the massacre that destroyed her Wayuu community. *Cultural Survival Quarterly*, 28(4): 41–43.
Coimbra, C.E.A. and Basta, P.C. (2007) The burden of tuberculosis in indigenous peoples in Amazonia, Brazil. *Transactions of the Royal Society of Tropical Medicine and Hygiene* 101(7): 635–636.
Coimbra Jr., C.E.A. and Garnelo, L. (2003) *Questoes de saúde reproductiva da mulher indígena no Brasil.* Working Paper no. 7. Porto Velho, Brazil: Escola Nacional de Saúde Pública, Universidad Federal de Rondonia.
Coimbra Jr., C.E.A., and Santos, R.V. (2000) Saúde, minorias e desigualdade: algumas-teias de inter-relações, com ênfase nos povos indígenas no Brasil. *Ciência and Saúde Colectiva* 5(1): 125–132.
Coimbra Jr., C.E.A. and Santos, R.V. (2004) Emerging health needs and epidemiological research in indigenous peoples in Brazil. In *Lost Paradises and the Ethics of Research and Publication*, F.M. Salzano and A.M. Hurtado (eds). Oxford, UK: Oxford University Press, 89–109.
Collins, A.E. (2002) Health ecology and environmental management in Mozambique. *Health and Place* 8(4): 263–272.
Collyns, D. (2011) Peru shaman murders investigated. *The Guardian Weekly* (6 October). See www.guardian.co.uk/world/2011/oct/06/peru-shaman-murders (accessed 14 October 2011).
Conklin B.A. and Graham, L.R. (1995) The shifting middle ground: Amazonian Indians and eco-politics. *American Anthropologist* 97(4): 695–710.
Denevan, W.M. (ed.). (1992) *The Native Population of the Americas in 1492.* Madison, WI: Wisconsin University Press.
DeWalt, B.R. (1998) The political ecology of population increase and malnutrition in Southern Honduras. In *Building a New Biocultural Synthesis: Political Economic Perspectives on Human Biology*, A. H. Goodman and T. L. Leatherman (eds). Ann Arbor, MI: University of Michigan Press 295–316.
Escobar, A.L., Coimbra Jr., C.E., Camacho, L.A., and Santos, R.V. (2004) Tuberculin reactivity and tuberculosis epidemiology in the Pakaanóva (Wari') Indians of Rondônia, South-Western Brazilian Amazon. *International Journal of Tuberculosis and Lung Disease* 8(1): 45–51.
Fearnside, P.M. (2008) The roles and movements of actors in the deforestation of Brazilian Amazonia. *Ecology and Society* 13(1): 23. See www.ecologyandsociety. org/vol13/iss1/art23 (accessed 18 June 2012).
Federación Shuar. (1976) *Federación de Centros Shuar: Solución Orginal a un Problema Actual.* Quito, Ecuador: Federación Shuar.
Flowers, N.M. (1994) Demographic crisis and recovery: a case study of the Xavánte of Pimentel Barbosa. *South American Indian* 25(4): 18–36.
Goicolea, I., San Sebastián, M., and Wulff, M. (2008) Women's reproductive right in the Amazon basin of Ecuador: challenges for transforming policy into practice. *Health and Human Rights Journal* 10(20): 91–103.

Gomes, M.P. (2000) *The Indians and Brazil.* Gainesville, FL: University Press of Florida.
Gracey, M. and King, M. (2009) Indigenous health part 1: determinants and disease patterns. *The Lancet* 374(9683): 65–75.
Greenhalgh, S. (1995) Anthropology theorizes reproduction: integrating practice, political economic, and feminist perspectives. In *Situating Fertility: Anthropology and Demographic Inquiry*, S. Greenhalgh (ed.). Cambridge, UK: Cambridge University Press: 3–28.
Guthman J. and Mansfield, B. (2008) Toward a political ecology of the body? Paper presented at Annual Meeting of the Association of American Geographers, Boston, MA, 15–19 April.
Hale, C.R. (2005) Neoliberal multiculturalism: the remaking of cultural rights and racial dominance in Central America. *Political and Legal Anthropology Review (PoLAR)* 28(1): 10–19.
Hames, R. and Kuzara, J. (2004) The nexus of Yanomamö growth, health, and demography. In *Lost Paradises and the Ethics of Research and Publication*, F. M. Salzano and A. M. Hurtado (eds). Oxford: Oxford University Press: 110–145.
Hanchette, C.L. (2008) The political ecology of lead poisoning in eastern North Carolina. *Health and Place* 14(2): 209–216.
Hern, W.M. (1992) Polygyny and fertility among the Shipibo of the Peruvian Amazon. *Population Studies* 46(1): 53–64.
Hern, W.M. (1994) Health and demography of Native Amazonians: historical perspective and current status. In *Amazonian Indians from Prehistory to the Present*, A. Roosevelt (ed.). Tucson, AZ: The University of Arizona Press: 123–150.
Holt, F.L. (2005) The Catch 22 of conservation: indigenous peoples, biologists, and cultural change. *Human Ecology* 33(2): 199–215.
de la Hoz, F., Perez, L., Wheeler, J.G., de Neira M., and Hall, A.J. (2005) Vaccine coverage with hepatitis B and other vaccines in the Colombian Amazon: do health worker knowledge and perception influence coverage? *Tropical Medicine and International Health* 10(4): 322–329.
Hughes, J. (2004) *Gender, Equity, and Indigenous Women's Health in the Americas.* Washington, DC: Pan-American Health Organization (PAHO).
Hurtado, A.M., Lambourne, C.A., James, P., Hill, K., Cheman, K., and Baca, K. (2005) Human rights, biomedical science, and infectious disease among South American indigenous groups. *Annual Review of Anthropology* 34: 639–665.
Izquierdo, C. (2005) When "health" is not enough: societal, individual and biomedical assessments of well-being among the Matsigenka of the Peruvian Amazon. *Social Science and Medicine* 61(4): 767–783.
Jokisch, B. D. and McSweeney, K. (2011) Assessing the potential of indigenous-run demographic/health surveys: the 2005 Shuar Survey, Ecuador. *Human Ecology* 39(5): 683–698.
Kennedy, D.P. and Perz, S.G. (2000) Who are Brazil's indígenas? Contributions of census data analysis to anthropological demography of indigenous populations. *Human Organization* 59(3): 311–324.
Kertzer, D.I. and Arel, D. (eds). (2002) *Census and Identity: The Politics of Race, Ethnicity, and Language in National Censuses.* Cambridge, UK: Cambridge University Press.
King, B. (2010) Political ecologies of health. *Progress in Human Geography* 34(1): 38–55.
King, M., Smith, A., and Gracey, M. (2009) Indigenous health part 2: the underlying causes of the health gap. *The Lancet* 374(9683): 76–85.
Kramer, K.L. and Greaves, R.D. (2007) Changing patterns of infant mortality and maternal fertility among Pumé foragers and horticulturalists. *American Anthropologist* 109(4): 713–726.

Lauer, M. (2005) Conflicts between indigenous politicians and conservationists in the Upper Orinoco-Casiquiare Biosphere Reserve. *Cultural Survival Quarterly* 28(4): 48–52.

Leatherman, T.L. and Thomas, R.B. (2001) Political ecology and constructions of environment in biological anthropology. In *New Directions in Anthropology and Environment*, C.L. Crumley (ed.). Walnut Creek, CA: AltaMira Press: 113–131.

Lee, D. and Fulford, T. (2000) The beast within: the imperial legacy of vaccination in history and literature. *Literature and History* 9(1): 1–23.

Levin, B.W. and Browner, C.H. (2005) The social production of health: critical contributions from evolutionary, biological, and cultural anthropology. *Social Science and Medicine* 61(4): 745–50.

McKinley, M. (2003) Planning other families: negotiating population and identity politics in the Peruvian Amazon. *Identities: Global Studies in Culture and Power* 10(1): 31–58.

McSweeney, K. (2005) Indigenous population growth in the lowland neotropics: social science insights for biodiversity conservation. *Conservation Biology* 19(5): 1375–1384.

McSweeney, K. and Arps, S. (2005a) A "demographic turnaround:" the rapid growth of indigenous populations in lowland Latin America. *Latin American Research Review* 40(1): 3–29.

McSweeney, K. and Arps, S. (2005b) Meta-analysis of demographic trends among indigenous populations in lowland Latin America. XXV IUSSP International Population Conference, Tours, France, 18–23 July.

McSweeney, K. and Pearson, Z. (2009) Waorani at the head of the table: towards inclusive conservation in Yasuní. *Environmental Research Letters* 4(3): 031001.

Mansfield, B. (2008) Health as a nature–society question. *Environment and Planning A* 40(5): 1015–1019.

Mansfield, B. (2011) Is fish health food or poison? Farmed fish and the material production of un/healthy nature. *Antipode* 43(2): 413–434.

Martí i Puig, S. (2010) The emergence of indigenous movements in Latin America and their impact on the Latin American political scene: interpretive tools at the local and global levels. *Latin American Perspectives* 37(6): 74–92.

Mayer, J. (1996) The political ecology of disease as one new focus for medical geography. *Progress in Human Geography* 20(4): 441–456.

Meireles, D.M. (1988) Sugestões para uma análise comparativa da fecundidade em populações indígenas. *Revista Brasileira de Estudos de População* 5(1): 1–20.

Middleton, E. (2010) A political ecology of healing. *Journal of Political Ecology* 17: 1–28.

Montenegro, R.A. and Stephens. C. (2006) Indigenous health in Latin America and the Caribbean. *The Lancet* 367(9525): 1859–1869.

Morales, P. (ed.). (2009) Political environments: development, dissent and the new extraction. Special Issue. *NACLA Report on the Americas* 42(5): 11–41.

Morales, W.Q. (2011) From revolution to revolution: Bolivia's national revolution and the "re-founding" revolution of Evo Morales. *The Latin Americanist* 55(1): 131–144.

Nawaz, H., Rahman, M.A., Graham, D., Katz, D.L., and Jekel, J.F. (2001) Health risk behaviors and health perceptions in the Peruvian Amazon. *American Journal of Tropical Medicine and Hygiene* 65(3): 252–256.

Nepstad, D., Schwartzman, S., Bamberger, B., Santilli, M., Ray, D., Shlesinger, P., Lefebvre, P., Alencar, A., Prinz, E., Fiske, G., and Rolla, A. (2006) Inhibition of Amazon deforestation and fire by parks and indigenous lands. *Conservation Biology* 20(1): 65–73.

Oldham J. (2006) Rethinking the link: a critical review of population–environment programs. Amherst, MA: Population and Development Program, Hampshire College, and the Political Economy Research Institute, University of Massachusetts.

Orta-Martínez, M. and Finer, M. (2010) Oil frontiers and indigenous resistance in the Peruvian Amazon. *Ecological Economics* 70: 207–218.

Pagliaro, H. and Junqueria, C. (2005) Fertility trends and cultural patterns of the Kamaiurá women, Upper Xingu, Central Brazil. XXV IUSSP International Population Conference. Tours, France, 18–23 July.

Pagliaro, H., Azevedo, M.M., and Santos, R.V. (eds). (2005) *Demografia dos Povos Indígenas no Brasil*. Rio de Janeiro and Campinas, Brazil: Editora Fiocruz /Associação Brasileira de Estudos Populacionais (ABEP).

Pagliaro, H., Mendonca, S., Carvalho, N., Santos de Macedo, E., and Baruzzi, R.G. (2008). Fecundidade e saúde reproductiva das mulheres Suyá (Kisêdje), Parque Indígena do Xingu, Brasil Central (1970–2007). Paper presented at the XVI Meeting of the Brazilian Population Association (ABEP), Minas Gerais, Brazil, 29 September–2 October.

PAHO. (2004a) Maternal and child mortality among the indigenous peoples of the Americas. *Bulletin of Pan-American Health Organization: Healing Our Spirit Worldwide* 2(May): 1, 3.

PAHO. (2004b) Health of the indigenous peoples initiative: data collection and disaggregation, 19–21 January 2004. Secretariat of the Permanent Forum on Indigenous Issues, United Nations. See www.un.org/esa/socdev/unpfii/documents/workshop_data_ilo.doc (accessed 7 July 2012).

Panelli, R. (2010) More-than-human social geographies: posthuman and other possibilities. *Progress in Human Geography* 34(1): 79–87.

Paulson, S. (2000) Cultural bodies in Bolivia's gendered environment. *International Journal of Sexuality and Gender Studies* 5(2): 125–140.

Peet, R., Robbins, P., and Watts, M.J. (2011) Global nature. In *Global Political Ecologies*, R. Peet, P. Robbins, and M.J. Watts (eds). Abingdon, UK: Routledge, 1–48.

Penna, T. (1984) Por que demografia indígena brasileira? See www.abep.nepo.unicamp.br/docs/anais/pdf/1984/T84V03A16.pdf (accessed 18 June 2012).

Perreault, T. (2001) Developing identities: Indigenous mobilization, rural livelihoods, and resource access in Ecuadorian Amazonia. *Ecumene* 8(4): 381–413.

Perz S.G., Warren, J., and Kennedy, D.P. (2008) Contributions of racial-ethnic reclassification and demographic processes to indigenous population resurgence: the case of Brazil. *Latin American Research Review* 43(2): 7–33.

Price, D. (1994) Notes on Nambiquara demography. *South American Indian Studies* 4(March): 63–76.

Radcliffe, S. and Pequeño, A. (2010) Ethnicity, development and gender: Tsáchila indigenous women in Ecuador. *Development and Change* 41(6): 983–1016.

Raich, J.R. (2004) Ecosystem approach to rapid health assessments among indigenous cultures in degraded tropical rainforest environments: case study of unexplained deaths among the Secoya of Ecuador. *EcoHealth* 1(1): 86–100.

Renfrew, D. (2011) The curse of wealth: political ecologies of Latin American neoliberalism. *Geography Compass* 5(8): 581–594.

Ribas, D.L.B., Sganzerla, A., Zorzatto, J.R., and Philippi, S.T. (2001) Nutrição e saúde infantil em uma comunidade indígena Teréna, Mato Grosso do Sul, Brasil. *Cadernos de Saúde Pública* 17(2): 323–331.

Richmond, C., Elliott, S.J., Matthews, R., and Elliott, B. (2005) The political ecology of health: perceptions of environment, economy, health and well-being among 'Namgis First Nation. *Health and Place* 11(4): 349–65.

Robbins, P. (2004) *Political Ecology: A Critical Introduction*. Malden, MA: Blackwell.

Rosengren, D. (2003) The collective self and the ethnopolitical movement: "rhizomes" and "taproots" in the Amazon. *Identities: Global Studies in Culture and Power* 10(2): 221–240.

Rubenstein, S. (2001) Colonialism, the Shuar Federation, and the Ecuadorian State. *Environment and Planning D: Society and Space* 19(3): 263–293.

Salzano, F.M. and Hurtado, A.M. (eds). (2004) *Lost Paradises and the Ethics of Research and Publication*. Oxford, UK: Oxford University Press.

Souza-Santos, R., de Oliveira, M.V.G., Escobar, A.L., Santos, R.V., and Coimbra, C.E.A. (2008) Spatial heterogeneity of malaria in Indian reserves of Southwestern Amazonia, Brazil. *International Journal of Health Geographics* 7: 55–65.

Stocks, A. (2005) Too much for too few: problems of indigenous land rights in Latin America. *Annual Review of Anthropology* 34: 85–104.

Stocks, A., McMahan, B., and Taber, P. (2007) Indigenous, colonist, and government impacts on Nicaragua's Bosawas Reserve. *Conservation Biology* 21(6): 1495–1505.

Teixeira, P. (2005) *Sateré-Mawé: Retrato de um Povo Indígena*. Manaus, Brazil: UNICEF.

Teixeira, P. and Brasil, M. (2005) Estudo demográfico dos Sateré-Mawé: um exemplo de censo participativo. In *Demografia dos Povos Indígenas no Brasil*, H. Pagliaro, M.M. Azevedo, and R. Ventura Santos (eds). Rio de Janeiro and Campinas, Brazil: Editora Fiocruz and Associação Brasileira de Estudos Populacionais (ABEP), 135–154.

Tiwi, W. (2005) Diagnóstico de situación de salud de las nacionalides Shuar y Achuar. Unpublished proposal for a health survey. Sucua, Ecuador: FICSH.

Tollefson, J. (2011) A struggle for power. *Nature* 479(9 November): 160–161.

UNICEF. (2003) *Ensuring the Rights of Indigenous Children*. Innocenti Digest no. 11. Florence, Italy: UNICEF Innocenti Research Centre.

UNICEF. (2007) *Diagnóstico de Salud de las Nacionalidades Shuar y Achuar*. Quito, Ecuador: UNICEF-Ecuador.

UNICEF-Ecuador. (2007) *The Right to a Name and Nationality*. Quito, Ecuador: UNICEF-Ecuador.

United Nations. (2008) *United Nations Declaration on the Rights of Indigenous Peoples*. New York, NY: United Nations.

United Nations. (2009) *State of the World's Indigenous Peoples*. New York, NY: United Nations, Department of Economic and Social Affairs.

Velasco, M. (2011) Contested territoriality: ethnic challenges to Colombia's territorial regimes. *Bulletin of Latin American Research* 30(2): 213–228.

Wainwright, J. and Bryan, J. (2009) Cartography, territory, property: postcolonial reflections on indigenous counter-mapping in Nicaragua and Belize. *Cultural Geographies* 16: 153–178.

Wilson, P. (2010) Indigenous leadership and the shifting politics of development in Ecuador's Amazon. In *Editing Eden: A Reconsideration of Identity, Politics, and Place in Amazonia*, F. Hutchins and P.C. Wilson (eds). Lincoln, NB: Board of Regents of the University of Nebraska, 218–245.

Witzig, R. and Ascencios, M.(1999) The road to indigenous extinction: case study of resource exportation, disease importation, and human rights violations against the Urarina in the Peruvian Amazon. *Health and Human Rights* 4(1): 60–881.

Yashar, D.J. (2005) *Contesting Citizenship in Latin America: The Rise of Indigenous Movements and the Postliberal Challenge*. Cambridge, UK: Cambridge University Press.

Zimmerman, B., Peres, C.A. Malcom, J.R., and Turner, T. (2001) Conservation and development alliances with the Kayapó of south-eastern Amazonia, a tropical forest indigenous people. *Environmental Conservation* 28(1): 10–22.

9 Tsetse and trypanosomiasis

Eradication, control, and coexistence in Africa

Paul F. McCord, Joseph P. Messina, and Carolyn A. Fahey

And it shall come to pass in that day,
That the Lord shall hiss for the fly
That is in the uttermost part of the rivers of Egypt ...
And they shall come, and shall rest all of them
In the rugged valleys, and in the holes of the rocks,
And upon all thorns, and upon all brambles.
(Isaiah vii: 18–19)

Introduction

African trypanosomiasis (AT), a neglected tropical disease, is a zoonotic, parasitic infection of wildlife, domesticated animals, and humans. Transmitted by the bite of the tsetse fly (genus *Glossina*), the causative agents are parasites of the *Trypanosoma brucei* species complex. Approximately 8.5 million km^2 in 37 sub-Saharan Africa countries, representing locations of ecological suitability,[1] are infested with tsetse (Allsopp 2001). The vast area occupied by tsetse results in approximately 70 million people facing exposure risk (World Health Organization 2006a). Two major epidemics occurred in the first half of the twentieth century, one between 1896 and 1906, and the other in 1920. By the mid-1960s, human African trypanosomiasis (HAT), also known as sleeping sickness, appeared to be under control. However, by the mid-1970s, HAT re-emerged due to a breakdown in surveillance and control programs compounded by drug resistance, genetic changes in the parasite, civil conflict, and anthropogenic and natural environmental changes (World Health Organization 2006a). In the mid-1990s it was estimated that at least 300,000 cases were underreported due to lack of surveillance capabilities, diagnostic expertise, and healthcare access (World Health Organization 2006a). In response to these limitations, the World Health Organization, with public–private partnerships, initiated a new surveillance and elimination program. During the implementation of this program, approximately 17,500 new cases were reported with an estimated cumulative rate of 50,000–70,000 cases (World Health Organization 2006b). The disease is also considered one of the most important economically debilitating diseases in sub-Saharan Africa, with African animal trypanosomiasis (AAT) reducing livestock productivity by 20–40 percent in tsetse areas (Hursey 2001). In this chapter, we seek to explore the social, political, and structural challenges to control of tsetse and trypanosomiasis. We suggest that historical precedent, modern eradication language, and global health organizations, influence, and, in fact, hinder control efforts. We examine the tortured history of control measures followed by a discussion of why the tsetse fly continues to trouble so many in spite of over a century's worth of active suppression. Finally, we close with a discussion of the contentious issue of tsetse fly control versus fly eradication; on this divisive topic we advocate for sustained *control* of the fly.

In Mombasa, Kenya in October 1999, the 25th International Scientific Council for Trypanosomiasis Research and Control (ISCTRC) recommended the creation of a pan-African initiative to address tsetse eradication. As a result, the Organization for African Unity established the Pan African Tsetse and Trypanosomiasis Eradication Campaign (PATTEC) in October 2001 (Schofield and Kabayo 2008). While the goal of a tsetse free Africa is clear, PATTEC does not suggest nor follow a specific method for arriving at such a goal. Rather, to eliminate the tsetse fly, a host of techniques to be implemented throughout sub-Saharan Africa are encouraged, including combinations of traps and insecticide-impregnated targets, insecticide-treated cattle, aerial and ground spraying of insecticides, and deployment of sterilized male tsetse, also known as the sterilized insect technique (SIT) (Schofield and Kabayo 2008). Several vector control methods, such as trapping techniques, have been used to suppress the tsetse fly for many years, while other methods, such as the aerial spraying method listed above, have a more recent history. However, vector control is only one part of the complicated history of tsetse control; animal host population and administrative control have also experienced periods of popularity. Despite the persistent control efforts, the tsetse fly continues to plague sub-Saharan Africa.

The tsetse fly throughout history and early practices of living with the vector

Sleeping sickness is far from a recent phenomenon. As early as 1374, the Arab writer al-Qualquashandi described the death of the King of Mali as the result of sleeping sickness (Knight 1971). In fact, the discovery of *Glossina* fossil impressions in Colorado by Theodore Dru Alison Cockerell in the early 1900s allowed researchers to conclude that tsetse flies existed in North America during the Miocene (Brues 1923). Brues (1923) argues that these flies possibly carried trypanosomes during this period, suggesting that tsetse may have contributed to the extinction of select large mammals that once inhabited North America. John Atkins, an English naval surgeon, more recently recognized the presence of sleeping sickness when he used the term "negro lethargy" in 1742 to describe slaves in western Africa, and in 1803 Dr Thomas Winterbottom commented that slave-dealers would not buy Africans with enlarged glands, perhaps the most physically identifiable trait of trypanosomiasis (Lambrecht 1964). Sleeping sickness undoubtedly played a large role in the slave trade, and in the process, promoted negative stereotypes. Browne (1953) provides an observation made by the late Milton J. Rosenau, once a professor of public health at Harvard University, where Dr Rosenau points out that:

> … the ravages of sleeping sickness were well known to the old slave traders and the presence of "lazy niggers" lying prostrate on wharves and decks with saliva drooling from their mouths, insensible to pain or emotion, was a familiar sight.
>
> (Browne 1953: 150)

African societies have long coexisted with trypanosomiasis. In John Ford's seminal work *The Role of the Trypanosomiases in African Ecology*, he contends that pre-colonial societies achieved resistance to the disease through limited, continuous exposure to the trypanosome (Ford 1971). Accordingly, protection was acquired by modifying the environment to regulate interactions between humans, domesticated animals, wild fauna, tsetse flies, and the trypanosome. Giblin (1990) reviews Ford's work, and presents alternative pre-colonial methods of responding to the tsetse fly. One such method is found in Kjekshus's *Ecology Control and Economic Development in East African History: the Case of Tanganyika, 1850–1950* (1977) in which avoidance of the tsetse fly was encouraged. Kjekshus's method therefore was one of evading the fly, while Ford found low levels of contact necessary in human coexistence with tsetse. Torday (1910) seems to agree with Kjekshus's isolationist approach when he describes the people of the Kasai Basin in the Belgian Congo where sleeping sickness was controlled by keeping populations away from the fly and through the practice of removing sick villagers to isolated forests.

Whether pre-colonial Africans coexisted with trypanosomiasis and its persistent vector through a process of limited but continuous exposure (as suggested by Ford) or if an isolationist approach was key to survival, the arrival of Europeans certainly disrupted the established cohabitation practices. In discussing the colonial administration of north-eastern Rhodesia, Vail (1977) lists gun control laws, game control policies, village amalgamation policies, hut taxes, and labor recruitment campaigns as a collection of policies that disrupted the pre-colonial coexistence of man and the fly. As early as 1908, David Bruce, the Scottish microbiologist who played the largest role in identifying the cause of sleeping sickness, noted the impact of colonists in spreading sleeping sickness:

> It cannot be forgotten that the introduction of sleeping sickness into Uganda was due to England's interference with existing conditions. The movement of large masses of men or animals from the conditions to which they have become adapted is always attended with danger. Civilized man presents the untutored savage … with what he calls the dignity of labour with one hand, while with the other he scatters abroad the seeds of tuberculosis, sleeping-sickness, and other pestilences which I need not enumerate.
>
> (Bruce 1908: 258–259)

In their influential work examining land degradation from the political ecology perspective, Blaikie and Brookfield (1987) discuss the colonial role in the spread of the tsetse fly and trypanosomiasis. Such an expansion in distribution occurred as the result of colonial policies annexing land and forcing indigenous groups to move to areas previously avoided. For a related discussion of the relationships between human health and frontier settlement, see Chapter 7. These forced relocations following the Europeans' arrival contributed in no small part to the first sleeping sickness epidemic (Vail 1977). Uganda and the Congo Basin experienced the worst of this epidemic where, from 1896–1906, 200,000 people died from the disease (World Health Organization 2006a).

Following Walker (2006) we approach these issues from a traditional political ecology framework and focus on the influence of power in shaping human–environment relations.

Tsetse and trypanosomiasis control: early colonial practices

Knight (1971) suggests that the earliest form of active trypanosomiasis control was population evacuation. This technique often involved moving entire villages to "safe areas." In 1908 North-Eastern Rhodesia, ruled by the British South Africa Company, pursued one such mass evacuation from the eastern bank of the Luapula River (Figure 9.1). All villages along the bank from Kabila to the Nsakaluba stream were moved to higher ground to avoid the tsetse population at the river's edge (Musambachime 1981). During this move, the abandoned villages were burned in an effort to discourage the natives from returning. Those who were sick were sent into quarantines, from which they rarely returned.

Often during these evacuation events, the villagers were only allowed to carry basic necessities, and rarely were their destinations adequately prepared for their arrival (Musambachime 1981). Once relocated, it was common for villages to be densely resettled in order to promote expedient collection of taxes by colonial officials and to allow for ease in medically examining the resettled people (Vail 1977). The crowded conditions spawned enormous overuse of land that contributed to soil erosion, and the abandonment of previously used land allowed tsetse habitat to regenerate, which created an environment ripe for the continued spread of sleeping sickness (Vail 1977). The practice of population relocation was therefore nothing short of disastrous. Musambachime (1981) presents an observation that, due to the disorderly nature of such relocations from the Luapula River, more people died from hunger than died of sleeping sickness.

In addition to fostering conditions disastrous to the resettled population's health, these programs also disrupted lives both spiritually and economically. Relocation meant the abandonment of ancestral resting places. According to Torday (1929: 168), such displacement was like "a tree cut off from its roots," as relocated people were moved to environments deprived of physical, cultural, and religious familiarity. Relocation as well as the restrictions placed on the resettled individuals' movement also disrupted rituals such as rain prayers and made pilgrimages to sacred locations impossible (Musambachime 1981). Cattle, the economic core for many African villages, often fared poorly in the newly settled areas due to concentrated conditions that facilitated the spread of the disease (Soff 1969). As villagers lost their cattle and other animals to AAT, it

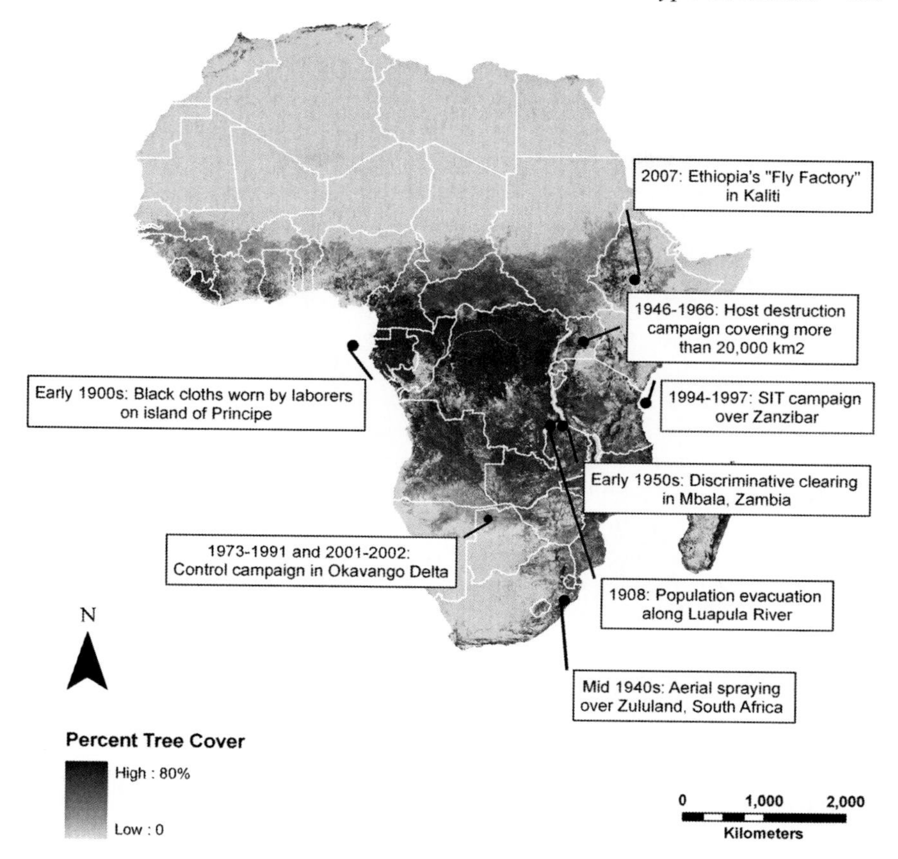

Figure 9.1 Location and timing of several campaigns against the tsetse fly across sub-Saharan Africa. The map displays the percentage of tree cover across the continent as the tsetse fly has a similar distribution (Cecchi *et al.* 2008).

Source: Tree-cover data from DeFries *et al.* (2000)

Note: Current political boundaries shown.

was not uncommon for livestock theft to increase, as occurred between tribal groupings in British East Africa at the turn of the twentieth century (Soff 1969). Additionally, population relocation further jeopardized economic prospects by eliminating the use of more fertile land and disrupting regional trading of items such as salt, palm oil, and fish (Musambachime 1981).

Population relocation programs used at the beginning of the twentieth century were plagued by hasty evacuations, poor preparation of resettlement destinations, and insufficient consideration of the resettled people's cultural and economic values. It is not surprising, therefore, that such programs led to adverse health, societal, and economic outcomes.

Like population evacuation, host destruction and habitat clearing were both popular colonial administrative solutions to tsetse. Bruce (1905) presents an early discussion of the role wild animals played as a reservoir of the disease. Bruce remarked that wild animals such as the buffalo and the wildebeest carried trypanosomes in their blood, but that the parasite did not seem to hurt these animals. On the other hand, when the parasite was introduced in domestic animals, AAT would result, often leading to death. The practice of game destruction as a means of eliminating both the parasite reservoir and a food source for the tsetse quickly followed Bruce's discovery.

Elimination of wild species was widely debated following the discovery of their role in harboring the parasite. At a meeting of the Royal African Society in 1913, Dr Warrington Yorke claimed that the only effective means of combating sleeping sickness was through the elimination of the reservoir of the virus (today known to be a unicellular parasitic protozoa). Dr Yorke was specifically addressing the situation in Rhodesia and Nyasaland where the population was troubled by the tsetse species G. morsitans. As the local population could not be moved away from the infested areas, Dr Yorke (1913: 27) commented that in attempting to destroy the reservoir of the virus, "it is obvious that the mere isolation of infected human beings is futile in view of the fact that the main reservoir of the virus is the blood of the big game."

In addition to ridding the population of the parasite's host, big game destruction was also promoted as a means of eliminating the fly's food supply, and thus causing the fly to disappear (Yorke 1913). Three years before Dr Yorke's call for the destruction of wild game, Alfred Sharpe, a British colonial administrator, warned the Royal African Society of the difficulties and potential unintended consequences of game destruction. He emphasized that certain conditions make areas suitable for the existence of tsetse flies, and the presence of big game makes little difference in their choice of habitat (Sharpe 1910). It was additionally cautioned that if the primary food source for the fly was removed, humans could be targeted as a secondary source, thus making sleeping sickness more prevalent (Sharpe 1910). Arguments were also made for more judicious slaughtering of game, acknowledging the successes of host destruction but calling for better knowledge of the relationship between the fly and wild animals to avoid ruthless mass killings (Swynnerton and Buxton 1938).

Active host destruction ultimately declined as a control technique by the 1940s due to its increasing social anathema and the rise of insecticides as a vector control option (Hargrove 2003; Cox 2004). However, during its period of popularity and even after insecticides became more broadly used, host destruction did offer effective, though modest, tsetse control. Jordan (1986) describes one such success that took place in Uganda from 1946–1966 (Figure 9.1). During this period, active hunting of wild animals substantially reduced the populations of two tsetse species from more than 20,000 km^2, primarily due to its intensive culling of even the most elusive animals (Jordan 1986). Host destruction is often jeopardized when the level of hunting is not sufficient to achieve the necessary degree of wild animal elimination. Reinvasion of the fly is common, and sustainable elimination of hosts is therefore required for successful tsetse control. Vail (1977) demonstrates how colonial policies limited the ability to sustain this necessary level of wildlife suppression in North-Eastern Rhodesia. During the first half of the twentieth century, the administration of North-Eastern Rhodesia, and later Northern Rhodesia, was not unlike many colonial regions of its time. Policies concerned with the capturing of resources and the promotion of European interests frequently increased the vulnerability of Africans and left them to cope with foreseeable and unforeseeable consequences of these policies. Such policies included gun control laws stemming from the Brussels Act of 1890, the conservation of wild animals from the London Convention of 1900, hut taxes, and labor recruitment programs (Vail 1977).

Throughout Central Africa, the British, fearing confrontation with the vastly more numerous local populations, implemented gun control laws (Vail 1977). While largely successful in sterilizing the chances of conflict, gun control, coupled with the protection of wildlife at sites such as the Mweru Marsh Game Reserve, increased tsetse food supplies, and simultaneously increased the number of reservoirs for the parasite. With infected wildlife existing in greater numbers and the means for checking their growth becoming increasingly limited, trypanosomiasis was able to sweep across North-Eastern Rhodesia with ease in the early 1900s. Furthermore, as early as 1898, the North-Eastern Rhodesia administration had been imposing a five shilling hut tax within its villages (Vail 1977). Unable to pay the tax due to limited employment opportunities, the men of North-Eastern Rhodesia often journeyed, and were recruited, to Southern Rhodesia where employment opportunities were more widespread, further reducing the available labor pool most capable of managing wildlife populations. During the colonial period, policies concerned with promoting European interests often ignored potential negative externalities. Such was the case in Rhodesia where concerns over access to firearms, faunal diversity, and the costs of government inadvertently led to increasing cases of trypanosomiasis by expanding the host population. This host population growth was often also accompanied by an expansion of tsetse habitat.

The discovery that many sleeping sickness cases existed along the shores of rivers and lakes led to discussions of tsetse habitat clearing. In Bruce (1908), the distribution of sleeping sickness cases was presented to the Royal African Society as occurring along the west coast of Africa, along the shores of Lake Tanganyika and Lake Victoria, in parts of western Uganda, and at Wadelai on the Nile River. These sites were shown to coincide with the existence of *G. palpalis*. With such reports clearly locating the presence of sleeping sickness to such confined areas inhabited by *G. palpalis*, bush clearing practices became more frequent and often were employed alongside game destruction to curb sleeping sickness cases caused by the more widely dispersed *G. morsitans* (Harcourt 1912).

Shrub and bush clearing occurred around Central and East African lakes in the early 1900s, and these practices were met with some success; however, the projects were often costly and required the movement of human populations (Harcourt 1912). Bruce (1908: 257) made clear the daunting task of habitat destruction, stating that "If we picture the hundreds of miles of lake and island shore, with huge trees and dense undergrowth up to the water's edge, we must come to the conclusion that the wholesale destruction of the fly is impossible." Soff (1969) highlighted the inefficiencies often found in bush clearing when recounting Ugandan Protectorate's Governor Hesketh Bell's belief that all bush harbored tsetse. Such views frequently led to total destruction of lakeshore vegetation when, in reality, only certain species of bush, based on physiological structure, provide habitat for the fly. Additionally, unless the land was populated after clearing or the bush kept from regenerating, the tsetse fly would return. Habitat destruction therefore typically included the encouragement of villagers to live closer together, as their routines of building, collecting firewood, and practicing agriculture would discourage the return of tsetse habitat (Science News-Letter 1927). The increase in agriculture that followed the process of concentrating villagers was promoted as an opportunity to increase the standard of living for those in the fly belts. This practice in turn was believed to have a positive cyclical effect: as the economic status of an area increased, so too would the public health (Gilks 1935).

Despite its inefficiencies, the practice of habitat destruction, when coupled with additional control techniques, has proven to be very effective at combating the tsetse fly (Hargrove 2003). Moreover, the practice of total habitat destruction, employed in the infancy of tsetse habitat removal, was discovered to be only one of several effective forms of habitat clearing, with the others (partial, selective, and discriminative clearing) being far less damaging. In the early 1950s for example, Mbala, Zambia (then Abercorn, Northern Rhodesia) was able to effectively control the tsetse fly using a discriminative clearing scheme where only the woody vegetation from a plant community was removed (Figure 9.1). This achievement resulted from clearing only 3 percent of the vegetation in a 725 km^2 area (Hargrove 2003).

Similar to the host destruction method, habitat clearing began to fall out of favor with the rise of insecticides (Cox 2004), and currently it is rarely deliberately used as a control technique. However, the anthropogenic impact of population growth throughout Africa represents a less deliberate form of habitat, as well as host, destruction (Bourn *et al.* 2001). Agricultural operations expand with growing populations and the hunting of game intensifies limiting the extent of natural tsetse habitat. Additionally, because the land is intensively and continuously used, agricultural expansion provides a sustainable form of control that, as stated earlier, is necessary to prevent habitat regrowth and reinvasion by the fly. Unfortunately, during the colonial period, continuous and intensive use of the land was often hampered by official policies (Vail 1977; Musambachime 1981). The regrettable result was the expansion of tsetse habitat during the early twentieth century and the continued pervasiveness of sleeping sickness cases.

It has already been demonstrated that man and tsetse coexisted for centuries in Africa prior to the arrival of Europeans. Kjekshus (1977), among others, suggests that tsetse were cleared from areas leading to tsetse-free zones. Such zones were created by the intensive use of land, which eliminated the fly's preferred habitat. Unfortunately, several of the European policies already mentioned as allowing reservoir host populations to increase also led to tsetse habitat expansion; the hut tax is one such policy. With the imposition of the North-Eastern Rhodesia hut tax in 1898, men moved to areas offering better employment opportunities (e.g., Southern Rhodesia), and, consequently, land management of the fields left behind declined (Vail 1977). Agricultural output accordingly decreased and the fields were allowed to revert to habitat conducive to the fly.

The conclusion of World War I provided another opportunity for the fly to reclaim lost habitat. In the early 1920s, the administration of Northern Rhodesia was eager to move Africans off from the most fertile lands in order to attract European settlers intent on growing tobacco. This practice resulted in approximately 3,500 square miles of inferior land set aside for African reserves, and 6,500 square miles of the most fertile land allocated to Europeans and those settlers expected to follow (Vail 1977). The collapse of the tobacco market in the 1920s discouraged settlers from moving to the "European" land, and after a fallow period the area was reclaimed by the tsetse fly. Sleeping sickness then became a greater risk for the Africans living in the nearby reserves and in one village a death rate of 66 per thousand was reported each year throughout the 1930s (Vail 1977).

Colonial administrators often oversaw the tsetse and trypanosomiasis control techniques used at the turn of the twentieth century. However, these same administrators commonly facilitated the failings of the control campaigns through policies that moved Africans into tsetse-infested zones, increased the number of host reservoirs, and encouraged the regrowth of tsetse habitat. In this way, the early tsetse and trypanosomiasis control methods largely failed due to poor planning and policy implementation. These failures were thought to be a problem of the past with the emergence of insecticides as a vector control technique in the 1940s; suddenly it was possible to swiftly control tsetse populations over large areas. Unfortunately, both old and new obstacles have limited the successes of the more contemporary control methods.

Tsetse and trypanosomiasis control: post-World War II practices

In 1874, dichlorodiphenyltrichloroethane (DDT) was first synthesized. It was not until 1936, however, that its insecticidal properties were realized, and not until the conclusion of World War II were insecticides widely used to control vector populations (Garnham 1967). With the use of insecticides, vector eradication finally seemed possible, and the process by which it could be achieved was often easier than previous control measures (Garnham 1967). In an article titled "DDT Can Wipe Out Plagues" (Science News-Letter 1945: 147), optimism was expressed that DDT could send disease-carrying insects to "join the dodo and the dinosaur," and thereby end "these particular plagues for all time."

Application of insecticides initially occurred most commonly in the form of ground spraying (Allsopp 2001). These operations typically included large teams of trained staff dispatched over the control area. The staff, consisting of control officers and laborers equipped with pressurized and non-pressurized sprayers carried on their backs, applied insecticides to vegetation frequented by tsetse. Spraying operations could only be successful if the control area was made uninhabitable for both the adult flies and the tsetse puparia buried in the soil (Jordan 1986). This was, and still is, achieved in one of two ways: through the use of residual insecticides that remain lethal long enough to control tsetse once they have emerged from their puparial case after roughly 22–25 days (Hargrove 2003), or through reapplication of a non-residual insecticide.

Currently, ground spraying is used infrequently due to its high costs, dependence on large numbers of well trained technicians, susceptibility to reinvasion, and regular dependence on residual insecticides (Hargrove 2003). In fact, the use of residual insecticides in West Africa in the late 1970s and early 1980s led to a decline in aquatic arthropod populations (FAO 1992). Other creatures that saw population numbers drop after sprayings included the plant-hoppers and silverfish in Zimbabwe at the beginning of the 1990s and the little bee-eaters during the 1980s (FAO 1992). Due in no small part to the devastating effects it had on non-target species, such indiscriminate, high-dose sprayings have largely been replaced by more selective, low-dose sprayings.

The first widespread aerial spraying campaign of tsetse habitat took place in the mid- and late 1940s with South African Air Force pilots flying over Zululand, South Africa (Figure 9.1; Science News-Letter 1947). Pilots applied DDT to an area of 100 square miles during the operation, and ground teams set off DDT grenades to target habitat missed by the aerial spraying. The campaign effectively controlled *G. pallidipes* in the sprayed area (Hargrove 2003), and optimism that the fly could be removed from the continent began to grow. Unfortunately, issues such as limited funding, environmental concerns, poor coordination between countries, and the ability and efficiency of the fly to reinvade cleared areas have limited the successes of control efforts (PATTEC 2001). The large-scale 1973–1991 control campaign over Botswana's Okavango Delta is one such example (Figure 9.1). During this operation, extensive and repeated aerial sprayings occurred over the vast wetland, but in order to entirely remove any opportunity for reinvasion, pilots also needed to spray the portion of the fly belt that extended into neighboring Namibia (Hargrove 2003). Unfortunately, permission was not granted for such cross-boundary flights, and a corridor of reinvasion was made available to the fly. The inability to cross political boundaries, however, was by no means the sole culprit for the operation's ultimate failure. The spray zones within the wetland were far too small, and movement between sprayings was not impeded to any great degree through the use of barriers, allowing the fly to avoid areas receiving aerial treatments (Hargrove 2003). Thus, despite nearly two decades of active suppression, inadequate management of the spraying campaign and poor coordination across political boundaries led to persistence of the fly in the Okavango Delta. In 2001, determined to eliminate the fly from the area, aerial spraying operations were renewed (Figure 9.1).

The second spraying operation, a campaign spanning 16,000 km^2, took place during two periods from 2001–2002 with full backing of Botswana's government (Kgori *et al.* 2006). The first treatment occurred in the northern half of the Okavango Delta from June to September 2001, and the second treatment in the southern half from May to August 2002. The sheer size of the separate spray regions was an improvement upon the 1973–1991 operation, as it was much more difficult for flies to seek refuge in untreated areas. The use of 12,000 deltamethrin-treated targets to create barriers separating spray zones also improved the second spraying campaign as fly movements were impeded and reinvasion made more difficult. Following the completion of this government-sponsored operation, no tsetse flies were caught in the area, allowing for claims of a tsetse-free Okavango Delta (Kgori *et al.* 2006).

The sterile insect technique (SIT) is another large-scale and internationally popular method for control. This technique, which relies on the use of radiation to sterilize male flies, has been advanced as a less destructive means of achieving control since only the target species is harmed (SIT, however, is used in conjunction with other techniques that may cause damage to non-target species). In the 1950s, the release of artificially sterilized male screw-worm flies over the island of Curaçao successfully eliminated the screw-worm from the island and led to the use of sterilized male flies to eliminate the pest from the southern United States (Simpson 1958). Elsewhere, SIT eradicated the melon fly from Okinawa and the Mediterranean fruit fly from Mexico, Chile, and southern Peru (Townson 2009). Such successes naturally spurred interest in the use of the sterilized male technique to confront the tsetse fly.

In Tanga, Tanzania in the 1970s, researchers supported by the US Agency for International Development set up a "fly factory" that produced thousands of sterile male tsetse flies per week. In this operation, unhatched male pupae were sterilized with small doses of cesium-137, and then released, sometimes 10,000 each week, into the wild where it was determined that the sterilized males led to a decrease in the tsetse population (Broad 1978). More recently, the island of Zanzibar was declared tsetse-free in 1997 following the release of nearly 8 million sterile male flies over the island from 1994–1997 (Figure 9.1; Nuclear News 1998). Other countries have also explored SIT, such as Ethiopia which recently spent roughly $12 million on a "fly factory" (Enserink 2007) expected to assist in tsetse eradication from 25,000 km^2 of Ethiopia's Southern Rift Valley by 2017 (Figure 9.1).

The costs of SIT are a barrier to implementation. In the successful use of sterile males over Zanzibar, nearly $6 million was spent from 1994–1997. This figure, while large, does not include the costs of establishing a facility to rear the sterile flies, nor does it include the costs of previous suppression work on the island (Molyneux 2001). Furthermore, Zanzibar, as an island, is an anomaly due to its natural barriers to reinvasion and presence of only one tsetse species. On the African mainland, SIT success is questionable with few natural barriers to reinvasion and the frequent presence of several tsetse species in the same area. The rearing of more than one sterile species, which would be required across much of Africa, for any SIT campaign would lead to a substantial increase in costs (Hargrove 2003). SIT is also only successful if the sterile males outnumber wild males ten to one (Enserink 2007), and in order to achieve this, traditional insecticidal techniques are still required. On the island of Zanzibar, the wild tsetse population had to be suppressed by 90 percent using traps and other techniques before SIT could achieve its goal. This result has caused some skeptics to suggest that continued use of the control technique that has achieved 90 percent control should be able to clear the remaining flies (Rogers and Randolph 2002). In this way, the economic burden of rearing sterile males would be avoided.

The aforementioned methods of ground spraying, aerial spraying, and SIT comprise the group of large-scale techniques. These were, and are, the methods that have given hope to the idea of Africa-wide tsetse eradication. However, since the 1970s there has been an ongoing decline in spending by African governments on tsetse and trypanosomiasis control (Hargrove 2000). This decline in funding has partly been the result of a reduction in donor support due to increasing donor concerns regarding the environmental consequences of large-scale control efforts and impatience resulting from witnessing only minimal improvements from large investments (Hargrove 2000). As donor and government support for control projects has waned, "community participation" has become a common phrase in aid projects (Catley and Leyland 2001). Under this approach, direct involvement from those benefitting from the aid program is encouraged. The African villager who has suffered the loss of cattle from trypanosomiasis therefore becomes responsible for controlling the problem rather than an aid agency of the government. This shift from government-centered or donor-centered control to local control has predictably brought about a shift from large-scale to small-scale control techniques (Hargrove 2000; Torr *et al.* 2005). The consequences of this shift have yet to be fully realized, but Torr *et al.* (2005) offer that in order to achieve elimination of tsetse, a return to large-scale campaigns must occur.

PATTEC plays the role of coordinating continent-wide tsetse eradication across political boundaries, providing technical guidance to member countries, and obtaining financial and material support when possible (PATTEC 2001). It promotes a variety of control methods, including aerial spraying and SIT, in order to attack tsetse in distinct fly belts across the continent (Kabayo 2002). In a direct appeal to the donor community, PATTEC (2001) offered that the environmental impact of all eradication campaigns will be considered before implementation, and that, despite the substantial initial cost of the large-scale efforts, eradication is a "once-and-for-all cost," while control costs recur indefinitely. Even with such announcements, the large initial cost of eradication may simply be too great for donors and governments given the long legacy of control failures. Consequently, the transition to cheaper, smaller, and more environmentally benign forms of control seems irreversible. These localized forms consist of three point-source methods of control: targets, traps, and insecticide-treated cattle.

The point-source practices

As mentioned above, the apparent success of the control effort in the Okavango Delta was largely due to the use of barriers to restrict reinvasion of sprayed areas. In that operation, tsetse attempts to reinvade the northern zone after treatment were prevented by a barrier of targets (screens sprayed with insecticides) set up between the two spray zones. This method of using targets and traps (the target's three-dimensional counterpart) to impede the fly's reinvasion efforts has frequently been used alongside large-scale control efforts (Kuzoe and Schofield 2004). More recently, traps and targets have been used alone due to the rise of community participation in control campaigns. These devices are effective when used alone (Hargrove 2003); however, their success hinges on their ability to provide an attractive visual and/or aromatic stimulant to lure in the fly.

The use of stimuli to attract tsetse to control devices has taken a variety of creative forms since the early 1900s. Several of these traps include the animal trap by Morris and Morris (Jordan 1986) and the "moving staircase" trap (Swynnerton 1933). However, the first successful use of visual stimuli to capture tsetse quite possibly took place on the island of Principe, off the west coast of Africa. Bulhões Maldonado, the estate manager of a cocoa and coffee plantation on the island, noticed that the tsetse on Principe were attracted to the backs of the plantation's laborers (Maldonado 1910). In an innovative strategy to limit the fly's presence in and around the plantation, Mr Maldonado ordered the laborers to wear black cloths covered with a "glutinous substance" on their backs (Figure 9.1). The black cloths effectively mimicked the daytime shadows that flies seek when locating resting sites (Steverding and Troscianko 2004). Mr Maldonado's strategy was successful: between April 1906 and the end of 1907, 133,778 tsetse were captured (Maldonado 1910), and in the process the use of visual stimuli to attract tsetse to control devices was born.

Regarding aromatic attractants, a progression of ideas similar to the development of visual stimuli occurred. These advances included the use of live animal baits in the 1930s (Kuzoe and Schofield 2004), which has progressed to the contemporary method of chemical attractants, such as acetone, octenol, and synthetic phenols that mimic host odors (Leak *et al.* 2008).

The third form of point-source control, the deployment of insecticide-treated cattle, has also witnessed a recent surge in popularity. In this method, also referred to as cattle dipping, cattle are commonly sprayed with pyrethroids (Hargrove 2003) along the parts of the body where the tsetse feed, typically the legs and belly (Torr *et al.* 2005). Using the host is an inexpensive and simple alternative to trapping and targeting devices, and the method finds allies in those advocating for smaller-scale, community-driven control efforts. With cattle dipping, often the livestock owner, rather than the government or a donor agency, is responsible for applying the insecticide and determining the frequency of applications (Torr *et al.* 2007). Thus, the individuals investing in the spraying of the livestock are also those receiving the direct benefit of control.

Like other tsetse control efforts the deployment of insecticide-treated cattle presents control efficiency challenges. The mere fact that cattle roam differentiates them from other point-source methods of control, since it cannot be ensured that the entire control area will be equally served. From a financial perspective, ability and willingness to pay may also become an issue when expecting stockowners to purchase the insecticides sprayed on the cattle. This financial burden was calculated to be roughly $0.20 per animal per year when just the legs and bellies of cattle are sprayed (Torr *et al.* 2007), a considerable obstacle in sub-Saharan Africa where it is not uncommon for 70 percent of the region's population, or more, to live on less than $2 per day (World Bank 2010).

Conclusions: control versus eradication

The complexity of tsetse infestation and trypanosomiasis, the options for suppressing the vector, and the past and present social, political, and economic challenges of the disease system have been discussed in this chapter. In the process, a number of factors, ranging from fly reinvasion, colonial programs, and environmental interests, have been highlighted as contributing to the fly's persistence across sub-Saharan Africa. To give adequate attention to these diverse factors, the political ecology approach was applied to offer a social interpretation for the presence and perpetuation of both the disease vector and the larger disease system (Mayer 2000).

Control efforts prior to World War II failed due to poorly designed policies, ineffective administration, lack of funding, and an inability to prevent reinvasion. Campaigns in the years following have likewise continued to suffer from limited funding and fly reinvasions; however, ideological differences centered on the end goal of tsetse suppression have further hindered the effectiveness of such campaigns. These differences, which are by no means unique to the tsetse fly (e.g., Gubler 1989 identified differing camps in the control of *Aedes aegypti* and its spread of dengue fever), have been of vector control versus eradication.

Continent-wide eradication was discussed earlier as the ultimate target of PATTEC, and an Africa free of the tsetse fly is undoubtedly a venerable goal. A one-time payment for eradication promoted in PATTEC (2001) is certainly appealing, and donors would naturally prefer projects with clear conclusions (eradication) rather than indefinite commitments (control). However, the current palate for tsetse control favors cheaper, more localized campaigns; thus, the large, "once-and-for-all" investments of eradication, as well as the need to conduct these campaigns at the national level and beyond leads to wavering interest in eradication from both donors and governments. Feldmann *et al.* (2005) lists more immediate concerns of governments, such as improvement of roads, schools, and medical services, which often take precedence over medium to long-term eradication campaigns against the fly. Eradication, in addition, requires an exceptional level of coordination that seems to be downplayed in the PATTEC (2001) document where it is suggested that eradication can be achieved by simply locating and wiping out isolated "Zanzibars" across the African continent.

Reinvasion and/or persistence of tsetse at low population densities has disappointed many who may have optimistically believed that the tsetse fly had been eradicated from a particular area. In Zululand, southeast Zimbabwe, and even the island of Principe, the tsetse fly eventually returned after it was believed that the original population had been eradicated (Hargrove 2003). The ever-present problem of reinvasion has limited the willingness of donors to give to a cause that has historically failed and has made generating the support of decision-makers increasingly difficult (Feldmann *et al.* 2005). It also calls into question whether continent-wide eradication is an achievable goal. It is clear that the knowledge and methods exist to achieve eradication in tsetse pockets, but conditions, whether conflict, corruption, or geography, in sub-Saharan Africa simply do not allow for continent-wide eradication. International organizations, private donors, and governments with conflicting goals limit the large investments needed to protect against reinvasion of cleared areas. Broad-scale campaigns versus campaigns with limited reach and programs with an emphasis on insecticides versus environmentally conscious campaigns are two such opposing views that reduce the efficiency of exogenous funding of efforts against the fly. Eradication also relies on cooperation within and among countries. During the 1960s and 1970s, Zimbabwe successfully cleared some areas of the fly; however, the country suffered extensive tsetse reinvasion during its push for independence (Jordan 1986). And without an equal desire for control from neighboring countries, campaigns carried out by individual governments succeed only within political boundaries, with reinvasion from across boundaries remaining a significant threat. PATTEC has a part to play in expanding control campaigns beyond political boundaries. However, the political, ideological, and socio-economic differences and instabilities that exist among the countries in sub-Saharan Africa make continent-wide eradication a seemingly unreachable goal. For these reasons, the approach against the tsetse fly must be one emphasizing sustained control, with participation across geographical scales.

Sustained control quells concerns from the donor community that tsetse suppression campaigns are too destructive to the environment, as the point-source techniques can be used to achieve this goal. Additionally, eradication efforts lead to large external debts for the country or countries taking on the ambitious task; on the other hand, sustained control can be achieved using methods far less financially burdensome, such as traps, targets, and trypanocides to treat herds in tsetse belts. Rogers and Randolph (2002) provide an estimate that the use of trypanocides is four times more cost-beneficial than any other tsetse or trypanosomiasis control method. Such methods allow the costs of control to be, at least partly, borne at the community level. And while eradication campaigns rarely leave infrastructure in place that could be used to control the fly if reinvasion occurs, the very nature of sustained control requires the continuous use of control infrastructure, such as traps and targets (Rogers and Randolph 2002). Therefore, sustained control avoids the large, ongoing (re)investments that accompany eradication campaigns when reinvasion inevitably occurs.

Tsetse and trypanosomiasis annually account for over $4.5 billion in losses across Africa (Oluwafemi 2009). This figure includes reduced milk and meat yields from cattle, the inability to use livestock for traction, and the limitation put on bringing additional land under cultivation. New control techniques such as the push–pull method (ICIPE 2010) and the identification of tsetse reservoirs (DeVisser *et al.* 2010) coupled with the continued use of point-source methods and trypanocides allow for the keeping of healthier livestock and the expansion of agricultural lands. These methods additionally suppress the fly in both an environmentally and economically responsible fashion. Through their sustained application, fly populations can be brought under control, and the burden of this neglected tropical disease made to weigh less heavily on the people of Africa.

Note

1 Tsetse rely on the presence of suitable habitat conditions for their survival. These conditions consist of proper land cover, climate, and food sources (Pollock 1982a, b). Preferred land cover types vary by tsetse species, with the *morsitans* group species preferring savanna and grassy woodland, the *palpalis* group species preferring mangrove swamps, lakeshores, and forests, and the *fusca* group species generally preferring thickly forested areas (Pollock 1982a). Climatically, depending on the species, temperatures between 19–28°C are preferred (Pollock 1982b). When temperatures rise above the preferred levels, tsetse seek shelter that helps to mitigate the higher temperatures (Leak 1999). As temperatures drop below the preferred levels, a "chill coma" sets in, which prevents tsetse from flying and eventually leads to starvation (Knight 1971; Terblanche *et al.* 2008). Soil moisture levels must be adequate for pupa development (Pollock 1982b).

Tsetse primarily feed on wild ungulates and ruminants, including the warthog, bushpig, kudu, and bushbuck (Jordan 1986). They take a blood meal from a host every two to three days (Schofield and Torr 2002). Tsetse also feed upon humans and livestock, including cattle, sheep, and goats.

References

Allsopp, R. (2001) Options for vector control against trypanosomiasis in Africa. *Trends in Parasitology* 17(1): 15–19.

Blaikie, P. and Brookfield, H. (eds). (1987) *Land Degradation and Society*. London, UK: Methuen & Co.

Bourn, D., Reid R., Rogers D., Snow B., and Wint, W. (eds). (2001) *Environmental Change and the Autonomous Control of Tsetse and Trypanosomosis in Sub-Saharan Africa: Case histories from Ethiopia, The Gambia, Kenya, Nigeria, and Zimbabwe*. Oxford, UK: Environmental Research Group Oxford Limited.

Broad, W.J. (1978) Taming the tsetse. *Science News* 114(7): 108–110.

Browne, W.J. (1953) Health as a factor in African development. *Phylon (1940–1956)* 14(2): 148–156.

Bruce, D. (1905) The advance in our knowledge of the causation and methods of prevention of stock diseases in South Africa during the last ten years. *Science* 22(558): 289–299.

Bruce, D. (1908) Sleeping-sickness in Africa. *Journal of the Royal African Society* 7(27): 249–260.

Brues, C.T. (1923) Ancient insects: fossils in amber and other deposits. *The Scientific Monthly* 17(4): 289–304.

Catley, A. and Leyland, T. (2001) Community participation and the delivery of veterinary services in Africa. *Preventative Veterinary Medicine* 49(1–2): 95–113.

Cecchi, G., Mattioli, R.C, Slingenbergh, J., and de la Rocque, S. (2008) *Standardizing Land Cover Mapping for Tsetse and Trypanosomiasis Decision Making*. PAAT Information Service Publications, PAAT Technical and Scientific Series no. 8. Rome, Italy: Food and Agriculture Organization of the United Nations; IAEA joint Division, International Atomic Energy Agency.

Cox, F. (2004) History of sleeping sickness (African trypanosomiasis). *Infectious Disease Clinics of North America* 18(2): 231–245.

DeFries, R., Hansen, M., Townshend, J.R.G., Janetos, A.C., and Loveland, T.R. (2000) 1 kilometer tree cover continuous fields, 1.0. Department of Geography, University of Maryland, College Park, MD. Coverage dates: 1 April 1992–1 April 1993. See www.glcf.umd.edu/data/treecover (accessed 10 February 2010).

DeVisser, M.H., Messina, J.P., Moore, N.J., Lusch, D.P., and Maitima, J. (2010) A dynamic distribution model of Glossina subgenus Morsitans: the identification of tsetse reservoirs and refugia. *Ecosphere* 1(1): 1–21.

Enserink, M. (2007) Welcome to Ethiopia's fly factory. *Science* 317(5836): 310–313.

FAO (1992) *Trypanosomiasis and Tsetse – Africa's Disease Challenge*. Rome, Italy: Food and Agriculture Organization of the United Nations.

Feldmann, U., Dyck, V.A., Mattioli, R.C., and Jannin, J. (2005) Potential impact of tsetse fly control involving the sterile insect technique. In *Sterile Insect Technique: Principles and Practice in Area-Wide Integrated Pest Management*, V.A. Dyck, J. Hendrichs, and A.S. Robinson (eds). Dordrecht, The Netherlands: Springer, 701–724.

Ford, J. (ed) (1971) *The Role of the Trypanosomiases in African Ecology: A Study of the Tsetse Fly Problem*. Oxford, UK: Clarendon Press.

Garnham, P.C.C. (1967) Importance of pesticides in preventive medicine. *Proceedings of the Royal Society of London B* 167(1007): 134–140.

Giblin, J. (1990) Trypanosomiasis control in African history: an evaded issue? *The Journal of African History* 31(1): 59–80.

Gilks, J.L. (1935) The relation of economic development to public health in rural Africa. *Journal of the Royal African Society* 34(134): 31–40.

Gubler, D.J. (1989) *Aedes aegypti* and *Aedes aegypti*-borne disease control in the 1990s: top down or bottom up. *American Journal of Tropical Medicine and Hygiene* 40(6): 571–578.

Harcourt, L. (1912) The crown colonies and protectorates and the colonial office. *Journal of the Society of Comparative Legislation* 13(1): 11–40.

Hargrove, J.W. (2000) A theoretical study of the invasion of cleared areas by tsetse flies (Diptera: Glossinidae). *Bulletin of Entomological Research* 90(3): 201–209.

Hargrove, J.W. (2003) *Tsetse Eradication: Sufficiency, Necessity, and Desirability*. Research report. Edingurgh, UK: University of Edinburgh, DFID Animal Health Programme, Centre for Tropical Veterinary Medicine. See www.dfid.gov.uk/r4d/PDF/Outputs/RLAHtsetse_Erad.pdf (accessed 18 June 2012).

Hursey, B.S. (2001) The program against African trypanosomiasis: aims, objectives and achievements. *Trends in Parasitology* 17(1): 2–3.

ICIPE (International Centre of Insect Physiology and Ecology). (2010) Tsetse research. See www.icipe.org/home/60.html?task=view (accessed 28 June 2010).

Jordan, A.M. (ed) (1986) *Trypanosomiasis Control and African Rural Development*. New York, NY: Longman Group Limited.

Kabayo, J.P. (2002) Aiming to eliminate tsetse from Africa. *Trends in Parasitology* 18(11): 473–475.

Kgori, P., Modo, S., and Torr, S. (2006) The use of aerial spraying to eliminate tsetse from the Okavango Delta of Botswana. *Acta Tropica* 99(2–3): 184–199.

Kjekshus, H. (ed) (1977) *Ecology Control and Economic Development in East African History: the case of Tanganyika, 1850–1950*. London, UK: Heinemann.

Knight, C.G. (1971) The ecology of African sleeping sickness. *Annals of the Association of American Geographers* 61(1): 23–44.

Kuzoe, F.A.S. and Schofield, C.J. (2004) *Strategic Review of Traps and Targets for Tsetse and African Trypanosomiasis Control*. Geneva, Switzerland: UNICEF/UNDP/World Bank/WHO, Special programme for research and training in tropical diseases (TDR).

Lambrecht, F.L. (1964) Aspects of evolution and ecology of tsetse flies and trypanosomiasis in prehistoric African environment. *The Journal of African History* 5(1): 1–24.

Leak, S.G.A. (ed) (1999) *Tsetse Biology and Ecology: Their Role in the Epidemiology and Control of Trypanosomosis*. Wallingford, UK: CABI Publishing with the International Livestock Research Institute.

Leak, S.G.A., Ejigu, D., and Vreysen, M.J.B. (2008) *Collection of Entomological Baseline Data for Tsetse Area-Wide Integrated Pest Management Programmes.* Rome, Italy: Food and Agriculture Organization of the United Nations. See www.fao.org/docrep/011/i0535e/i0535e00.htm (accessed 23 May 2010).

Maldonado, B. (1910) English abstract of Portuguese texts of 1906 and 1909. *Sleeping Sickness Bureau Bulletin* 2: 26.

Mayer, J.D. (2000) Geography, ecology and emerging infectious diseases. *Social Science and Medicine* 50(7–8): 937–952.

Molyneux, D.H. (2001) Sterile insect release and trypanosomiasis control: a plea for realism. *Trends in Parasitology* 17(9): 413–414.

Musambachime, M.C. (1981) The social and economic effects of sleeping sickness in Mweru-Luapula 1906–1922. *African Economic History* (10): 151–173.

Nuclear News (1998) Tsetse fly eliminated on Zanzibar. *Nuclear News* 1: 56–61.

Oluwafemi, R.A. (2009) The impact of African animal trypanosomisis and tsetse fly on the livelihood and well-being of cattle and their owners in the BICOT study area of Nigeria. *The Internet Journal of Veterinary Medicine* 5(2). See www.ispub.com/journal/the-internet-journal-of-veterinary-medicine/volume-5-number-2/the-impact-of-african-animal-trypanosomosis-and-tsetse-fly-on-the-livelihood-and-well-being-of-cattle-and-their-owners-in-the-bicot-study-area-of-nigeria.html (accessed 7 July 2012).

PATTEC (Pan African Tsetse and Trypanosomosis Eradication Campaign). (2001) *Plan of Action: Enhancing Africa's Prosperity.* Addis Ababa, Ethiopia: Organization of African Unity. See www.africa-union.org/Structure_of_the_Commission/Pattec/PATTECAction_Plan_English.pdf (accessed 18 June 2012).

Pollock, J.N. (1982a) *Training Manual for Tsetse Control Personnel. Volume 1: Tsetse Biology, Systematics and Distribution, Techniques.* Rome, Italy: Food and Agriculture Organization of the United Nations. See www.fao.org/docrep/009/p5178e/p5178e00.htm (accessed 2 March 2010).

Pollock, J.N. (1982b) *Training Manual for Tsetse Control Personnel. Volume 2: Ecology and Behaviour of Tsetse. Rome.* Rome, Italy: Food and Agriculture Organization of the United Nations. See www.fao.org/docrep/009/p5444e/p5444e00.htm (accessed 2 March 2010).

Rogers, D.J. and Randolph, S.E. (2002) A response to the aim of eradicating tsetse from Africa. *Trends in Parasitology* 18(12): 534–536.

Schofield, C.J. and Kabayo, J.P. (2008) Trypanosomiasis vector control in Africa and Latin America. *Parasites and Vectors* 1(24): 1–9.

Schofield, C.J. and Torr, S.J. (2002) A comparison of the feeding behaviour of tsetse and stable flies. *Medical and Veterinary Entomology* 16(2): 177–185.

Science News-Letter. (1927) Fight tsetse fly. *The Science News-Letter* 11(310): 179–180.

Science News-Letter. (1945) DDT can wipe out plagues. *The Science News-Letter* 48(10): 147.

Science News-Letter. (1947) DDT war on African flies. *The Science News-Letter* 52(20): 317.

Sharpe, A. (1910) Recent progress in Nyasaland. *Journal of the Royal African Society* 9(36): 337–348.

Simpson, H.R. (1958) The effect of sterilised males on a natural tsetse fly population. *Biometrics* 14(2): 159–173.

Soff, H.G. (1969) Sleeping sickness in the Lake Victoria region of British East Africa, 1900–1915. *African Historical Studies* 2(2): 255–268.

Steverding, D. and Troscianko, T. (2004) On the role of blue shadows in the visual behaviour of tsetse flies. *Proceedings of the Royal Society B* 271(Supplement 3): S16–S17.

See http://rspb.royalsocietypublishing.org/content/271/Suppl_3/S16.short (accessed 17 January 2010).

Swynnerton, C.F.M. (1933) Some traps for tsetse-flies. *Bulletin of Entomological Research* 24(1): 69–102.

Swynnerton, C.F.M. and Buxton, P.A. (1938) Tsetse-flies of East Africa. *Journal of the Royal African Society* 37(146): 92–94.

Terblanche, J.S., Clusella-Trullas, S., Deere, J.A., and Chown, S.L. (2008) Thermal tolerance in a south-east African population of the tsetse fly *Glossina pallidipes* (Diptera, Glossinidae): implications for forecasting climate change impacts. *Journal of Insect Physiology* 54(1): 114–127.

Torday, E. (1910) Land and peoples of the Kasai Basin. *The Geographical Journal* 36(1): 26–53. See www.jstor.org/pss/1777651 (accessed 29 November 2009).

Torday, E. (1929) The morality of African races. *International Journal of Ethics* 39(2): 167–176.

Torr, S.J., Hargrove, J.W., and Vale, G.A. (2005) Towards a rational policy for dealing with tsetse. *Trends in Parasitology* 21(11): 537–541.

Torr, S.J., Maudlin, I., and Vale, G.A. (2007) Less is more: restricted application of insecticide to cattle to improve the cost and efficacy of tsetse control. *Medical and Veterinary Entomology* 21(1): 53–64.

Townson, H. (2009) SIT for African malaria vectors: epilogue. *Malaria Journal* 8(Supplement 2): S10.

Vail, L. (1977) Ecology and history: the example of eastern Zambia. *Journal of Southern African Studies* 3(2): 129–155. See www.jstor.org/pss/263633 (accessed 21 January 2010).

Walker, P.A. (2006) Political ecology: where is the policy? *Progress in Human Geography* 30(3): 382–395.

World Bank. (2010) Data: poverty. See http://data.worldbank.org/topic/poverty (accessed 8 May 2010).

World Health Organization. (2006a) African trypanosomiasis (sleeping sickness). See www.who.int/mediacentre/factsheets/fs259/en (accessed 16 March 2010).

World Health Organization. (2006b) *Weekly Epidemiological Report*. Geneva, Switzerland: World Health Organization.

Yorke W. (1913) The relation of big game to sleeping sickness. *Journal of the Royal African Society* 13(49): 23–32.

10 Geographies of HIV and marginalization

A case study of HIV/AIDS risk among Mayan communities in western Belize

Cynthia Pope

Introduction

Research and scholarship within the social sciences on human immunodeficiency virus (HIV) have transitioned from focusing exclusively on risk groups within the early days of learning about the virus, to individual risk behaviors, and then finally to the role social environments play in encouraging or hindering HIV transmission. One of the more recent research frameworks, referred to as "social ecology," highlights a growing awareness of the role of social, structural/institutional, and environmental influences on HIV risk behaviors (Friedman *et al.* 2009; Jones and Moon 1993; Kearns 1993; Latkin *et al.* 2010; Latkin and Knowlton 2005; Parker *et al.* 2000; Pope *et al.* 2008; Tobin *et al.* 2010). These approaches are particularly salient for the field of health geography because they have delved into more nuanced ways of understanding how places constitute, as well as contain, social relations and physical resources.

Building upon this tradition within medical geography and health geography, geographers Cummins *et al.* (2007) present a foundational argument for this chapter as they discuss how places can be construed in health geography research as "relational." This framework sheds light on how rural Mayan communities could be at increased risk for HIV and helps to reveal why HIV prevention and intervention programs in these communities have taken so long to institute. I advocate in this chapter against traditional social ecological views for two reasons. First, a tendency remains to describe rural communities as cohesive, and static units. Second, ethnicity is presented as a binding identity that dictates attitudes, practices, behaviors, and opinions. Instead, I argue that power structures are created, reified, and can also change at different scales and times within these communities, a viewpoint that corresponds to Cummins *et al.*'s (2007) "relational" approach to health and place. This relational view of place differs from a more conventional view in a number of important ways, including understandings of multiscaled definitions of place, the role of socio-relational distance as opposed to physical distance, the mobility of individuals between places, changes in paths of access to certain resources over space and time, fluid versus static definitions of area, and the importance of social power relations and cultural meaning in determining territorial divisions (Cummins *et al.* 2007: 1827). The authors thus argue for a concentration on the "*processes* and *interactions* occurring between people and places and over time which may be important for health" (Cummins *et al.* 2007: 1828; emphasis original). Further, they argue for a better understanding of the position of places relative to each other, as well as the flows of capital, culture, and people between places. Porous international and inter-community borders mean that movement is an important element of these relationships and of the social and disease landscapes of rural communities.

In this chapter I explore ideas of physical and symbolic locations of rural populations, historical segregation, cultural norms, and social structures in Belize that lead to indigenous marginalization and health risk. Belize's current adult (age 15–49) HIV rate is approximately 2.7 percent, which ranks it as the second highest prevalence rate in the Western Hemisphere and 27th in the world (CIA 2011; UNAIDS and WHO 2010). While the country has received significant funding from organizations such as the Global Fund for HIV, AIDS, and Tuberculosis, the issues of rurality and ethnicity have meant that some populations who may be at risk for HIV are yet to be integrated entirely into HIV prevention, education, and intervention programs. The goal of this chapter is to use the lens of geography to analyze some reasons that rural populations, primarily those of Mayan descent, may continue to be at risk. The theoretical objective of this chapter is to explore the traditional exclusion of Mayans in HIV/AIDS discourse and policies in Belize. The structural invisibility of these individuals is couched in colonial and post-colonial history whereby coastal populations have been the focus and thus "privileged" in national HIV policy. Of course, ironically, to be privileged in HIV policies and thus distribution of funds is to be the population considered most at risk. While the coastal Garifuna, Afro-Belizean, and Creole populations are the focus of most HIV projects funded from abroad, the Mayan populations have been essentially considered outside the traditional purview of the virus. My contention is that historical models of social, economic, geographic, and political disenfranchisement of these communities have laid the foundation for increased HIV transmission. In this context, HIV risk is not a reflection of skin color or racial heritage, but rather of structural factors that I work to uncover in this chapter.

In order to accomplish this, I concentrate upon triangulating information from various written sources while relying on a variety of qualitative methods conducted in the Cayo Region, Caye Caulker, and southern Belize since 2005 addressing economic dependency and women's risk for HIV. The geographic isolation of many of these Mayan communities and their lack of participation in crucial sectors, such as health, government, and national education agendas, is noteworthy. Mayan communities are kept at the margin of HIV outreach efforts and attempts at national integration. Meanwhile their image is at the heart of recent development efforts, particularly in the tourist sector. I reveal in this case study of Belize that ethnicity, and HIV/AIDS risk and policy is the symbolism of ethnicity may increase certain communities' risk for HIV, due to stereotypes, physical distance, and impoverishment that began in colonial times and continues to the present day.

Social context of indigenous communities in Belize

For such a small territory (23,000 km^2) with a small population (321,000), Belize is very diverse ethnically. Mestizos (those of European and indigenous ancestry) comprise approximately 49 percent of the population, Creole (Belizeans of African descent) comprise 25 percent, Maya 11 percent, Garifuna (ancestors of Carib "Indians" and Europeans) comprise 6 percent, and nearly 10 percent of the population is classified as "other," which includes east Indian, Middle Eastern, Chinese, and Europeans (CIA 2011). While English is the official language, Spanish is spoken as a first language by 46 percent of residents, Creole by 33 percent, Mayan dialects by 9 percent, English by 4 percent, Garifuna (also known as Carib) by 3 percent, German 3 percent, and other (such as Chinese) by 1 percent (CIA 2011; Statistical Institute of Belize 2010). While the geopolitical image of Belize may have an English-speaking face, the society appears to be becoming more like its Central American neighbors than its Caribbean ones. For example, the most recent data show that more citizens speak Spanish as a first language than Creole, and much more than English. The prevalence of Christianity in Belize is instrumental to understanding the dominant culture, and the country even has an official prayer calling on Jesus Christ to help in its development (Belize Tourism Board 2011). Christianity is not indigenous to Mayan communities, creating an additional layer of difference between the indigenous and the descendants of the settlers who established political norms and economic hierarchies in the country. Roman Catholics make up 50 percent of the population, Protestant denominations 27 percent, "other" religions (including Mayan and African religions) 14 percent, and 9 percent of the population identifies no religious affiliation (CIA 2011; Statistical Institute of Belize 2010).

Pre-Mayan and Mayan groups first populated Belize, and they reached their highest numbers during the Classic Period of 300–900 AD. The country was colonized in the 1500s by the Spanish, and then through treaty became part of the British Empire from 1862 until 1981 (Sullivan 2003). The British legacy is reflected in the official language, the system of government, membership in CARICOM (Caribbean Common Market—islands colonized by Britain, the education system, the legal system, and the social hierarchy. Discourses of ethnicity and definitions of "Belizean" have tended to favor the white population and the coastal populations, many of whom are descendants of Africans. Although a new capital city was established in the center of the country in 1970, the previous capital of Belize City, on the Caribbean coast, is still the center of many government offices, the country's main hospital, and the headquarters of HIV testing centers and policy creation, such as the National AIDS Commission.

It is one of my contentions in this chapter that the British legacy is one of geographic marginalization of Mayans in the country. The highest poverty rates correspond to the two districts with the most Mayan residents (Toledo in the south is 57.6 percent Mayan, and Cayo in the west is 41.0 percent; National AIDS Commission 2006). The Mayans in the south (Kek'chi and Mopan) were native to the area and then fled into neighboring Guatemala when foreign loggers started working on their lands without consent (Mayan People of Southern Belize *et al.* 1997; Shoman 2000). They have since returned. The Yucatec Mayans arrived to the north and west of the country in the 1800s fleeing the Caste Wars on the Yucatan Peninsula. There has been no area of Belize that was not at one time under Mayan control (Sharer and Traxler 2005). Trade has always linked highland and lowland Mayan groups, and transborder movement (between Belize and Guatemala, in particular, and more recently between Belize and the United States) is a feature in these communities.

Mayans have never held significant political or economic power in Belize. Even if one looks at one of the most important symbols of any nation, the flag, Mayans are absent. The flag portrays a shield with two images—a black male agricultural worker and a white one. The image promoted is, ironically, one of agricultural exploitation and harkens back to a colonial racialized history where black and white men shaped the landscape.

Despite their absence on the flag, Mayan history, cosmology, and even traditional healing (Arvigo and Balick 1993) are being used to sell Belize as a tourism destination. Belize is still defining its post-colonial nationhood and identifying an economic niche in a highly competitive and globalized world (Santos 2000). Ironically, while Mayans have been marginalized from social services, they have been integrated into the most recent attempt for the Belizean development. "*Mundo Maya*" (Mayan World) and "*La Ruta Maya*" (The Mayan Route) are tourism programs based on the geographic legacy of Mayan civilization in Guatemala, Mexico, Belize, Honduras, Nicaragua, and El Salvador. Even the new capital built in 1970, Belmopan (*Bel*ize+*Mopan* Mayan) has been designed to resemble a traditional Mayan city,

Figure 10.1 Flag of Belize.

although few Mayans occupy positions of power in the national government nor do many (if any) live in the city or the surrounding areas. This fissure was true even last decade when Sutherland wrote:

> While the tourist industry promotes with fanfare La Ruta Maya and the celebration of ancient sites in the Yucatan, Belize, and Guatemala, the Maya quietly carry on with their lives. It is always somewhat jarring when great play is given to the Maya as an ancient people while the living Maya are ignored or trampled in the rush toward archaeology …
>
> (Sutherland 1993: 313)

However, promoting international tourism has the perhaps unintended consequence of increased risk for HIV transmission through new sexual networks and sometimes the introduction of drug use into those new tourism communities. Padilla *et al.* (2010) call for greater attention to the context of sexual vulnerability in tourism and call the areas of increased tourism growth "ecologies of heightened vulnerability." These ecologies play out in this case study in the interplay between tourists and rural Mayan communities in Cayo and Toledo districts and in the northern cayes. Additionally, the cruise ship industry leads to quick interactions between foreigners and Belizeans, which sets the stage for transactional sex (Sutherland 1998). As happens in other tourist destinations in the Caribbean, people may migrate to these areas for work. Large-scale tourism, however, while most prevalent on the cayes, is developing in the southern part of the country with new resort construction. These tourism nodes where the majority of international tourists visit, such as the cayes, are increasingly characterized by ecologies of vulnerability as the low wages combined with opportunities to gain money through sexual activities may lead to increased exposure to sexually transmitted infections. Thus, one of the challenges for Belize is to increase its income while not increasing HIV transmission because the disease can be transferred from the tourism nodes to rural villages, as in other Caribbean and Central American countries. For example, it is not uncommon to have infections occur in the tourism nodes and then the infected individual returns to his or her home community, often leading to infection in that location (Kempadoo 1999).

Another mode of disenfranchisement includes a growing lack of access to resources. The Toledo Association of Mayans has noted that they have always felt on the economic margins of development plans and have had to fight for resources they feel are historically theirs (Mayan People of Southern Belize *et al.* 1997). For example, foreign loggers and foreign evangelists are encroaching in southern Belize, breaking chains of indigenous knowledge systems, and pushing Mayans to cities for employment (Steinberg 2002, 2005). In one case, even though the government granted reservation status to the Mayan in the south, it was taken away without the village's knowledge, and an Asian firm was granted logging concessions (Steinberg 2002).

To better gauge ecologies of risk in Mayan communities it is necessary to deconstruct how the image of Mayans is created. For example, Wilk (1997) notes an anachronistic portrayal of Mayans in Belize. Many anthropologists depict "traditional" societies as living outside the boundaries of contemporary time and activities, for example working in subsistence agriculture and outside of a capitalist economy rather than in urban settings and relying on indigenous plant-based medicine rather than allopathic medicine. Thus, Mayan societies represent how a society "used to be" and are often infantilized. However, recent work on Belize shows that Mayans, through a city-state form of government, have always integrated ideas, languages, and norms from other Mayan groups, and more recently, other Central American populations (Sharer and Traxler 2005). They have adapted to cash economies, new languages, and new agricultural and construction technologies. Most Mayans do not live separate from modern society, nor have they "collapsed," as many textbooks report. As Shoman states, "the Maya were never finally 'conquered.' They have continued to recreate their civilization with accommodations and adjustments reflecting their particular environment" (Shoman 2000: 3). Indeed, Mayans are still living in the areas that they have inhabited for hundreds, and in some cases thousands, of years.

However, the notion of the "Pristine Myth" still dominates. This concept took the post-conquest history of the Americas and molded the image of indigenous to live in small-scale subsistence societies that rarely modify the landscape (Denevan 1992). For example, Sullivan, through analyzing postcards and photographs, writes that "The Orient might be decadent lasciviousness, but the New Word is ageless virginal passivity" (Sullivan 2003: 321). If Mayans are depicted as "the imagined Indian" (i.e. passive and pure; Sullivan 2003) or as elements of a pristine landscape (Denevan 1992), is an assumption of Mayan asexuality running through health discourses? Could this assumption be not only because of lack of sexuality (and maybe an overemphasis on the assumption of African sexuality in Belize), but also because HIV has been portrayed as a disease of sin, a moral disease (Allen 2000), a "disease of modernity" (Bancroft 2001: 89)? Thus, if policymakers do not envision Mayans as modern, these cultures, and thus their communities, may be (unwittingly perhaps) framed as immune to this new global virus, which is imbued with images of sexuality, marginalization, drug use, and sex work.

It therefore appears that Mayans in Belize represent a paradox. On one hand, their image as exotic is integrated into tourism campaigns. On the other hand, they may be considered so unique that they are outside the realm of allopathic/Western healthcare delivery and disease risk, including HIV. While Belize is becoming a "modern" nation and shedding its colonial past, the needs of the indigenous are not at the forefront.

HIV prevention, transmission, and risk in Belize

The first case of HIV in Belize was diagnosed in 1986, and the country now has the highest HIV prevalence rate in Central America. The rate is nearly twice that of neighboring Guatemala and eight times the rate in Mexico. Cohen (2006) appropriately labels Central America as the site of an overlooked epidemic. UNAIDS and WHO report that the median prevalence rate is approximately 2.1 percent.[1] Several sources cite the tremendous under-reporting of stigmatized health conditions, such as HIV, in the region (Jaramillo and Gough 2005;

Wheeler *et al.* 2001). Despite the high prevalence rates, Belize's HIV situation continues to be understudied in academic literature (PAHO 2010).

Policy-makers and statisticians have noted that the data in Belize are skewed geographically (Jaramillo and Gough 2005). If one were to believe the official HIV statistics then the Caribbean coast of the country would host almost all of the HIV cases. While it appears that the Garifuna population on the coast is hit particularly hard (Friedman 2007; N. Ken, medical doctor and HIV expert, 2005, personal communication), high HIV rates may be experienced throughout the country. For example, individuals who seek HIV testing often provide a Belize City address instead of their true district of residence, due to a fear of lack of confidentiality and stigma. Even when I attended a support group meeting for people living with HIV/AIDS (PLWHA) in Cayo district, I met many more than the five or so people reported by the Ministry of Health as testing positive for HIV in the district at the time (C. Orozco, co-founder of UNIBAM, Belize City, Belize, 2007, personal communication). Additionally, most people with HIV probably never get tested (C. Dominguez, Women's Department officer, Cayo District, 2006, 2007, personal communication).

While voluntary testing and counseling sites exist in each of the six districts, most testing is conducted in Belize City. Thus, the data, and ensuing programs, are skewed toward urban and Afro-Belizean and Creole populations. Moreover, a recent billboard campaign focuses on girls and HIV risk in Belize City (see Figure 10.2). The campaign is forward-thinking as it links economic dependence, sexuality, and structural issues to HIV risk; however, the campaign is also notable in that the targeted population is Afro-Belizean and urban.

Figure 10.2 Billboard in Belize City showing that girls may need to engage in sex for money to pay for school tuition.

Belize receives funding from the Global Fund for HIV, TB, and Malaria for distribution of antiretroviral (ARV) drugs. The most recent World Health Organization data estimate that 45 percent of those needing ARV therapy receive it (UNAIDS and WHO 2010). The medication is distributed at eight sites throughout the country, but the confounding factors of stigma and inaccessibility to these health centers (due to such things as distance, hours of operation, poor roads, lack of transportation) may lead to the low rate of coverage (UNAIDS and WHO 2010). Additionally, the remoteness of Mayan communities, as well as their small populations, could lead to more stigmatization over HIV testing and diagnosis than urban dwellers. For example, in a rural community it is much easier to see who is visiting a health clinic than in a crowded urban area, and the social connections are much closer than in urban areas. These two issues make confidentiality a concern, whether one is getting tested for HIV or any other condition. If one wishes to ensure a higher level of confidentiality, one must travel quite a distance, which may be monetarily or temporally prohibitive.

HIV studies and policy-makers in Belize note that conclusions about the major mode of HIV transmission in Belize are difficult to make due to incomplete reporting of information (Belize Women's Department 2004; Cohen 2006; Jaramillo and Gough 2005). However, it appears the pattern of transmission is primarily heterosexual in nature, and the numbers of men and women infected is at about a 1:1 ratio (UNAIDS 2009). Several groups in Belize have been identified as being particularly vulnerable to HIV infection, including persons living in poverty, commercial sex workers (Kane 1993; Kinsler *et al.* 2004), persons living with sexually transmitted infections (STIs), youth (Goldberg 2006; Vergara 2006), men who have sex with men, members of the uniformed services and incarcerated populations (Cohen 2006), and mobile and migrant populations (Bronfman *et al.* 2002). Mobility is an important factor in the Cayo and Toledo regions where many people go back and forth across the Belize–Guatemala border to visit relatives, regional markets, commercial sex workers, and for employment. Also contributing to the country's high HIV prevalence rate is a culture in which many men have more than one sexual partner. Additionally, high rates of sexual activity occur among teenagers, often without any method of contraception or disease prevention (Goldberg 2006). The low income of the country's residents, one-third of whom have earn less than US$2,000 annually, also contributes to the spread of HIV. Thus, the confounding factors of rurality and poverty make Mayan communities vulnerable to HIV.

Language disenfranchisement is also a factor in the disease landscape of Mayan communities. Anthropologist McClaurin (1996) employs Elmendorf's (1983) term "linguistic tyranny" to describe her inability to access Mayan women for her research in Toledo District. Since men speak English and/or Spanish, and are thus able to interact with those outside of Mayan-speaking communities, the majority of women are excluded from knowledge systems. Thus, if one does not speak Mopan or Kek'chi Mayan (most people do not), it is difficult if not impossible to communicate with women. This linguistic barrier keeps social workers and healthcare workers either outside the community or needing to filter the information through men in order to communicate with women. Gendered language constraints make it a challenge to access certain types of information on structural vulnerabilities to disease. One type of information that is difficult to tease out is household violence. Much research has shown that domestic violence is linked to women's economic subordination and thus linked to HIV risk. When one cannot control an income, or in this case lives in isolation, one is dependent on a male spouse or partner for cash income. The issue of abuse and non-monogamy, and having to submit to these, are common for women throughout Belize, and particularly in isolated indigenous communities (McClusky 2001). The Belizean government has recognized partner violence as a major issue in HIV vulnerability and has begun working to create policies to address it (PAHO 2010). In 1993 Belize adopted a Domestic Violence Act making domestic violence punishable by law. However, according to the Women's Department, domestic violence has not necessarily waned. For example, the Cayo Women's Department instituted a program with police officers to respond to domestic violence as a crime and not treat it as a "family problem" (C. Dominguez, Women's Department officer, Cayo District, 2006, 2007, personal communication). Judging from McClusky's (2001) observations and my interviews with practitioners at Belize Family Life Association, the Women's Department, and ProBelize (an international non-governmental organization), the prevalence of violence may be high due to language barriers in reporting to police, lack of phones to call police, and gender norms. Women in abusive relationships may contract HIV from a partner who has become infected from sexual dalliances and transmits it to her through sex. If she requests a condom, she is often subject to abuse (or at the very least suspicion of being unfaithful). Even if she is worried about infection, she often has sex because of the economic necessity of a male partner (for a related discussion, see Chapter 13). However, one researcher observed the beginnings of a "young woman's revolt" whereby some Belizean women are choosing not to get married and instead becoming professionals. This development may alter women's risk for HIV as they may be able to choose to live alone and thereby reduce occasions of sexual acquiescence to a male partner (McClusky 2001).

Case studies of marginality and risk: HIV risk in San Antonio and Cristo Rey, Cayo district

In order to outline some of the dimensions shaping vulnerabilities to HIV among Mayan communities, as well as relational views of place, this section provides case studies from San Antonio and Cristo Rey in Cayo District. The villages of San Antonio and Cristo Rey, both in Cayo District, present coherent examples of a framework that prioritizes a "relational" approach to place and health attributes (see Figure 10.1). While I argue that Mayan communities may be at risk for HIV, in this section I highlight how two of these communities in Western Belize are resisting marginalization, which have the potential to shape the disease landscape in the future.

Cayo borders Guatemala, and the two areas are much closer culturally and historically to each other than Cayo is with the rest of Belize. However, many of the same structural conditions throughout Belize are also experienced in

Cayo. For example, NGO workers and Women's Department officials have highlighted the high rate of violence against women (abuse, incest, and rape); the lack of economic opportunity for men and women; traditional gender norms that encourage young adolescent boys to cross into Guatemala to lose their virginity; and women crossing into Belize to have sex for money or to meet a "sugar daddy," a wealthy man, to support them (E. Castellanos, president, Tikkun Olam non-governmental organization, 2010, personal communication; C. Dominguez, Women's Department officer, Cayo District, 2006, 2007, personal communication). While injected drugs do not appear to be the most pressing issue in the Cayo region, infected needles could become a transmission route in the future given that Belize is one of the corridors through which drugs pass from South America and Mexico into the United States. In addition, this is the district with the second highest number of Mayan residents and is also the second poorest in the country. High poverty rates are often linked to high-risk activities, such as engaging in sex work to meet economic needs. Further, many Mayan villages have no access to healthcare or social services. Thus, knowledge about HIV risk and its cultural contexts may be limited to a small group of residents, healthcare workers, NGO workers, and educators. A local cultural activist found that HIV prevention campaigns often do not reach the Cayo Mayan communities since the virus cannot be seen, and those with the infection often appear healthy. However, the fear is such that PLWHAs often leave their home villages for fear of discrimination (Emily Tzul, outreach coordinator, Mopan Maya Community, Cayo District, 2010, personal communication).

Expanding on the notion of Mayans and disease vulnerability, the president of the NGO Tikkun Olam Belize wrote this in an email to me about her experiences in the area:

> Belize has Mayan presence although sometimes I think that governments and donors and politicians forget that fact. Every week there are countless meetings with big name donors and they talk a lot about statistical facts and figures. Mayan women and young girls are never counted. The amount of statistical data in relation to Mayan women and HIV is in the neighborhood of nil—although there might be an iota there is not even the intent of focus on this issue. When I worked as an outreach officer for a USAID NGO several years ago, we would go into the Belize military camps to facilitate sexual and reproductive health (SRH) trainings and education sessions with one of our target populations which was uniformed populations (including Belize Defense Forces [BDF] soldiers, police, and firefighters). On several occasions when I went to these military bases, I would go into the neighboring Mayan villages. I noticed there were many young mothers with multiple children from multiple fathers—the fathers were all different BDF soldiers. These Mayan women and their risks for HIV infection from multiple partners are not being taking into consideration in any quantitative or qualitative research that I am aware of as I cannot track any useful or relevant information in regards to Belize Mayan women and anything related to SRH. It is only in the recent years that we see more involvement and openness from the Mayan

> community in politics and in the fearless defense for their land rights against the government of Belize. My hope is that young Mayan women will have access to vital information that can help them achieve economic empowerment and the education they need to be successful in whatever field of work so they can live healthy lives making informed decisions about their SRH.
>
> (E. Castellanos, president, Tikkun Olam, 2011, personal communication)

The villages of San Antonio and Cristo Rey, both in Cayo District, present coherent examples of a framework that prioritizes a "relational" approach to place and health attributes. While I argue that Mayan communities may be at risk for HIV, in this section I highlight how two of these communities in western Belize are resisting marginalization, which have the potential to shape the disease landscape in the future.

San Antonio is home to roughly 2000 people and is located several kilometers from the main town in Cayo District, San Ignacio, and a few kilometers from the Guatemala border. It is Yucatec Mayan and Mestizo (primarily from Guatemala and El Salvador), and the first population census was conducted by a Peace Corps worker in 2006–2007. The village is accessible by a road that is paved for a few miles and then turns into dirt and gravel. It serves as a stopping point on tourist treks to Mayan ruins and also a stopping point for Guatemalans coming into Belize on market days or seeking healthcare from indigenous healers. Cristo Rey is smaller and is on the same road to San Ignacio. These two villages are connected through family relations, travel routes, and a school. Both suffer a high rate of unemployment and a low level of school completion. Despite this, both villages have initiated progressive social programs, particularly in health.

The social context of San Antonio reifies the importance of Cummins *et al.*'s (2007: 1827) description of a "relational" understanding of place whereby territorial divisions and infrastructure are imbued with social power relations and cultural meanings. Foreign encroachment is evident in San Antonio through the access that the road gives them (generally considered beneficial in terms of access to tourists and access to a larger town) and, particularly important for HIV transmission, evangelism. While Catholic churches have been a mainstay in this region (and historically have usurped indigenous Mayan religions), the encroachment of evangelical churches on villages has far-reaching implications. In San Antonio one Catholic church existed, but since the late 1990s three Pentecostal churches have been built, recruiting heavily from the impoverished population by offering people clothes or rice for attending services. As religious groups run most schools in Belize, sexual abstinence is typically the foundation of health education courses (if any are taught). An interesting anecdote comes from one of the study participants who mentioned that her daughter's Catholic primary school had a large mural of Jesus with the words "abstinence until marriage" underneath it. She said it was odd that a school that opposes sex education would draw young children's attention to sex before they were thinking about it on their own.

Indeed, in many ways the school system in Belize still reflects the Spanish and British colonial legacies whereby the Church was to be the main provider of education (Lewis 2000). Given that secondary schools are only attended by roughly half of the population, most people receive messages about sexuality and gender roles in elementary school. Despite the traditional infusion of religiosity in the schools, girls continue to become pregnant at a young age, and the Ministry of Health cites high STI rates (Pan American Health Organization 2010). Unfortunately, the number of HIV cases is not known at such a local scale. I contend that the confounding factors of poverty and abstinence education may have created high-risk behaviors and environments for HIV.

However, local organizations are trying to create spaces of economic opportunity that may reduce HIV risk. Addressing economic dependency, the San Antonio Women's Association shows a promising project to curtail HIV. While its main purpose is to raise money to build a new elementary school (the government will not provide one), their recently established crafts group creates a way for women to earn their own money. This opportunity has the potential for women to leave abusive relationships where women's economic dependency on their partners is the main reason for staying. This change could also curtail the rate of domestic violence since women become the primary breadwinners in the household. Indeed, many international studies have shown that women's economic independence lowers their HIV risk rates (Greig and Koopman 2003; Pan American Health Organization 2010; Türman 2003). Thus, by using traditional organization techniques, women's groups are going outside colonial systems of power to positively transform their community spaces.

Cristo Rey ("Christ the King") primary school has responded in a different way to HIV risk. The primary school principal, a respected Mayan healer from San Antonio, integrated teaching in the Yucatec Mayan language, which had been prohibited by the national government from colonial times until recently. He also has been using traditional herbs found in the nearby forest to treat patients for various ailments and he stated that he is trying to develop the proper combination of treatments for autoimmune diseases, like HIV/AIDS (D. S. Canto, naturalist and school principal, San Antonio, Belize, 2005, personal communication). Many local people, including Guatemalans, see him as a type of primary physician. Along with a local non-governmental organization, the principal decided to resist the trend to keep HIV and its transmission routes hidden and highly stigmatized. Instead, he showed that there is room to resist the silence surrounding HIV in Mayan communities (Bolland 2004). One important project was organizing the primary school students into a parade with each student holding a hand-drawn placard addressing stigma and discrimination, two barriers to seeking HIV tests (Andrewin and Chien 2008; Pope and Shoultz 2010).

This is an excellent example demonstrating how this community is dynamic and that people have the ability to change a potentially stigmatized space into a safer space, where taboo issues of sexuality are discussed openly. By marching through the village, these children and their instructors brought messages to most of the households in a traditional way that information is passed in rural communities—through discussions and word of mouth (Cropley 2004).

The case studies of San Antonio and Cristo Rey demonstrate why a "relational" approach to geography and health is so useful in this context. For example, the "constellation of connections" between people and place is clearly shown. Not

Figure 10.3 Students in an anti-HIV stigma parade in Cristo Rey.

only is the network between people living in Cristo Rey and neighboring villages evident in how they can change the education level about HIV/AIDS in the area, but it shows that the area is "imbued with social power relations and cultural meaning" and it is important to note that individuals in this village, like other rural Mayan villages, often travel to cities for work, healthcare, education, and market days (Cummins *et al.* 2007: 1827). Oftentimes, such as in Cristo Rey, relatives are located in bordering Guatemala, and thus the spaces of vulnerability for HIV or other transmissible diseases occupies a much larger space than official political maps indicate. This implies the necessity of cross-border and multinational HIV prevention cooperation that target the movement zones between countries. For example, in other parts of Central America, truck routes have been a primary mode of disease transmission because of the transactional sex work that takes place in the truck stops (Schifter 2001). Is this something that also occurs at official truck border crossings between Belize and Honduras, Guatemala, and Mexico? Do disease transmission opportunities also occur in the unofficial land and water crossings between villages on either side of the official borders? At least one Belizean non-governmental organization is working to identify the border HIV transmission possibilities between Mexico and Belize. Perhaps once this is done there will be more official cooperation between neighboring countries in the field of public health. From a more theoretical perspective, border crossing behaviors expand our notion of vulnerable spaces from a cohesive regional or district perspective with defined political boundaries to a more functional level of space that recognizes familial connections, social connections defined by a common history and language, and economic activities.

Conclusion: returning to issues of marginalization and HIV risk in rural Belize

Belize presents an interesting scenario about post-colonial nation-building, cultural norms, and geographies of HIV risk. In this chapter I have drawn from multiple qualitative methods utilized in two communities in western Belize to situate Cummins *et al.*'s (2007) relational approach between health and space. I rely on recent HIV/AIDS risk literature to inform the ways in which ethnicity, power, and movement contextualize risk, and how these risky places can also be transformed into safe ones. Geographers are particularly well-placed to expand on traditional social ecology approaches to understand of how ethnicity, place, movement, and place-making become mutually constitutive processes that create and/or hinder HIV risky environments. Communities are not bound by physical geography; rather social and economic relations may lead individuals to work in tourism areas, to cities for education and services, and the closest population centers for market days. Thus, in this chapter I presented a case study of how historical structural marginality contributes to an ecology of HIV vulnerability and how this marginality is reflected or resisted by indigenous communities.

As I have argued throughout this chapter, marginalization of Mayan villages is due not only to their ethnicity and language, but also to their geographic distance, which influences their economic and political access. Part of this Belizean case study is the evaluation and promotion of "indigenous culture" according to the priorities and interests of the ruling, non-indigenous groups. Historically, colonial governments saw the indigenous as a roadblock to modernization, and the government continues to break agreements brokered to grant indigenous their reservations (Steinberg 2002, 2005). Their towns and villages are rural, and social services are difficult to access. From the viewpoint of historical and geographic disenfranchisement, it is not surprising that Mayans have not been included in the targeted population for HIV prevention and intervention. Indeed, language factors, cultural norms, gendered opportunities, and historical friction between indigenous groups and "outsiders," have led to national health programs that do not cater to these communities.

These factors lead me to ask about the role of location; is it distance from former capital and current coastal HIV/AIDS "hotspots" that makes Mayan communities invisible? In other words, is it the symbolic or the physical distance that makes Mayans appear to be so "other" that they are not considered at risk, or is it the other historic factors that lead to these communities not being targeted for intervention? Conversely, by assuming that Garifuna and African cultures are cultures at risk, does this not reify notions of racialized risk in lieu of behavioral or structural risk? These questions encompass the various notions that Cummins *et al.* (2007) bring forth in a "relational" view of health and its physical context.

However, there are individuals and small groups that recognize and hoping to alter this indigenous geography of risk. These groups are staffed by Belizeans, instead of relying on foreign NGOs for funding and staffing. For example, the Ministry of Health, the Women's Department, Belize Family Life Association, the National AIDS Committee, and UNIBAM are taking steps toward using more localized knowledge systems in HIV prevention (National AIDS Commission 2010). If they can increase the funding available to fight HIV in the country, and if they acknowledge that a potential epidemic extends beyond the Caribbean coast, then there is a chance that Mayan communities could be integrated successfully into culturally appropriate prevention and treatment programs.

While I use Belize as a case study for this chapter, the findings have broader implications for theory and policy application. I demonstrate a fruitful way in which social scientists and healthcare practitioners and policy-makers must transcend the narrow confines of "social determinants" of health to include cultural and symbolic aspects in a suitable and holistic manner. I hope to have shown in this chapter that a relational approach to health and place is useful in this case study, and, additionally, that this case study of rurality, ethnicity, and national development adds evidence to this approach. To answer Cummins *et al.*'s (2007) call for expanded research, I highlight the ways in which Mayans relate to how places are produced and maintained and what this indicates for the health of individuals.

I add to this relational perspective the various ways in which differently scaled notions of power plays out *vis-à-vis* HIV risk. While I couch most of this chapter in terms of geopolitical power, I also demonstrate that intra-community and gendered power relations affect risk. This case highlights that new political ecological spaces have been molded and "safe spaces" are being created in Cristo Rey. Indeed, this spatialized and relational approach to risk can be applied to other places where indigenous are stigmatized with the goal of rendering the indigenous visible in national planning, unfettered from stereotypes, and with full rights and access to healthcare and conditions for well-being.

Acknowledgments

I would like to thank the Belizean researchers, policymakers, and HIV/AIDS advocates who helped on various aspects of this project, particularly Elisa Castellanos, president of Tikkun Olam Belize. Other individuals who were particularly helpful for data collection and helping with the framework include Mercedes Cuz, Claudia Domínguez, Bart and Suze Mickler, Assad Shoman, Emeli Tzul, and members of UNIBAM (United Belize Advocacy Movement, www.unibam.org/index.php).

Note

1 Note that the statistical analysis of data changed in 2008. For example, through 2007, UNAIDS and WHO estimated a high rate of HIV in Belize around 5 percent. By 2008 that changed to about 3 percent, and is currently averaged at 2.1 percent (UNAIDS and WHO 2010).

References

Allen, P.L. (2000) *The Wages of Sin: Sex and Disease, Past and Present.* Chicago, IL: The University of Chicago Press.

Andrewin, A. and Chien, L.-Y. (2008) Stigmatization of patients with HIV/AIDS among doctors and nurses in Belize. *AIDS Patient Care and STDs* 22(11): 897–906.

Arvigo, R. and Balick, M.J. (1993) *Rainforest Remedies: One Hundred Healing Herbs of Belize.* Twin Lakes, WI: Lotus Press.

Bancroft, A. (2001) Globalisation and HIV/AIDS: inequality and the boundaries of a symbolic epidemic. *Health, Risk and Society* 3(1): 89–98.

Belize Tourism Board. (2010) About Belize. See www.travelbelize.org/about-belize (accessed 7 July 2012).

Belize Women's Department. (2004) *Belize's Report on the Implementation of the Beijing Platform for Action (1995) and the Outcome of the Twenty-Third Special Session of the General Assembly (2000).* Belmopan, Belize: Ministry of Human Development, Government of Belize.

Bolland, O. N. (2004) *Colonialism and Resistance in Belize: Essays in Historical Sociology.* Mona, Jamaica: University of West Indies Press.

Bronfman, M.N., Leyva, R., Negroni, M.J., and Rueda, C.M. (2002) Mobile populations and HIV/AIDS in Central America and Mexico: research for action. *AIDS* 16(Supplement 3): S42–S49.

CIA (Central Intelligence Agency). (2011) World factbook—Belize, 2011. See https://www.cia.gov/library/publications/the-world-factbook/geos/bh.html (accessed 4 October 2010).

Cohen, J. (2006) The new world of global health. *Science* 311(5758): 162–167.

Cropley, L. (2004) The effect of health education interventions on child malaria treatment-seeking practices among mothers in rural refugee villages in Belize, Central America. *Health Promotion International* 19(4): 445–452.

Cummins, S., Curtis, S., Diez-Roux, A.V., and Macintyre, S. (2007) Understanding and representing "place" in health research: a relational approach. *Social Science and Medicine* 65: 1825–1838.

Denevan, W. (1992) The pristine myth: the landscape of the Americas in 1492. *Annals of the Association of American Geographers* 82(3): 369–385.

Elmendorf, M.L. (1983) *Nine Mayan Women: A Village Faces Change.* Rochester, VT: Schenkman Books.

Friedman, S.A. (2007) Educating for health: UNICEF Belize, a small partner gets the big picture. *AIDSLink* 103. See www.globalhealth.org/publications/article.php3?id=1672 (accessed 4 October 2010).

Friedman, S., Cooper, H.L.F., and Osbome, A.H., (2009) Structural and social contexts of HIV risk among African Americans. *American Journal of Public Health* 99(6): 1002–1008.

Goldberg (2006) Cultural factors contribute to high HIV prevalence in Belize. *KPBS News*, 21 February. See http://dailyreports.kff.org/Daily-Reports/2006/February/22/dr00035542.aspx (accessed 18 June 2010).

Greig, F. and Koopman, C. (2003) Multilevel analysis of women's empowerment and HIV prevention: quantitative survey results from a preliminary study in Botswana. *AIDS and Behavior* (7)2: 195–208.

Jaramillo, R. and Gough, E. (2005) *Belize National Specialist, Dangriga Composite Policy Index—2006.* Belize City, Belize: National AIDS Committee, 3.

Jones, K. and Moon, G. (1993) Medical geography: taking space seriously. *Progress in Human Geography* 17: 515–524.

Kane, S. (1993) Prostitution and the military: planning AIDS intervention in Belize. *Social Science and Medicine* 36(7): 965–979.

Kearns, R. (1993) Place and health: toward a reformed medical geography. *Professional Geographer* 45: 139–147.

Kempadoo, K. (1999) Continuities and change. In *Sun, Sex and Gold: Tourism and Sex Work in the Caribbean*, K. Kempadoo (ed.). Lanham, MD: Rowman & Littlefield, 3–36.

Kinsler, J., Sneed, C.D., Morisky, D.E., and Ang, A. (2004) Evaluation of a school-based intervention for HIV/AIDS prevention among Belizean adolescents. *Health Education Research* 19(6): 730–738.

Latkin, C.A. and A.R. Knowlton. (2005) Micro-social structural approaches to HIV prevention: a social ecological perspective. *AIDS Care* 17(Supplement 1): S102–S113.

Latkin C.A., Kuramoto, S.J., Davey-Rothwell, M.A., and Tobin, K.E. (2010) Social norms, social networks, and HIV risk behavior among injection drug users. *AIDS Behavior* 14(5): 1169–1181.

McClaurin, I. (1996) *Women of Belize: Gender and Change in Central America*. New Brunswick, NJ: Rutgers University Press.

McClusky, L.J. (2001) *Here, Our Culture Is Hard: Stories of Domestic Violence from a Mayan Community in Belize*. Austin, TX: University of Texas Press.

Mayan People of Southern Belize, Toledo Maya Cultural Council, Toledo Alcaldes Association, and A.D. Nystrom. (1997) *Maya Atlas: The Struggle to Preserve Maya Land in Southern Belize*. Berkeley, CA: Toledo Maya Cultural Council and Toledo Alcaldes Association.

National AIDS Commission. (2006) *Responding to HIV/AIDS in Belize*. Belize City, Belize: National AIDS Commission.

National AIDS Commission. (2010) *One Response E-Bulletin* 1(1), 9 July.

Padilla, M.B., Guillermo-Ramos, V., Bouris, A., and Reyes, A.M. (2010) HIV/AIDS and tourism in the Caribbean: an ecological systems perspective. *American Journal of Public Health* 100(1): 70–77.

Pan American Health Organization (PAHO). (2010) *HIV and Violence Against Women in Belize*. Final Report, March. See www.nacbelize.org/dms20/dm_new.asp (accessed 4 October 2010).

Pope, C.K. and G. Shoultz. (2010) An interdisciplinary approach to HIV/AIDS stigma and discrimination in Belize: the roles of geography and ethnicity. *GeoJournal* (April, doi:10.1007/s10708–010–9360-z).

Pope, C.K., R.T. White, and R. Malow. (2008) Global convergences. In *HIV/AIDS: Global Frontiers in Prevention/Intervention*, C. Pope, R.T. White, and R. Malow (eds). New York, NY: Routledge.

Santos, C. (2000) Statement to the United Nations General Assembly—Twenty-fourth special session. Geneva, Switzerland: United Nations. See www.un.org/socialsummit/speeches/296bel.htm (accessed 4 October 2010).

Schifter, J. (2001) *Latino Truck Driver Trade: Sex and HIV in Central America*. Binghampton, NY: Haworth Press.

Sharer, R. and Traxler, L. (2005) *The Ancient Maya*, 6th Edition. Stanford, CA: Stanford University Press.

Shoman, A. (2000) *Thirteen Chapters of a History of Belize*. Belize City, Belize: Angelus Press Limited.

Statistical Office of Belize. (2010). General statistics. Belmopan, Belize: Statistical Office. See www.statisticsbelize.org.bz (accessed 4 October 2010).

Steinberg, M.K. (2002) The second conquest: religious conversion and the erosion of the cultural ecological core among the Mopan Maya. *Journal of Cultural Geography* 20(1): 91–105.

Steinberg, M.K. (2005) Mahogany (*Swietenia macrophylla*) in the Maya lowlands: implications for past land use and environmental change? *Journal of Latin American Geography* 4(1): 127–134.

Sullivan, C. (2003) Photographic images of the Mayan Indian: classic times to the Zapatistas. *The Journal of American Culture* 26(3): 313–328.

Sutherland, A. (1998) *The Making of Belize: Globalization in the Margins*, Westport, CT: Bergin & Garvey.

Tobin, K.E., Davey-Rothwell, M., and Latkin, C.A. (2010) Social-level correlates of shooting gallery attendance: a focus on networks and norms. *AIDS Behavior* 14(5): 1149–1158.

Treichler, P.A. (1987) AIDS, homophobia and biomedical discourse: an epidemic of signification. *Cultural Studies* 1(3): 263–305.

Türman, T. (2003) Gender and HIV/AIDS. *International Journal of Gynecology and Obstetrics* 82(3): 411–418.

UNAIDS and WHO. (2010) Epidemiological fact sheets on HIV/AIDS and sexually transmitted infections. Belize: World Health Organization, Pan American Health Organization, and UNAIDS. (Geneva, Switzerland: World Health Organization.) See www.unaids.org/en/dataanalysis/epidemiology/epidemiologicalfactsheets (accessed 4 October 2010).

Vergara, M. (2006) HIV/AIDS prevention among vulnerable youth in Central America and the Caribbean. See http://www.iasociety.org/Abstracts/A2199637.aspx (accessed 18 June 2012).

Wheeler, D.A., Arathoon, E.G., Pitts, M., Cedillos, R.A., Bu, T.E., Porras, G.D., Herrera, G., and Sosa, N. R., (2001) Availability of HIV Care in Central America. *Journal of the American Medical Association* 286: 853–860.

Wilk, R.R. (1997) *Household Ecology: Economic Change and Domestic Life among the Kekchi Mayan in Belize*. DeKalb, IL: Northern Illinois University.

11 The mosquito state

How technology, capital, and state practice mediate the ecologies of public health

Paul Robbins and Jacob C. Miller

In 2003, the mosquito acquired new significance in the southwestern United States. The arrival of West Nile virus (WNV) and its first associated human deaths ushered in a rereading of the mosquito from an itchy nuisance to a potentially life-threatening hazard. Mundane objects now required attention like never before. Swimming pools, irrigation canals, ditches, clogged gutters, and abandoned tires all became potential sources of a mobile public health hazard: the mosquito vector. In the state of Arizona, WNV went from a largely unanticipated epidemic situation to an endemic one in short order, where expectation of ongoing disease control quickly became a part of government obligations (Robbins *et al.* 2008).

This sustained hazard focused the energies of state officials and captured the concern of a transfixed public over the decade. In the wake of the disease, a wide variety of strategies were implemented by state agencies to address the problem. Sophisticated surveillance apparatuses were installed across dense urban areas. Notifications were issued in order to assuage public concern regarding the clouds of white fog projected from county pick-up trucks. Public health specialists roamed the streets disseminating information. Airplanes flew over residential neighborhoods looking for breeding hotspots. Predator fish were introduced to devour larvae. Investments in new technology were made by agencies increasingly burdened with training field agents and technicians, maintaining equipment and relations with industry, and organizing the logistics of trucks, sprayers, and databases.

The case and death rates in Arizona have risen and fallen in the years since 2003, as shown in Table 11.1, and while reported cases have fallen steadily nationally, those in Arizona have not declined overall. In 2010, seven years into the outbreak, the CDC reported 167 West Nile virus human infections in the state, resulting in 15 deaths, 26 percent of the reported deaths in the US that year (United States Center for Disease Control 2011a). Though the hazard has not come to exceed the severity of some other health risks—21 children drowned in Arizona in 2010 (Arizona Republic 2011)—the management of mosquitoes has become a state imperative.

As a result, sets of highly diverse strategies have been innovated by local and regional authorities. The resulting patchwork of mosquito control efforts varies widely in their concentration on larviciding versus adulticiding, information

Table 11.1 Reported West Nile virus activity in the state of Arizona and the United States, 2003–2010

Year	*Arizona cases*	*Arizona deaths*	*AZ cases per mill*	*US cases*	*US deaths*	*US cases per mill*
2003	13	1	2.0	9862	264	32.1
2004	391	16	60.2	2539	100	8.2
2005	113	5	17.4	3000	119	9.8
2006	150	11	23.1	4269	177	13.9
2007	97	6	14.9	3630	124	11.8
2008	114	7	17.5	1356	44	4.4
2009	20	0	3.1	720	32	2.3
2010	167	15	25.7	1021	57	3.3

Source: United States Center for Disease Control (2011b)

versus state control, and choices of technology and strategy. This raises fundamental questions regarding the role of the state in protecting public health. What priorities and metrics lead to the selection of technologies designed to address vector borne disease and protect public health? Conversely, how do technological choices direct the local and regional state commitments to certain vector knowledges and practices? When and why do state strategies differ?

In this chapter, we explore the relationship between state priorities and technological choices in the state of Arizona in the wake of WNV outbreak since 2003. We specifically draw upon comparative examination of the activities and attitudes within three county management agencies: Maricopa County Vector Control (a county environmental service), Yuma County Pest Abatement District (a pest abatement district), and Pinal County Environmental Health Services (a county health service). Using interviews with professional mosquito managers and examination of agency budgets and inventories, we seek to address two questions: (i) how state agents become committed to specific modes of management, and (ii) the extent to which those modes of management crystallize around institutional practices geared towards material technologies and investments. In doing so, we hope to better understand how both state and capital are implicated in complex networks that involve human and non-human actors—such as backyards, abandoned tires, irrigation canals, clogged drains, abandoned pools, adult mosquitoes, mosquito larvae, trucks, computers, chemical companies, and capital—and what these might mean for public health.

The situation in southern Arizona provides an excellent test case for state responses to novel disease risks. By confronting a mosquito hazard that was heretofore unknown in the state, relative to more well-developed mosquito abatement traditions in states like Florida (Patterson 2009), the evolution of Arizona institutions demonstrates how state institutions adapt and develop health capacities "on the fly," in the midst of an emerging problem, a model for any number of yet unanticipated disease problems that demand innovative adaptation (Miller *et al.* 1998). So, too, the central state mandate from the Arizona Department of Health

Services is not to dictate or over-rule county programs or initiatives. As such, the autonomy of local jurisdictions allows separate, parallel, and divergent strategies to form, presenting analysts with the opportunity to comparatively assess what influences state priorities, within a similar overarching ecological context (Shaw *et al.* 2010). Additionally, southern Arizona presents a situation where relative aridity (less than 500 mm of annual rainfall) has not fully hindered the development of mosquito populations. This is a result of the anthropogenic sources of mosquito breeding sites including both irrigated agricultural zones and urban areas, where landscaping, pools, and drainage provide ample habitat. This situation is also far from unique. Urban infrastructure is a key variable in determining mosquito abundance worldwide (Deichmeister *et al.* 2011), even while rural agricultural land covers can also maintain mosquito populations (Bowden *et al.* 2011). Rural–urban variability in southern Arizona provides further insight into the comparative hazard profiles of emerging anthropogenic ecologies, therefore.

The research described here, addressing this region through comparative county analysis, is based upon interviews with lead managers of Maricopa County Vector Control, Yuma County Pest Abatement District and Pinal County Environmental Health Services conducted in fall 2009, followed by email queries for all managers, a formal survey of management priorities at all three regional offices, an examination of recent agency budgets, an examination of lists of current equipment, and an interview with a lead manager at the statewide Arizona Department of Health Services.

Using the results of these to characterize the strategies in each county, and observing these as path-dependent outcomes rooted in contextual political ecologies, we conclude that, as the state struggles to engage the complexity of disease vector insects, it necessarily simplifies the natural world, selecting technologies of surveillance and control. Simultaneously, the proliferation of control technologies comes to direct state priorities and activities. These engagements are both mediated, moreover, by flows of capital that traverse state operations. The resulting *ad hoc* innovations of local-level bureaucracies tend to make them specialized in their outlook and programs, prone to capture by capitalized interests, and somewhat isolated from public knowledge and feedback. This problematic intersection of institutional habits, political economy and mosquito ecology presents problems for health management under conditions of rapid change.

Mosquito ecology and management technologies in southern Arizona

The case of the mosquito in the arid US southwest is paradigmatic of state learning amid ecological complexity, since it is a place where insect problems are increasing rapidly and confronting historically under-prepared management institutions. Despite the aridity of southern Arizona, the region has long been the center of mosquito-borne problems and epidemics (Dobyns 1976; Kessell 1976). Only through the transformation of water tables, loss of wetlands, and pioneering use of dichlorodiphenyltrichloroethane (DDT) did mosquito populations decrease in the mid-twentieth century (Russell 2001: 154). Mosquito populations rebounded in the late twentieth century, however, with the cessation of DDT fogging (Reiter and Gubler 1997) and the proliferation of new habitats: neglected swimming pools, unmaintained fountains, shade trees, solid waste, and restored wetlands (Karpiscak *et al.* 2004). By the late 1990s mosquitoes were on the rise throughout the region, though mosquito borne diseases were far from the policy agenda of managers.

Since 2003, this state of affairs has been fundamentally challenged. Specifically, the region has seen the increase in several species, including *Aedes aegypti* and *Culex quinquefasciatus* in residential areas. Surveys in and around the region have revealed that *Ae. aegypti*, the mosquito responsible for transmitting dengue, has colonized many cities and towns (Merrill *et al.* 2005) and viable eggs have been found in and around 46 percent of surveyed residences (Botz 2002). *C. quinquefasciatus*, found in both wetland and residential areas, continues to feed extensively in the region on humans and birds, presenting a serious West Nile risk (Zinser 2004; Zinser *et al.* 2004). West Nile virus arrived in Arizona in 2003, and in 2004 Maricopa County experienced an unprecedented outbreak of over 350 cases.

State response has been dramatic, but has proceeded through several, necessarily differentiated strategies, specifically including adulticiding of mature mosquitoes through aerial fogging, larviciding at water-based breeding sites, and provision of public information and behavior control. The heterogeneity of technological options is borne in part by the breeding cycle of the mosquito itself.

Though the habits of disease vectoring mosquitoes vary significantly by species or physiotype (Briegel 2003), the overall population cycle is generally similar. Adult females fly in pursuit of blood-meals rich in iron and other nutrients to support reproduction, incidentally contracting diseases (obtaining WNV, for example, by biting infected birds) and often transmitting them to humans in the process. These females lay "rafts" of several hundred eggs on water surfaces. The specific characteristics of water bodies may vary, though stagnant pooling with persistence over several days is ideal. Water bodies need not be very large to provide habitat. Retention basins of any size, such as from irrigated farming culverts to buckets and birdbaths, can become the wellspring. In these contexts, eggs hatch into larvae, which develop near the water's surface, clinging through surface tension and obtaining air through a snorkel-like siphon. Emerging through a pupal stage, individuals then leave the water and take flight as immature adults (Spielman *et al.* 2001).

The central problem posed by this reproductive cycle is that the insect's behaviors and habitats vary over its life course, with habits and conditions that are alternately amenable or invulnerable to differing management techniques. Abatement strategies therefore correspond to specific phases in the mosquito's life cycle.

Measures taken to obstruct the link between the adult mosquito and the next batch of eggs, for example, specifically target the reduction of widely dispersed water sources, ranging from bird feeders and discarded tires to bends in washes and arroyos. Vector control agencies address these, therefore, in part through public outreach campaigns and by distributing information regarding the spatially and environmentally diverse nature of mosquito breeding locations. Agencies seek to generate among the public a self-disciplining population of citizens who diligently clean out clogged gutters, drain stagnant swimming pools, and replace water in birdbaths (Shaw *et al.* 2010: 376).

Once mosquito eggs are laid, however, other institutional practices come to the fore: specifically monitoring sites and addressing larval development. When water bodies are located and are found to have larvae, a site might be logged in a GIS database and continually monitored. Sites might then be treated using chemical larvicides or oily surfactants that break surface water tension and effectively drown larval populations clinging to the water's surface. Chemical briquettes are deployed from backpack applicators or distributed from aerial vehicles. This approach also has a long history (Darling 1912).

Where and when larvae develop into flying and biting adult mosquitoes that may carry disease, a differing set of practices emerges. Vector control agencies supply information to the public regarding the simple measures that can be taken to reduce mosquito bites (e.g. wearing long sleeves or installing screens on house windows). Monitoring moves to direct trap counts of flying mosquitoes in suspect and sampled areas, utilized to check population densities and the presence of specific diseases.

Where a threat is identified by an agency, either through numbers of trapped mosquitoes, presence of a positive disease test case, or local complaints, ultra-low volume (ULV) truck-mounted adulticide (pesticide) foggers are directed to areas of concern. These machines project a blanket of adulticide fog aimed at potentially infected adult populations.

Significantly, each of these approaches requires differential investment in equipment and training provided by private vendors. A large number of companies are therefore deeply integrated into agency decision-making and strategies, providing sprayers, foggers, larvicidal and adulticidal chemicals, as well as GIS software and GPS hardware. These companies are typically national, albeit independently-owned, and field teams of technicians maintain ongoing contact with state, county, and municipal client agencies.

The development and maintenance of these relationships is facilitated by the State of Arizona Department of Health Services, whose annual disease conference and workshop dedicates a full day to promotion and interaction with private companies and where venders distribute promotional pamphlets and sometimes provide live field demonstrations. All field technicians who work with chemicals are required to fulfill six hours of continued education each year, which the meeting fulfills. This institutional requirement brings managers and technicians into an arena where technology and equipment is marketed and advertised, forming a somewhat captive audience for the vector control industry.

This pattern of relationships between state actors and private vendors is not unprecedented in United States vector control. In states with long histories of mosquito control, typically those where real estate and settlement development have depended heavily on mosquito nuisance control like Florida, such close relationships are long-standing (Patterson 2009; Robbins and Marks 2011). Even so, the relatively disaggregated scale at which this occurs in Arizona is notable, as is the absence of established norms to govern these relationships, or a public venue to adjudicate technological priorities. The result is a situation where firms hold significant power over the direction and selection of equipment and overall abatement strategies.

In the end then, the differentiated and morphologically diverse moments in mosquito life cycles present state managers with divergent socio-environmental problems, and demand differential commitments to public communication (breeding site control versus risk behavior education), forms of chemicals (larvicide versus adulticide), and forms and applications of equipment and software (foggers versus backpacks; GIS to monitor water sites or to track treatments and trap counts). Each such commitment requires interactions with differing actors in the private sector (homeowners versus pesticide companies) and differing demands of staff time and limited resources. Given socio-ecological differences, how might we predict the *technological* choices of agencies to influence or be influenced by habits in governance and the exigencies of *state power*? A brief review of the question stresses the historical tyranny of state simplification but also the critical role of past technology choices in directing future decisions.

Technologies, states, and insects

By way of definition, we follow Langdon Winner (1977: 8–12) in understanding the term "technology" to refer to the sphere of "rational" and "efficiency-seeking" methods in human society (here following Ellul 1964). This construct includes both physical *apparatus* of technical performance, like backpack sprayers and GIS databases, as well as "purposive, rational, and step by step" *techniques*, like the skills and procedures used to trap and categorize insects, as well as the knowledge to operate a truck-mounted fogger.

We simultaneously embrace the term "state," following Michael Mann (1984) and others, to refer to a politically sovereign institution, organized through a central authority, with both responsibility for, and monopoly power over, a defined and bounded territory, and the people and things contained within. Here, we include the specific agencies charged with managing vector-borne diseases and mosquito populations, as well as the bureaucratic and jurisdiction conditions and motivations to which these respond. So too, however, we acknowledge recent insights in political geography that suggest that the state does not act as a coherent organism and is only realized through the day-to-day practices of ordinary functionaries, including individual vector control agents and health information officials: the "everyday state" (Painter 2006).

So understood, the relationship between technology and state power and practice has long been a matter of concern. It has received increased attention in recent years, however, as apparently intractable hazards seem to proliferate through modernization, economic growth and urbanization (Beck 1992). Each strategy, technology, and activity a bureaucracy undertakes forces it to develop expertise, knowledge, and investment in a mode of governance. These impacts hold implications for the state's capacity to learn, adapt, and solve problems. In the context of accelerated economic and environmental change, these tendencies have been suggested to be problematic (Robbins *et al.* 2008). This difficulty is especially true for vector control, where the dynamic ecologies of animals (e.g. rabid dogs), insects (mosquitoes carrying WNV), and dirt (e.g. soil containing spores of Valley Fever fungus, *Coccidioides immitis*) intersect with state power.

The state has been observed to consistently simplify such ecologies, leading to problematic technological choices. As James Scott argues in *Seeing Like a State* (1998), the modern state works to control environments and populations through administrative tools allied with instrumental logic and scientific reason. These tools order complexity using inevitably reductive categories and gridded cartographies that ultimately fail in achieving the state's goals precisely as a result of their totalizing "tunnel vision." The hallmark of these state practices is technical knowledge or "*techne*," the privileged optic of operation, "based on logical deduction from self-evident principles" (Scott 1998: 317). Conversely, Scott (1998: 320) describes "*metis*" as an operating register that is not universal, but always "contextual and particular," a kind of localized knowledge at risk of disappearing as the capitalist state expands its grasp.

Other scholars of state and technology proceed from a reverse direction, asking whether technological commitments produce state outcomes. In *Autonomous Technology* (1977), Langdon Winner argues that the political field in advanced industrial societies is fundamentally shaped by sophisticated technological systems; a "technological imperative" informs political choices (Winner 1977: 251). In this scenario, technology is more than passively used to achieve state ends, but returns to the political arena as a directing force itself. While Winner acknowledges that the need for a specific technology is not free from "stimulation and manipulation" (Winner 1977: 248), as in advertising, his primary explanation of techno-politics rests with the possibility of *autonomous technology*. A test of whether or not a technological system has become autonomous is by determining if the original goals of an agency, for example, are *reverse adapted* by the adoption of a technology in a way that suits the needs of the technology itself.

These two models of state and technology influence the discussion of mosquitoes and public health in the literature. Eric Carter's (2008) work on mosquito control and state apparatus in Argentina in the late nineteenth century and early twentieth century, for example, shows how malaria control provided an only recently consolidated national state an opportunity to extend its power into previously unincorporated provinces, "building and maintaining the bureaucratic structures of state science" (Carter 2008: 280–281). At the same time, however, Carter stresses that the habits of public health practitioners began to possess their own momentum, as they became deeply invested in the techniques and technologies they deployed.

Theory and history, therefore, point to dialectics between technology and state power. What implications does this dialectic have for public health? How does technology shape state vector control?

Contexts of state ecology: the study region in Arizona

For purposes of this analysis, we compare the differential contexts of three subregions: the agricultural west (Yuma County), the burgeoning urban "Valley of the Sun" (Maricopa County), and the sprawling suburban uplands of the central East (Pinal County).

Table 11.2 Differing social and ecological indices in study counties

County	*Yuma*	*Maricopa*	*Pinal*
Population: 2010a	195,751	3,817,117	375,770
Population density: 2010 (per sq. mile)a	35.5	414.8	70.0
Population growth rate: 2000–2010 (percent)[a]	22.3	24.2	109.4
Workforce in farming, fishing and forestry: 2005–2009 (percent)[b]	5.7	0.2	1.9
Agricultural sales: 2007 (thousands $)[c]	959,968	813,491	799,811
Total housing units[d]	87,850	1,639,279	159,222
Housing units that received a foreclosure filing in March, 2011[e]	1/438	1/141	1/99
Mean precipitation in inches, 2000–2010, 12 month period ending in January[f]	4.27	8.21	10.11
Mean temperature, degrees F, 12-month period ending in January[f]	73.48	71.26	69.56

Sources:
a United States Census Bureau (2011a);
b American Community Survey five-year estimate, US Census Bureau (2011b);
c United States Department of Agriculture (2011);
d United States Census Bureau (2011b);
e RealtyTrac (2011);
f Prism data, WRCC, accessed with WestMap; PRISM Climate Group (2011)

As shown in Table 11.2, in many respects, these regions are similar in conditions relevant to mosquito breeding and survival. All are water-scarce, depend heavily on monsoonal moisture for most of the average annual rainfall, and have summer temperatures warm enough to maintain large seasonal mosquito populations. All have winter temperatures low enough to deter overwintering of mosquitoes. Most critically, the mosquito season is one that is delimited in all three counties by the timing and spacing of the southwest monsoon, in which summer low pressure draws moisture from the south, typically creating a distinctive rainy season between July and September, during which more than a third of the region's precipitation (mean annual ~300–400 mm) occurs. The entire region is therefore prone to wide inter-annual precipitation variability, as well as spatial unevenness, causing some areas to experience flooding while others remain bone dry (Arriaga-Ramirez and Cavazos 2010; Lizarraga-Celaya *et al.* 2010). Key differences are also in evidence, however. The climate in Yuma County is notably drier than Maricopa and Pinal Counties. Mean annual temperatures are lower in Pinal County, moreover, owing to its relatively higher elevation.

It is critical to acknowledge, however, that these climate differences are often immaterial since mosquitoes depend heavily on highly specified micro-climates for reproduction. Recent comparative modeling has demonstrated that in already water-stressed southwest areas, warming or drying trends may have negligible impact on mosquitoes, owing to dependence on permanent anthropogenic breeding sites (Morin and Comrie 2010). Overwintering of mosquito eggs has been similarly associated with urban housing (Botz 2002) and mosquito survival depends on air-conditioning and evaporative coolers, outdoor vegetation cover, and access to piped water (Hayden *et al.* 2010). Thus, the material specificities of micro-scale anthropogenic ecologies (e.g. standing water, drains, or abandoned pools) provide an enormous counterweight to any ambient temperature and rainfall conditions.

In this regard, the subtle socio-ecologies of the subregions become more relevant. Yuma's preponderance of agriculture poses a specific environmental context for management, stressing large, contiguous, irrigated landscapes, spread over dispersed human populations. Conversely, the dense urban areas of Maricopa County present patchy, artificially landscaped features and urban drainage, determined by the decision-making of nearly 4 million people, whose aesthetic and economic decisions govern more than 1.5 million housing units. The fast-growing but widespread suburban populations of Pinal County represent a very different challenge, notably including a 21 percent rate of residential abandonment in the 2010 census, with 1 in 99 homes receiving a foreclosure filing in March 2011. Abandonment of private property leads to neglected mosquito breeding areas and poses a serious challenge for managers. Though it is difficult to know the precise impact of such conditions, they represent a stark contrast to the irrigated fence lines of Yuma.

These three counties therefore share similar controlling climatic conditions, but highly differentiated socio-ecological contexts for mosquito managers. Within these divergent socio-ecologies, state actors are forced to make commitments of limited resources to differing strategies and technologies in order to govern a health threat with little regional precedent. This comparative context makes it is possible to ask, therefore, why differing state institutions select specific insect-control technologies and how those technologies impinge both on regional health as well as state capacity and function.

Mosquito state(s): institutional practices and material technologies

The three agencies examined below are representative of highly differentiated contexts for decision-making. As noted previously, Maricopa County Vector Control (MCVC) is situated in a dense, wealthy, and fast-growing urban corridor. Yuma County Pest Abatement District (YCPAD) has jurisdiction over a more rural region at the edge of the intensely cultivated Imperial Valley. The area of operation for Pinal County Environmental Health Services (PCEHS) is a largely suburban corridor spanning the space between Phoenix and Tucson. Each agency operates with different kinds of expertise as well as different resources that frame their manifestations of the mosquito state. Table 11.3 shows the response of agency directors to a request to rate agency activities in terms of importance to their operations.

As shown in Table 11.3, many activities are universal to mosquito control, including monitoring, larviciding, and responding to complaints. But important differences are also evident. Where adulticide fogging is critical in Maricopa

Table 11.3 Responses of managers in three counties when asked to rate the importance of mosquito abatement techniques

	MCVC	*YCPAD*	*PCEHS*
Monitoring traps	M	M	M
Larviciding	M	M	M
Responding to complaints	M	M	M
Adulticide fogging	M		
Driving around and looking for potential breeding sites		M	M
Maintaining contact with VCI vendors		L	L
Talking to home builders about home and yard design		L	
Using GIS			L
Giving educational presentations at home owners associations, trailer parks, etc.		L	M
Driving around and talking to people about vector control	L	L	M

Note: L represents "least important" or "unimportant," and M represents "most important."

County, it is less so in other counties. Maintaining contact with vendors and using GIS technology are least important in Pinal County, and such interactions are far more important elsewhere. Public communication and educational efforts, conversely, are most important in Pinal and far less important in other counties. A brief comparison of the commitments of each agency helps explain what internal logics support these divergent practices.

Maricopa County Vector Control

Maricopa County is one of the fastest growing urban areas in the country. It also has the largest budget for vector control at $2.2 million in 2009. Changes in the suburban landscape have made many parts of the city subject to mosquito infestation. While a variety of techniques are used—following an integrated vector management paradigm—our interviews reveal differential preferences for control borne out in the specifics of day-to-day operations.

While MCVC does include educational information on their website that encourages the public to act in ways that reduce mosquito breeding sites, their daily practices focus on surveillance and mobilization. MCVC focuses the bulk of their time, energy and resources on monitoring and destroying the adult mosquito. In 2005, shortly following the arrival of WNV and the first WNV-caused human death, this agency received over two million dollars in emergency funds from the state government to address the issue. These funds were used to add staff and equipment, adulticide foggers and trucks. Spread out like a grid over the city, a system of CO_2 mosquito traps is monitored weekly to check the number of specific mosquito species and to test for WNV.

While MCVC responds to public complaints, the "vector fleet" is mobilized only in instances of an exceedingly high number of mosquitoes found in a single trap or if a positive case of WNV is found. Then, a warning goes out to the neighborhood and a driver is assigned to fog a grid-square corresponding to an acute signal in the system. Each truck is equipped with a heavy-duty ULV adulticide fogger with automatic flow-rate adjustment and a GPS linked to the mainframe GIS system back at headquarters. Following the WNV outbreak, the size of this fleet increased from 6 trucks to 23. When asked what the most important pieces of equipment are for MCVC's operation, the responding manager asserted: (i) trucks, (ii) adulticide foggers, and (iii) larvicide applicator backpacks. Our discussion of these techniques suggested that the use of this expensive equipment has been embedded in the way that MCVC understands effective management strategy. The manager drew on the institutional memory of the agency and elaborated on the self-evident common sense of MCVC's approach:

> Since the beginning … when they used the thermo-foggers in the 1960s and even earlier … we have used these technologies to get the chemical out there and treat the adult mosquito population … It is really pretty simple. You have an engine and air compressor that delivers the chemical. It hasn't changed that much and probably will always be that easy.
>
> (Maricopa County Vector Control official, September 2009)

Thus, in the resource rich environment of Maricopa County, the Vector Control agency has developed a strategy of *rapid response and mobilization* that deploys an arsenal of vector control equipment: trucks, traps, and foggers. These techno-institutional practices revolve around managing an extensive GIS-supported grid of traps and the deployment of trucks for the application of adulticide fog. MCVC was the only agency that reported adulticide application as one of the "most important" abatement strategies.

Simultaneously, this approach largely eschews depending on the public for information, especially complaints, or indeed acknowledging local knowledge in any form. Indeed, interviews with managers suggest that the removal of human judgment in dealing with mosquito problems is essential. One respondent described a future in which the vector response system is fully automated, with CO_2 traps automatically triggering self-guided GPS-driven foggers that would treat effected neighborhoods with machine-calibrated doses. This notion of technological currency is a common trope. As a MCVC agent explained:

> We need the durable, top of the line models with powerful engines because of the large areas covered … when new technology comes along, everything gets upgraded. We try and stay in touch with the latest technology, just like anything else.
>
> (Maricopa County Vector Control official, September 2009)

This technological imperative necessitates, and is further propelled by, relationships with private vendors, which are a critical part of MCVC staff certification and inculcation. These vendors sell equipment but also have a well-developed and extensive outreach apparatus to demonstrate and train public employees. Upon purchase of any equipment, companies send representatives to agencies for instruction on proper installation, use and maintenance. From this point, the agency will maintain contact with the company through mail-lists and product brochures, as well as periodic visits from sales representatives. Annually, the "Univar guy" or "Clarke guy" stops by for an update to further solidify customer loyalty. One company offers a "Mosquito University" workshop that has become the first line of education on mosquitoes and technology, which many Arizona state and county employees receive, especially Maricopa county personnel.

Indeed, since the WNV outbreak, MCVC's response has been to increase efforts in this direction precisely by adding more equipment and staff training. This response creates a close and interdependent relationship between MCVC officials and vendors. For instance, MCVC is currently using a prototype model that a company has lent them for a trial period to test the performance of a new electric motor model in the Arizona heat. The MCVC manager has been impressed so far and says that they might "switch over to that model" in the next round of upgrades. A pattern of adulticide emphasis, chemical treatment, and capital-intensive solutions enmesh officials in relationships that are heavily mediated by capital.

Yuma County Pest Abatement District

In southwest Arizona, a rural environment has also supported the emergence of vector species. YCPAD governs a landscape that includes fields, ranchettes and horse pastures, in which irrigation systems have become ideal mosquito breeding sites. Here, larvaciding is the critical part of agency strategy, though adulticiding receives some emphasis. The agency stresses intervention at the breeding locations that prevent larvae from becoming adult mosquitoes. "We try and be [*sic*] proactive and prevent the problem before people are bitten" (Yuma County Pest Abatement District official, October 2009).

The concomitant technological selections, therefore, stress monitoring activities. Like MCVC, YCPAD uses CO_2 traps and has flocks of sentinel chickens used for disease monitoring but a stress on monitoring leads to a different regime of state practice. Specifically, the staff clock many road miles in their work, literally driving, talking to people, and responding to complaints in an effort to identify potential problem areas and log them into their GIS database. As opposed to the Cartesian grid that hovers atop of Maricopa County, in Yuma County mosquitoes are known and controlled by "just keeping your eyes open, being observant and talking to people." Perhaps their most emblematic strategy involves hiring airplanes to fly over neighborhoods in search of abandoned swimming pools. As a result, technological commitments differ from those in Maricopa. When asked what the most important pieces of equipment are, the responding manager hesitated and carefully considered his response: "It depends on how you think about it. If you're thinking about managing the problem, it is surveillance." The surveillance program is "the building block that everything else stands on. If you don't look, you don't find," he explained, stressing the necessity of "getting a measure of what's going on out there" (Yuma County Pest Abatement District official, October 2009). Only the YCPAD and the PCEHS reported that "driving around and looking for breeding sites" was one of the "most important" abatement strategies (Table 11.3).

In part, these preferences reflect very different demographic conditions. Rural Yuma County is more sparely populated than dense Maricopa County. The selection of aerial surveillance, unique in the three counties, further reflects the agrarian history of the area and the availability of crop-dusting light aircraft. But techno-institutional history plays a key role as well. Unlike MCVC, which formulated its post-WNV strategy within an experienced agency that had traditionally mobilized adulticide applications, and where such technologies are historically established, YCPAD was formed only in 1992 and its manager reports that he "had to learn a great deal about mosquito management." The first source of education, in this case, was Arizona Department of Health Services (ADHS), rather than industry. Notably, the YCPAD agency manager reported that the state's annual conference serves as an "important source of ideas, resources." YCPAD also has a "close working relationship with the folks at ADHS," who encourage specific practices, providing materials like Altosid larvicide (Yuma County Pest Abatement District official, October 2009).

Though interest in technology is in evidence—the Yuma manager was so impressed with a specific sprayer model at a recent showcase that they switched to that model—a differential pattern of allegiance to vendors is in evidence. YCPAD responded that "maintaining contact with vendors" was one of the "least important" agency activities (Table 11.3). Similarly, a lack of enthusiasm for advanced models was expressed. The much smaller YCPAD is content with the mid-grade GIS system they use, for example. The manager knew from the start in 1992 that "it made all the sense in the world" to incorporate GIS; the YPCAD has restrained from upgrading and the model they use now is simpler and more appropriate for their purposes as opposed to what he described as "exceedingly complex systems" used by agencies like MCVC (Yuma County Pest Abatement District official, October 2009).

Thus, while the YCPAD uses equipment similar to the MCVC, their overall approach to mosquito management is different. The pesticide budget alone for the MCVC in 2009 was around 11 times larger than YCPAD's entire mosquito control budget for the year. The much smaller YPCAD has adopted an approach that stresses airplanes, GIS support and direct communication with residents in an effort to locate and attack the insect in the larval stage. Furthermore, like MCVC, YCPAD depends on varying technological investments, like ULV sprayers, but its role in the overall strategy is mediated by an emphasis on a techno-institutional practice of *dispersed surveillance*. Adulticide application becomes a last resort and is accompanied by a managerial focus on horizontal communication in the community and preventative intervention. Thus, while similar technology is *adopted* by different agencies, those technologies are *adapted* for specific use by managers and technicians in very different ways.

Pinal County Environmental Health Services

Of the three agencies interviewed, the PCEHS is the least connected to high investment technology and the vector control industry. The main strategies of PCEHS were reported to be (i) surveillance, (ii) public education, and (iii) source reduction (managing breeding sites). These imperatives make the central task for management contacting home-owners associations (HOAs) and doing outreach to trailer parks and recreational vehicle (RV) parks. The current manager reports that her strong communication skills were central to her being hired.

While the PCEHS does have an electric motor ULV fogger and uses it 20–30 times a year, this manager asserts that "fogging is not the answer and never has been. It is a Band-Aid on the larger problem." Her list of the key equipment includes: (i) an electric truck-mounted fogger, (ii) "our eyes to look for larvae," and (iii) "our mouths to speak to the public, to get the word out." In terms of monitoring, the PCEHS manager reports anticipating where the future problem areas would be by analyzing the surrounding environment. "Just by looking at geographic areas, I would ask myself, 'if I were a mosquito, where would I be?'" Her problem continues to be "how to catch it before people get sick" (Pinal County Environmental Services official, November 2009).

The goal of outreach, by comparison, is not solely to bolster monitoring, as it is in Yuma County, or to destroy adult mosquitoes as in Maricopa County, but to change perceptions and behaviors. Extensive presentations at RV parks and community organizations are intended to inculcate a knowledge and responsibility for the problem within homeowner communities and collective institutions, like HOAs, who hold appreciably greater power over individual residents than health departments. As she explained, "you just walk up to people and talk about the issue." Thus, the techno-institutional practices of this agency are organized around *generating self-disciplining behavior in the community* in hopes of reducing the mosquito population through everyday maintenance of private property (Pinal County Environmental Services official, November 2009).

As noted, Pinal County has experienced rapid urban growth and has been hit especially hard with the collapse of the housing market. This was cited by the manager as a main problem in mosquito control. Part of their operation includes locating vacant properties where backyard swimming pools provide breeding sites. Many times, the owners of the property are difficult to reach, causing a dilemma for public health management. A concomitant challenge comes from engaging developers and builders to cooperate and alter landscape designs. The PCEHS manager cited malfunctioning or improperly built "dry wells" as a contributing factor. This further engagement seeks to bring public health to the attention of development capital and induce the industry to comply with preventative design measures, which builders report to be "unnecessary" expenditures (Pinal County Environmental Services official, November 2009).

Discussion: mosquito knowledge, machines, and capital

These results suggest that the public health practices emerge from distinct socio-ecological contexts, but also from uneven flows of capital and knowledge. Simplifications are surely in evidence, as where Yuma County managers reduce the inevitable complexities of mosquito management to "surveillance," but each simplification is unique. No single state logic prevails. So too, there is evidence of reverse adaptation, as where the structure of operations in Maricopa County necessarily come to resemble the technical needs of a system where monitoring traps trigger automatic responses. But each such adaptation also appears locally derived and sensitive to divergent socio-ecology. As a result, and as depicted in Figure 11.1, the "Mosquito State" is better described as multiple "mosquito states."

The activities and logics of these agencies therefore highlight a series of issues that exceed "state simplification" and "reverse adaptation". It is evident that: (i) a state that *adopts* specific simplifications and technologies is a product of state-capital interaction, not a pre-existing entity; (ii) technics in state practice are *adapted* with uneven effects owing to the knowledges and logics that exist within local context; and (iii) *reverse adaptation* of agency practice leads towards specialization of agency behavior around these technics.

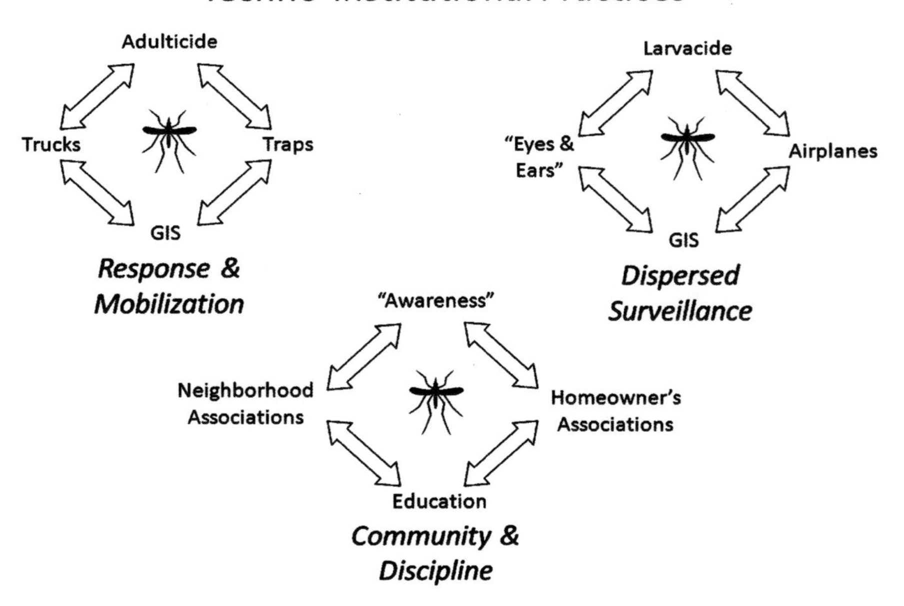

Figure 11.1 Three mosquito states of southern Arizona.

Adoption

According to product brochures distributed at the annual ADHS meeting in 2009, Altosid larvicide is "A simple solution to a complex problem." As we have seen, this effectively mandatory meeting is a key platform for industry, which seeks to establish and maintain relationships with state agencies whose job it is to protect public health. While the arrival of WNV presented a significant threat and challenge to the state, it also presented an opportunity for private capital. However, in the case of PCEHS and to an extent the YCPAD, it is clear that abatement strategies need not be entirely dependent on expensive or mechanical technologies. While PCEHS does use a fogger 20–30 times a year, their main strategies are not geared towards the constant use and perpetual replacement of material technologies. It is clear therefore that there is no universal compulsion to adopt simplified logics and capital-intensive solutions. Rather, the state must be *made* into a consumer of products through the cultivation of relationships. The industry develops agency knowledge by investing in mechanisms that strive to explain to state actors *why* they should adopt certain strategies over others. As Langdon Winner (1977: 248) puts it, "A need becomes a need in substantial part because a megatechnical system external to the person [or agency] needed that need to be needed."

Adaptation

The behavior of these institutions also shows how such relationships are unevenly articulated. For example, both MCVC and YCPAD have adopted geospatial technologies, but use them in entirely different ways. Operating across urban Phoenix, MCVC uses GIS to monitor mosquito traps, distributed across a grid, to activate a near-automated fogging response. The use of GIS in Yuma County, conversely, records "horizontal" knowledge of dispersed breeding sites, developed through "driving around" and finding breeding spaces in the more rural environment. Similarly, PCEHS uses adulticide foggers but in an operation that differs dramatically from the "vector fleet" in Maricopa County. PCEHS did suggest that using these technologies is "somewhat important," but this practice is seen as a last resort and its deployment is adapted to institutional practices of public education and outreach. On the other hand, for MCVC fogging technology is a "most important" activity, around which it has organized its operations.

What these examples show is that state operations utilize specific technologies according to highly localized conditions. While capital seeks to encourage a knowledge regime centered on the *techne* of universal control, it does not find, and cannot always produce, the state that it desires. A dialectical process of technological (in)determination operates within the mosquito state apparatus: vector capital works to ensure the adoption of certain technologies by state agencies, while these technologies are differentially adapted into unique socio-spatial practices. In this sense, the state is always local (Marston 2005).

Reverse adaptation

Finally, the local outcomes in each case do not appear haphazard, and reflect path dependency. The specificities of each outcome reflect an emergent relationship between capital interests, technological imperatives, and site-specific political cultures. Technological choices, ideas about mosquito ecology, and relationships with the public become mutually reinforcing.

In the case of MCVC, following the 2003 WNV outbreak, initial plans were formed to reinforce techniques that were already long-standing (trucks, traps, sprayers), but now with an expanded budget allowing them to increase the size of the vector fleet and hire more staff. These practices only become further integrated with the everyday institutional practices of the agency by training field technicians in the use of machines, monitoring the traps, and maintaining field equipment and databases. Thus, specific vector control technologies were *adopted* by the agency, *adapted* to their unique socio-spatial imaginaries of vector ecology, and, as a result, the agency itself became simplified as it was *reverse adapted* to the system of technics it had created.

In the case of PCEHS, an historic stress on local suburban political conditions, centered on the autonomy of HOAs, is coupled with a resource-poor management environment to produce a technology focused on of self-governance. These commitments come to be embodied in hiring personnel who stress public interaction.

Similarly, Yuma's commitment to monitoring initially reflects local conditions (extensive rural land uses) and available technology (airplanes), but over time internally develops a logic of surveillance to which the system is adapted, naturalizing a spatial model of mosquito behavior and control that suits the system it has created and geared towards a specific phase of mosquito reproduction. The results are institutionally *specialized* technics (Figure 11.2).

Specialization, capture, and isolation: problems of the mosquito state

What are the implications for public health when institutions become specialized around their technics? According to what logic and for whose benefit? Is this how we want health management to function? This assessment suggests three areas of concern in this regard. First, *specialization* is a condition to which these institutional systems appear to drift. Such processes of specialization, which appear strongly systemic, are potentially problematic if the complexity of mosquito ecology calls for institutions that are supple, diverse, and multipurposed. Efforts to restructure the architecture of such institutions away from over-specialization will necessarily need to address the way agency learning leads to a potential impoverishment of knowledge diversity. Future research must necessarily address how learning actually occurs in mosquito management agencies and how it might be diversified.

Second, the provisioning of technological solutions through the private sector is potentially dangerous, precisely because agencies adapt their practices and

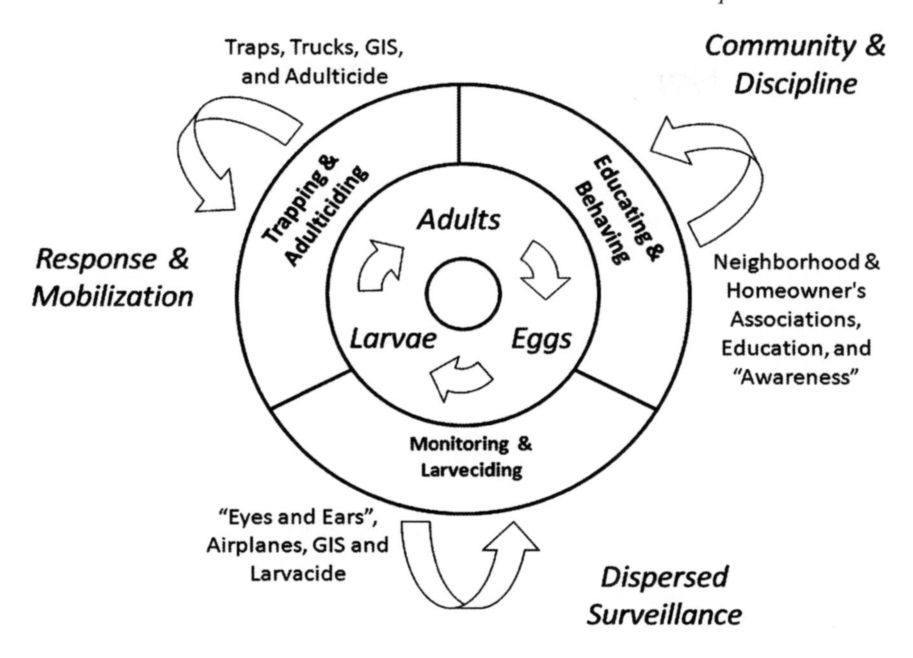

Figure 11.2 The reinforcing habits of mosquito states, centered on specific technics, geared to reproductive phases of the mosquito.

knowledges to the demands of their equipment. Merely by directing the selection of mosquito control equipment, vector capital effects agency *capture*. Even without intending to truncate public options or control agency priorities, the provisioning of foggers, insecticides, and GIS monitoring systems creates organizational capacity designed to do just that. So, too, where vector capital is most involved, the industry's general bias in favor of pesticide-heavy approaches (see Service 1995) will inevitably be most prevalent. Further research on the political economy of private contractors in state health operations is therefore essential.

Finally, it is clear that some technological choices create differential porousness to public knowledge, participation, and consultation. Most obviously in the case of Maricopa County, the voice of the public is marginalized as a result of specific technological commitments. The vector fleet now goes where the machine tells it to go, when triggered by traps and tests, with relative indifference to local *metis* outside the institution. The outcome is an undesirable state of agency *isolation* from the broader community in which it sits. Future research might then ask how public knowledge is acknowledged in institutions of vector control, and how do technological choices facilitate, mediate, or eliminate them?

To be clear, we have not concluded here that any one form of mosquito management or any specific agency is necessarily more or less effective in pursuing health outcomes. Rates of WNV in human populations in each of these counties are not significantly lower now than in the mid-2000s, but neither have they risen drastically: Yuma reported no cases in 2010; Pinal reported 17; and Maricopa County reported 129, all similar to 2006 rates (Arizona Department of Health Services 2011). Deaths attributed to WNV continue without decline state-wide, but have not ballooned. Institutional diversity, arguably, at some level may reflect the *genius loci* of the environments (ecological and political) in which agencies evolve and to which they are adapted. One size does not fit all.

Nor have we argued that any specific practice, on its own, is superior or inferior to another. During disease outbreak, adulticiding is a practical and necessary measure. Controlling hydrology on homeowner properties during the monsoon season is prerequisite to controlling populations. GIS databases showing past outbreaks and disease-vulnerable areas are critical tools for future disease prevention. A large toolbox is a strong one.

We have argued, however, that certain dangers are inherent in the interactions of state agencies, management technologies, and private interests: *specialization*, *capture*, and *isolation*. Meaningful control of WNV has remained thoroughly elusive in Arizona, moreover, which invites continued scrutiny of agency practices and habits. By recognizing some of the systemic tendencies in vector control, we may begin to think differently about what is driving health agency decisions. Rather than treating these organizational cultures as either fully optimal or wholly irrational, therefore, we might come to see these agency contexts merely as "frameworks for rational action" (following W.R. Scott 2008: 217). As an intervention in the way that we conceptualize vector control and public health strategies, therefore, this approach may help in developing effective abatement strategies in places like Arizona where institutional learning is ongoing, and climate and development coincide to create unprecedented vector habitats and ever-new threats to public health.

References

Arizona Department of Health Services. (2011) West Nile virus: Arizona data and maps. See www.azdhs.gov/phs/oids/westnile/data.htm (accessed 3 October 2011).

Arizona Republic. (2011) 2010 Arizona child drowning victims. See www.azcentral.com/news/articles/2011/05/28/20110528arizona-child-drownings-victims.html#ixzz1ZeGSf1O7 (accessed 3 October 2011).

Arriaga-Ramirez, S. and Cavazos, T. (2010) Regional trends of daily precipitation indices in northwest Mexico and southwest United States. *Journal of Geophysical Research-Atmospheres* 115.

Beck, U. (1992) *Risk Society: Towards a New Modernity*. London, UK: Sage Publications.

Botz, J.T. (2002) Survey of *Aedes aegypti* eggs in and around homes in Tucson, Arizona. *Journal of the American Mosquito Control Association* 18(1): 63–64.

Bowden, S.E., Magori, K., and Drake, J.M. (2011) Regional differences in the association between land cover and West Nile virus disease incidence in humans in the United States. *American Journal of Tropical Medicine and Hygiene* 84(2): 234–238.

Briegel, H. (2003) Physiological bases of mosquito ecology. *Journal of Vector Ecology* 28(1): 1–11.

Carter, E. (2008) State visions, landscape and disease: discovering malaria in Argentina, 1890–1920. *Geoforum* 39: 278–293.

Darling, S.T. (1912) A mosquito larvacide disinfectant and the methods of its standardization. *American Journal of Public Health* 2(2): 89–93.

Deichmeister, J.M. and Telang, A. (2011) Abundance of West Nile virus mosquito vectors in relation to climate and landscape variables. *Journal of Vector Ecology* 36(1): 75–85.

Dobyns, H.F. (1976) *Spanish Colonial Tucson: A Demographic History*. Tucson, AZ: University of Arizona Press.

Ellul, J. (1964) *The Technological Society*. New York, NY: Alfred A. Knopf.

Hayden, M.H., Uejio, C.K., Walker, K., Ramberg, F., Moreno, R., Rosales, C., Gameros, M., Mearns, L.O., Zielinski-Gutierrez, E., and Janes, C.R. (2010) Microclimate and human factors in the divergent ecology of *Aedes aegypti* along the Arizona, US/Sonora, MX border. *Ecohealth* 7(1): 64–77.

Karpiscak, M.M., Kingsley, K.J., Wass, R.D., Amalfi, F.A., Friel, J., Stewart, A.M., Tabora, J., and Zauderer, J. (2004) Constructed wetland technology and mosquito populations in Arizona. *Journal of Arid Environments* 56(4): 681–707.

Kessell, J.L. (1976) *Friars, Soldiers, and Reformers*. Tucson, AZ: University of Arizona Press.

Lizarraga-Celaya, C., Watts, C.J., Rodríguez, J.C., Garatuza-Payán, J., Scott, R.L., and Sáiz-Hernández, J. (2010) Spatio-temporal variations in surface characteristics over the North American monsoon region. *Journal of Arid Environments* 74(5): 540–548.

Mann, M. (1984) The autonomous power of the state: its origins, mechanisms, and results. *European Journal of Sociology* 25: 185–213.

Marston, S.A., Jones III, J.P., and Woodward, K.(2005) Human geography without scale. *Transactions of the Institute Of British Geographers* 30(4): 416–432.

Merrill, S.A., Ramberg, F.B., and Hagedorn, H.H. (2005) Phylogeography and population structure of *Aedes aegypti* in Arizona. *American Journal of Tropical Medicine and Hygiene* 72(3): 304–310.

Miller, A.J, Smith, J.L., and Buchanan, R.L. (1998) Factors affecting the emergence of new pathogens and research strategies leading to their control. *Journal of Food Safety* 18(4): 243–263.

Morin, C.W. and Comrie, A.C. (2010) Modeled response of the West Nile virus vector *Culex quinquefasciatus* to changing climate using the dynamic mosquito simulation model. *International Journal of Biometeorology* 54(5): 517–529.

Painter, J. (2006) Prosaic geographies of stateness. *Political Geography* 25(7): 752–774.

Patterson, G. (2009) *The Mosquito Crusades: A History of the American Anti-mosquito Movement from the Reed Commision to the First Earth Day*. New Brunswick, NJ: Rutgers University Press.

PRISM Climate Group. (2011) WestMap climate analysis & mapping toolbox. See http://cefa.dri.edu/Westmap/Westmap_home.php (accessed 6 October 2011).

RealtyTrac. (2011) Foreclosures, MLS listings and home values. See www.realtytrac.com (accessed 6 October 2011).

Reiter, P. and Gubler, D.J. (1997) Surveillance and control of urban dengue vectors. In *Dengue and Dengue Hemorrhagic Fever*, D.J. Gubler and G. Kuno (eds). New York, NY: CAB International, 425–462.

Robbins, P. and Marks, B. (2011) The political economy of mosquito control in the United States. Annual Meeting of the Association of American Geographers, Seattle, WA, 12–16 April.

Robbins, P., Farnsworth, R., and Jones, J.P. (2008) Insects and institutions: managing emergent hazards in the US southwest. *Journal of Environmental Policy and Planning* 10(1): 95–112.

Russell, E. (2001) *War and Nature: Fighting Humans and Insects with Chemicals from World War I to Silent Spring*. Cambridge, UK: Cambridge University Press.

Scott, J. (1998) *Seeing Like a State: How Certain Schemes to Improve the Human Condition Have Failed.* New Haven, CT: Yale University Press.

Scott, W.R. (2008) *Institutions and Organizations.* Los Angeles, CA: Sage.

Service, M.W. (1995) Can we control mosquitoes without pesticides: a summary. *Journal of the American Mosquito Control Association* 11(2): 290–293.

Shaw, I., Robbins, P., and Jones III, J.P. (2010) A bug's life and the spatial ontologies of mosquito management. *Annals of the Association of American Geographers* 100(2): 373–392

Spielman, A. and D'Antonio, M. (2001) *Mosquito: A Natural History of our Most Persistent and Deadly Foe.* New York, NY: Hyperion.

United States Census Bureau. (2011a) Fact finder. See http://factfinder2.census.gov/faces/tableservices/jsf/pages/productview.xhtml?pid=DEC_10_SF1_H1&prodType=table (accessed 7 October 2011).

United States Census Bureau. (2011b) State and county quickfacts. See http://quickfacts.census.gov/qfd/states/04/04013.html (accessed 7 October 2011).

United States Center for Disease Control. (2011a) Final 2010 West Nile virus human infections in the United States. See www.cdc.gov/ncidod/dvbid/westnile/surv&controlcasecount10_detailed.htm (accessed 2 October 2011).

United States Center for Disease Control. (2011b) Statistics, surveillance, and control archive. See www.cdc.gov/ncidod/dvbid/westnile/surv&control_archive.htm (accessed 2 October 2011).

United States Department of Agriculture. (2011) 2007 census of agriculture. See www.agcensus.usda.gov/Publications/Historical_Publications/index.php (accessed 6 October 2011).

Winner, L. (1977) *Autonomous Technology: Technics-out-of-control as a Theme in Political Thought.* Cambridge, MA: MIT Press.

Zinser, M. (2004). *Culex quinquefasciatus* host choices in residential, urban Tucson and at a constructed wetland. PhD thesis, Department of Entomology, University of Arizona, Tucson, AZ. See http://gradworks.umi.com/14/24/1424606.html (accessed 18 June 2012).

Zinser, M., Ramberg, F., and Willott, E. (2004). *Culex quinquefasciatus* (Diptera: Culicidae) as a potential West Nile virus vector in Tucson, Arizona: blood meal analysis indicates feeding on both humans and birds. *Journal of Insect Science* 4: 20.

Part IV
Health vulnerabilities

12 Exposure to heat stress in urban environments

Olga Wilhelmi, Alex de Sherbinin, and Mary Hayden

Extreme heat, exacerbated by the urban heat island (UHI) effect, is a leading cause of weather-related mortality in the United States and many other countries. Vulnerability to heat stress and patterns of heat-related morbidity and mortality are highly differentiated by age group and socio-economic status. Lower income populations often live in more built-up areas under more crowded conditions, and have less access to air conditioning or health services, making heat stress in urban areas a significant environmental justice issue. Current patterns of heat stress as a result of extreme heat events and the UHI effect are spatially varied at the city level, and can be shown to affect lower income populations and older age groups disproportionately. This chapter summarizes the literature on urban heat stress and associated morbidity and mortality, addresses aspects of urban ecology and the extent of built areas that promote UHI formation, and then addresses the social and political dimensions of exposure to heat stress. The chapter's concluding section addresses modifications to the urban environment and population's adaptive capacity through short-term response actions and long-term adaptation strategies that could help to reduce heat stress. We conclude that the inadequate appreciation of spatially differentiated factors contributing to extreme heat vulnerability limits the understanding of health risks and reduces the ability to prevent adverse heat health outcomes.

The health impacts of extreme heat

Impacts of heat stress on human health have been observed on all continents and within countries at all levels of income (Hajat and Kosatsky 2010). In the United States, mortality from heat stress accounts for more fatalities than any other weather hazard. From 1979–2003, exposure to extreme heat caused 8,015 deaths (CDC 2009). Future exposure of urban populations to extreme heat is projected to increase; as global warming continues, researchers anticipate increases in the severity, frequency, and duration of extreme heat events (Meehl and Tebaldi 2004; IPCC 2007; Diffenbaugh and Ashfaq 2010; Ganguly *et al.* 2009) in addition to seasonal warming of summer-month temperatures (Battisti and Naylor 2009). The effects of extreme heat on human health may vary significantly among geographic regions and demographic groups. However, general trends such as rapidly growing urban populations, aging of urban residents, the amplifying effect of UHIs and climate change (McMichael *et al.* 2003; Hayhoe *et al.* 2004) contribute to the increased health risks of urban populations. These trends require effective means of risk reduction in present and future climates. The potential dangers of climate change for future human health have been highlighted in a number of papers and reports (Patz *et al.* 2005; Gosling *et al.* 2009), making estimating and predicting the effects of a changing climate on human health a major public health priority (Frumkin *et al.* 2008; Kinney *et al.* 2008).

Heat-related morbidity and mortality may occur when the daily temperature exceeds a temperature range normal for a given climate and the local setting (Patz *et al.* 2005). Heat can also be a major health hazard in warmer subtropical and tropical climates, where high temperatures are typical throughout the summer season thus creating a heat-hazardous environment even without discrete episodes of heat waves (Harlan *et al.* 2006; Hajat *et al.* 2006; Hajat and Kosatsky 2010). Exposure to abnormally high temperatures during the night, an important characteristic of the UHI effect (as noted earlier), is a unique health risk factor. Earlier studies showed that significant numbers of heat-related mortality cases were attributed to high nighttime temperatures (Karl and Knight 1997; Meehl and Tebaldi 2004), mainly because of the inability of vulnerable urban residents to cool down and recover from the daytime heat.

Heat-related illness is caused by the inability of the human body to regulate its normal internal temperature either because of excess heat production or decreased transfer of heat to the environment (Jardine 2007; Ebi and Meehl 2007). Medical conditions resulting from prolonged exposure to heat can produce various degrees of severity and include heat-related edema, rash, cramps, exhaustion, and stroke. Heat stroke, the most serious of the heat-related illnesses, can cause shock, brain damage, internal organ failure, coma, and death. Heat stroke occurs when the core body temperature rises above 41°C (CDC 2009). Heat stroke can be life-threatening; Jardine (2007) reported a 12 percent death rate among adult patients diagnosed with heat stroke. In addition to heat stroke, many heat-related deaths are associated with cardiovascular and respiratory diseases (Kilbourne 1997). Helman and Habal (2009) describe two forms of heat stroke. The first is a "classic" heat stroke that usually occurs during extreme heat events or heat waves, especially in urban areas. It affects vulnerable residents who are unable to control their environment and water intake, sedentary elderly people, persons with pre-existing chronic illnesses, and very young children with immature thermoregulatory system. The second form is "exertional" heat stroke, which typically affects younger people and results from strenuous physical activity during very hot weather (Helman and Habal 2009).

Exposure to extreme heat has been a public health concern for over four decades (e.g., Schuman 1972). However, a number of deadly heat waves in US cities during the past two decades and in many European cities in the first decade of the twenty-first century (especially those in 2003, 2006, and 2010) have generated an increased public awareness and an expansion in heat-health research (for reviews see Gosling *et al.* 2009; Hajat and Kosatsky 2010). Despite the growing body of knowledge about urban vulnerabilities to extreme heat, the complex interplay of ecological, social, political, and medical factors contributing to heat stress makes it challenging to prevent heat-related mortality. Complexity is related to factors such as differential thresholds for negative health outcomes across geographic regions, urban–rural gradients and even among neighborhoods within a city.

There is no standard definition of a heat wave (Robinson 2001). Many existing definitions of heat waves or extreme heat events attempt to represent the interaction between the thermal environment and human body, which may result in negative health outcomes for the exposed population (Robinson 2001; Meehl and Tebaldi 2004). Meteorological definitions of a heat wave are usually intended to indicate an overall public health concern associated with heat stress and do not account for differential responses to heat among vulnerable urban residents along the UHI gradient. In the United States, for example, the National Weather Service defines a heat wave as a "period of abnormally and uncomfortably hot and unusually humid weather. Typically a heat wave lasts two or more days" (NWS 2009). Worldwide, definitions of extreme heat can range from "heat days" (Huang *et al.* 2010) to "excessive heat events" (EPA 2006) with a variety of thresholds. In their definition of excessive heat events, the US Environmental Protection Agency (EPA 2006) emphasizes that meteorological thresholds depend not only on the geographic location but also on a time of year. Kovats and Hajat (2008: 49) state that "The challenge lies in determining at which point the weather conditions become sufficiently hazardous to human health in a given population to warrant intervention."

Many epidemiological studies have used simple meteorological measures (e.g., daily mean, maximum, and minimum temperature, and daily maximum apparent temperature) to establish associations between temperature and heat-related mortality (Gosling *et al.* 2009; Hajat and Kosatsky 2010). Events, such as three consecutive nights of very hot temperatures, have been linked to negative health outcomes (Kalkstein and Smoyer 1993; Karl and Knight 1997) and thus have been applied in climate model simulations to analyze future heat waves (Meehl and Tebaldi 2004; Ganguly *et al.* 2009; McCarthy *et al.* 2010). Although most studies to date show vulnerability to heat stress in temperate climates (Davis *et al.* 2004), subtropical and tropical regions may show similar sensitivity as location-specific temperatures rise as a result of a changing climate (Patz *et al.* 2005). A few studies have investigated temperature–mortality relationships in subtropical cities in Europe, Asia and South America (Hajat and Kosatsky 2010; Huang *et al.* 2010; Guest *et al.* 1999; Romero-Lankao *et al.* forthcoming). A review of the heat–mortality relationship in 64 cities throughout the world (Hajat and Kosatsky 2010) showed that temperature–mortality thresholds were generally higher in cities with higher average summertime temperatures, which suggests physiological adaptation to heat.

While it is critical to establish mortality temperature thresholds that can be used for extreme heat monitoring, forecasting, and public health advisories, other environmental and social variables need to be considered in order to provide more targeted interventions, especially in places with high relative temperatures and for populations at risk. Research frameworks on extreme heat vulnerability (Wilhelmi and Hayden 2010) and a number of case studies (Uejio *et al.* 2011; Harlan *et al.* 2006; Johnson *et al.* 2009) demonstrated that both urban land cover (a factor in UHI formation) and demographic characteristics contribute to health outcomes. A multicity analysis across the globe (Hajat and Kosatsky 2010) illustrated that heat-related mortality occurred in a range of geographic settings and in all countries with all levels of income. However, in high-income countries most heat-related deaths were associated with underlying causes of cardiovascular and respiratory disease. Age distribution, population density, and gross domestic product (GDP) played an important role in the heterogeneous distribution of heat-health outcomes. And of course the conditions that lead to *exposure* to extreme heat (i.e., dense urban settings and the UHI, poor air circulation, lack of access to air conditioning) may in themselves be related to socio-economic differences at the urban scale.

Research on heat impacts on human health has expanded rapidly since the early 1990s in various academic disciplines (Kinney *et al.* 2008; Gosling *et al.* 2009). But it is only more recently that researchers have begun to explicitly incorporate information about social vulnerability in their studies (Smoyer 1998; Guest *et al.* 1999; Chan *et al.* 2001; Harlan *et al.* 2006; Gosling *et al.* 2009). Societal vulnerability often determines the severity of impacts of a heat hazard on an individual or a group and therefore needs to be considered with the same degree of importance that has been devoted to understanding the physical processes (i.e., heat waves or the UHI effect). A number of case studies have used epidemiological and statistical techniques to address the relationships among heat waves, heat-related morbidity, and mortality and to identify vulnerable groups of people (Smoyer 1998; Chan *et al.* 2001). These case studies demonstrate that certain groups in the population (e.g., elderly, very young, obese individuals, people using certain medications, socially isolated individuals, the poor, the mentally ill, those without air conditioning, outdoor workers) are disproportionately affected by exposure to extreme heat (Curriero *et al.* 2002; O'Neill *et al.* 2003; Medina-Ramon *et al.* 2006; O'Neill *et al.* 2005; Kalkstein and Greene 2007).

In large, sprawling cities with a highly variable socio-economic fabric, infrastructure and housing types, vulnerability is expected to be even more complex. The relative importance of individual and household heat-health risk factors has been investigated in several US cities (e.g., Smoyer 1998; Uejio *et al.* 2011). These integrated neighborhood-level studies showed that urban land surface characteristics associated with UHI formation together with socio-demographic characteristics can be associated with heat-wave morbidity or mortality information to identify first-order vulnerability indicators and highlight zones of elevated vulnerability within urban areas. Recent advances in geospatial methods and analysis tools allow for spatially explicit characterizations of heat-related vulnerabilities even in seemingly homogeneous urban environments (Wilhelmi *et al.* 2004). These case studies of local-level urban vulnerability to extreme heat show the importance of interdisciplinary approaches to analyzing and predicting heat-health outcomes. But the lack of such case studies in many countries, data gaps, different methods in coding heat-related morbidity and mortality creates challenges for international comparison of heat-health vulnerabilities, especially for assessments at the global level.

Ecological dimensions

Although the UHI effect has been measured using a variety of metrics, the fundamental definition is the same for all studies. The UHI effect represents the relatively higher temperatures found in urban areas compared to surrounding rural areas, especially at night. The relative differentials between urban and rural temperatures vary by season and climatic zone, as discussed below. There are several scales of analysis for studying the ecological dimensions of heat islands. On the micro scale, there are ecological, environmental, and geographical (e.g., topography, proximity to water bodies) factors that determine whether a neighborhood has higher or lower daytime or nighttime temperatures than surrounding neighborhoods. At the meso-scale, there are morphological factors that affect air flow into cities from surrounding areas. And on the regional and global scale, there are geographic and climatic factors that affect the production of UHIs.

A number of ecological and environmental factors affect urban climatology (McCarthy *et al.* 2010; Kuttler 2008; Oke 1987). One such factor is the replacement of natural land cover with sealed surfaces that have a strong three dimensional structure, including the reduction of the land area covered by vegetation, loss of surface water bodies, reduction of emissions of long-wave energy from the surface by street canyons (streets bordered by tall buildings), and disruption of airflow by buildings. A second factor is the release of waste heat from heating and cooling systems.

These factors affect radiation budgets and thermal properties such as evapotranspiration, water storage, and atmospheric exchange. Thermal properties, in turn, are determined by land use, the structure of buildings, and the overall extent of built-up areas. Vegetated surfaces associated with parks or tree-lined roadways tend to be cooler because shade reduces the amount of solar radiation that hits underlying surfaces, and because they have a lower heat capacity (Zhou and Shepherd 2009). Because water makes up a large proportion of biomass, at higher temperatures the water transpires through the leaf stomata generating a latent heat flux that cools surface temperatures.

By contrast with vegetated areas, paved surfaces and built structures tend to absorb shortwave radiation during the day and release long-wave radiation with increasing intensity in the afternoon and evening (Kuttler 2008). The proportion of shortwave energy that is absorbed versus reflected depends on surface spectral properties (i.e. albedo). Once shortwave energy is absorbed, the sensible heat flux from urban surfaces (e.g. roads, sidewalks, and buildings) depends on the thermal properties of the materials with which they are constructed (their density, heat capacity, and thermal conductivity) (Kuttler 2008). These factors, plus the angle of incidence of incoming radiation and the ways it is absorbed or reflected off surfaces in the street canyon, only to be absorbed and rereleased by other surfaces, all affect surface, canopy, and boundary-layer temperature conditions. The urban-canyon effect also reduces air flow, trapping heat near the surface. Finally, in areas with sealed surfaces rainfall tends to be quickly channeled to the underground sewer system, and the latent heat flux associated with evaporation is greatly reduced. Thus, more energy is available for long-wave emission and sensible heat flux from sealed surfaces.

Use of thermal-band data from satellite imagery is one way of investigating variations in heat intensity and their relationship to surface types within cities.[1] Using Landsat 7's thermal infrared band, Landsat-derived land cover data, and census blocks as the unit of analysis, Cheung (2002) found that for Washington, DC, 54 percent of the variance in surface temperature could be predicted by percent impervious surface, normalized difference vegetation index (an indicator of vegetation cover), and percent water bodies. Temperatures increase over impervious surfaces and decrease over vegetated areas and water bodies. Zhou and Shepherd (2009) found a similar relationship for Atlanta, and, in a study of 38 US cities, Imhoff *et al.* (2010) found that impervious surface area explains 70 percent of the variance in land-surface temperatures. In an assessment of night time surface temperature of Delhi, India (Javed *et al.* 2009), areas with greater vegetation cover were found to have lower mean surface temperatures, and changes in mean surface temperature between 2001 and 2005 were mainly due to vegetation loss owing to land cover change. The central business district, characterized by densely built-up areas, had a temperature differential of greater than 4°C compared with the suburbs.

Turning to urban form, the morphology of the city affects air flows into and out of the city. Wind speeds tend to be lower than in rural areas owing to surface roughness, which in turn depends on building (and vegetation) heights. As Kuttler (2008) points out, a number of common features can act as conduits for cooler air to reach the city and thus attenuate the effects of the UHI. These include highways and wide boulevards (which themselves may emit longwave radiation), railway tracks, public parks, and rivers and urban water bodies. The degree to which these corridors act to cool a city will depend on their orientation relative to prevailing winds. It is worth noting that these pathways can both help to cool the city and bring cleaner air, which is important for human health.

Size of the city is also an important aspect of morphology. Cities with larger built-up areas tend to have higher UHIs (Imhoff *et al.* 2010). Population size has often been used as a proxy for geographic extent (Kuttler 2008), but with the advent of urban land cover and urban extent data, such as the MODIS map of Global Urban Extent (Schneider *et al.* 2009) and the Global Rural–Urban Mapping Project (CIESIN *et al.* 2011), it is now possible to obtain more information on the actual extent of individual cities and conurbations. Cheung (2002) found that in Washington, DC from 1951 to 2001 the number of degree days above 35°C increased substantially, a period during which the metropolitan area rapidly expanded. Zhang *et al.* (2010) also found an increase in temperatures for the Shanghai region of 0.75°C per decade since 1976 owing in part to expansion of the urban agglomeration.

Though all large cities exhibit the formation of UHIs (Buyantuyev and Wu 2010), the geographic location of cities in terms of climatic zone, proximity to coast, and elevation all impact the degree of temperature difference between rural and urban areas during summer months. It should be added, however, that this is a relatively under-researched area, and comparative research is complicated by the adoption of different definitions and measurement methods for ascertaining the UHI effect (Alcoforado and Andrade 2008). A meta-analysis (Wienert and Kuttler 2005) of latitudinal differences in UHI included 150 studies that used a common metric, UHI*max*, which is the highest daily warming of an urban area relative to surrounding rural areas under the weather conditions most favorable for UHI development (cloud-free and low winds). Controlling for population size, energy use, topographic features, altitude, and energy balance, they found that the UHI*max* increased with latitude: cities in a tropical cluster had UHIs of 4°C on average, whereas those in subtropical and mid-latitude clusters had UHIs of 5.0°C and 6.1°C, respectively. Much of the difference is explained by anthropogenic heat production and radiation balance, which increases with latitude. Within the United States, Imhoff *et al.* (2010) found that the summer daytime UHI is largest for biomes dominated by temperate broadleaf and mixed forests, which exhibited an 8°C average differential between urban and rural areas, compared to a 6°C differential for grasslands and a –1°C differential in desert biomes, where daytime surface temperatures are actually hotter outside built up areas.

Beyond climate zones, other geographical factors can also exacerbate heat buildup. Mexico City and Santiago de Chile, for example, are both surrounded by mountain chains that serve to reduce air flow and increase temperatures and pollution levels. The UHI effects of cities in arid and semi-arid regions tend to increase during the dry season when canopy cover is reduced and drier soil conditions in rural areas increase the albedo of surfaces (Buyantuyev and Wu 2010). According to Nasrallah *et al.* (1990), similarities in landscape characteristics between urban and rural areas in hyperarid regions such as Kuwait lead to poorly defined UHIs. Yet climate modeling that includes urban land-surface effects by McCarthy *et al.* (2010) finds the opposite: that arid regions exhibit high UHI effects. They suggest that this trend is a function of high incident shortwave radiation at the surface and low soil moisture resulting in high daytime energy gains, but low surface-heat capacity and rapid cooling rates of the natural land cover at night, yielding large nocturnal temperature differences between urban and rural areas. Consistent with the findings of Wienert and Kuttler (2005), Roth (2007) reviewed a wide range of UHI studies in tropical and sub-tropical climates, and found that UHI intensities are generally lower compared to those of temperate cities with comparable population. They also show a seasonal variation with higher intensities during the dry season.

Emmanuel and Johansson (2006) conducted a micro-level study of the UHI in Colombo, Sri Lanka and found that maximum daytime temperatures tended to decrease with increasing building height/width ratio and proximity to the sea, with the temperature differences between sunlit and shaded urban surfaces reaching 20°C. The shade in urban canyons can be important in reducing insolation, but higher height/width ratios also reduce wind speeds and can block coastal breezes. The relative importance of these two conflicting factors depends on their proximity to the oceans. Zhang *et al.*'s (2010) study of Shanghai, China, which has a subtropical monsoon climate with distinct seasonal temperature variations, found the UHI to be most pronounced in the autumn, which corresponds to periods of less cloudiness and low wind speed, and weakest in the summer, when the air is convectively unstable and characterized by enhanced mixing over larger areas.

There are few if any comparative studies that look at the differential UHI effect in coastal vs. non-coastal cities within the same climate zone. Coastal cities, given their proximity to the relatively cooler ocean during summer months, suffer less from extreme temperatures than land-locked cities with continental climates, but they can also be expected to have high humidity levels that reduce the body's cooling ability. Since UHIs are measured as a relative difference between urban and rural areas, and both rural and urban settings would tend to be under the moderating influence of sea breezes, the actual development of the UHIs would also presumably be less than in interior areas.

The UHI effect is different than absolute temperatures experienced in urban areas. Obviously urban areas in cool temperate climates suffer fewer extreme temperature events over 35°C. But it is expected that with climate change, urban areas in sub-tropical and warm temperate and temperate climates will experience greater frequency of heat waves (Kuttler 2008). Indeed, McCarthy *et al.* (2010) predict that a combination of climate effects and a doubling of urban population will greatly exacerbate the UHI effect in the Middle East, central Asia, and east and west Africa, resulting in increases in night-time temperatures (T*min*) of +5 to +6.5°C with a doubling of atmospheric CO_2.

The social and political dimensions of heat stress

The risk of heat stress, which is associated with a differential exposure associated with urban ecology and climate change, is not borne equally by all members of society (UNDP 2008). As discussed earlier, certain indicators of heat vulnerability, such as age, education, social isolation, race and ethnicity, lack of air conditioners, and chronic deceases are commonly identified in case studies on excess deaths during extreme heat events (O'Neill and Ebi 2009). While physiological factors play a role in heat health outcomes, socio-economically disadvantaged populations tend to be disproportionately affected by heat stress. This raises issues of environmental justice. The concept of environmental justice was born out of a recognition that in many parts of the world there is an unequal spatial distribution of environmental burdens and benefits. It emerged in 1979 when a lawsuit was filed to prevent the siting of a landfill in a predominantly African-American neighborhood in Houston, Texas (Lee 2002). The United States Environmental Protection Agency defines environmental justice as "the fair treatment and meaningful involvement of all people regardless of race, color, national origin, or income with respect to the development, implementation, and enforcement of environmental laws, regulations, and policies" (EPA undated).[2] Traditionally, environmental justice has focused on socio-economically disadvantaged populations where harmful exposure to environmental toxins has resulted in adverse health outcomes. These exposures, often more pronounced in urban areas, consist of a variety of environmental hazards including toxic waste, air pollution, and extreme heat. In fact, the effect of interaction of certain air pollutants (e.g., ozone) and high temperature on human health have been noted in several studies (Filleul *et al.* 2006; Semenza *et al.* 2008).

Because the UHI phenomenon exacerbates the effects of heat waves, it is no surprise that those most vulnerable to extreme heat events often live in the poorest urban neighborhoods where studies have shown that environmental heat mitigators such as green spaces and overall vegetation cover are inequitably distributed (White-Newsome *et al.* 2005; Harlan *et al.* 2006; Ruddell *et al.* 2010).[3] Other vulnerability factors include lower socio-economic status, lower educational levels, and age, all of which have been observed to increase the risk of negative health outcomes in response to exposure to extreme heat (Klinenberg 2002). The resulting vulnerability or inability to shield oneself from the 'by-products' of industrialization is all too prevalent. Living in substandard housing coupled with inefficient or non-existent means of cooling one's personal environment subjects those with lower socio-economic status to negative health outcomes associated with exposure to extreme heat (Harlan *et al.* 2008).

Furthermore, as longevity and urban populations increase in developed countries, we can expect more segments of the population to be exposed to extreme heat. Among those most affected by extreme heat are the elderly because the ability to regulate body temperature and physiologically adapt to the heat lessens with age (Luber and McGeehin 2008). During one of the hottest summers on record in Europe, the elderly, defined as those above 65 years of age, experienced excess mortality as a result of the August 2003 heat wave. In a study by Le Tertre *et al.* (2006), an estimated 3,096 deaths were attributed to the August 2003 heat wave in nine French cities. In a case study of the week of 8–13 August 2003 in three French cities severely affected by the heat wave (Paris, Tours, and Orleans), risk factors for mortality among the elderly were examined. Results indicated that vulnerability to heat was associated with lack of mobility, lower socio-economic status (higher mortality among manual laborers), and social isolation, defined as not participating in any social, cultural, religious, or recreational activities. The specter of environmental justice is raised in the findings, as those living in older homes with no thermal insulation were found to be at highest risk. Other risk factors included living on the top floor of a building located in a UHI (Vandentorren *et al.* 2006).

For all of France, excess deaths (defined as deaths that would not have otherwise occurred) in the August 2003 heat wave exceeded 15,000 and were most frequent among the elderly living alone or in nursing homes, particularly among women who were divorced, single, or widowed (Fouillet *et al.* 2006). Excess deaths during the same time period in Italy (Bologna, Milan, Rome, and Turin) were highest in those over 75 years of age; mortality was also attributed to lower socio-economic status and levels of education. In this instance, socio-economic status was a notable factor in contributing to excess mortality because those who can afford to leave Italian cities in the hot summer months did so (Michelozzi *et al.* 2005).

A similar picture of excess mortality emerges in the British Isles where mortality in London during the August 2003 heat wave was 59 percent higher than expected. Again, the age group most affected was the elderly, defined there as those above 75 years of age (Johnson *et al.* 2009). Although there was documented excess mortality during the heat wave of 2003 in the Netherlands, a more nuanced picture indicates that temperatures were not as high as in other European nations nor did the majority of deaths occur in heavily populated urban areas (Garssen *et al.* 2005). However, Pascal *et al.* (2005) note that temperature thresholds for what constitutes extreme heat vary country by country and often, city by city. For example, extreme heat thresholds range from 41°C in Andalusia to 27.5°C in Belgium. Also of interest during the August 2003 time period is that most countries also noted air pollution indices that were above normal, exacerbating the negative health impacts of the heat wave.

In the US, heat disproportionately affects those in lower socio-economic brackets. This effect occurs not only for those living in heavily populated urban areas where UHIs are strong, but also for those working outdoors as manual laborers and in agricultural settings as migrant laborers. Farm laborers experience prolonged exposure to the sun, often 10 hours or more per day, and have little access to shade (Fougères 2007). In urban areas, Harlan *et al.* (2008) determined from their study of the 2005 heat wave in Phoenix, Arizona, that affluent whites were more likely to live in vegetated neighborhoods with reduced UHI characteristics, again highlighting the vulnerability of those with lower socio-economic status. The Chicago heat wave of 1995 resulted in 700 excess deaths, and a case study documented risk factors that included having an existing medical illness, living alone, and not having access to air conditioning (Semenza *et al.* 1996). The latter factor highlights the risk associated with poverty. A more nuanced question is not only whether there is air conditioning available, but rather whether those who have air conditioning can afford to use it to the extent necessary to provide adequate cooling.

Based on his analysis of the 1995 heat wave in Chicago, Klinenberg (2002) determined that those with lower socio-economic status were not affected equally. His study showed the importance of a more detailed documentation of risk factors and the need to go beyond recording excess mortality if reasonable interventions are to be addressed to reduce risk. Klinenberg theorized that a lack of social isolation was a key protective factor in survival because those with strong ties to family, friends, or neighbors were assured that someone would check up on them in the event of a heat wave. This hypothesis was corroborated in the 2003 heat wave in France, where social isolation was considered to be a major factor contributing to excess mortality.

Intersection between social dimensions and ecological dimensions

The ecological theory of resilience (Holling 1973, 1986) offers a means of addressing complex interactions within the ecological dimensions of a dynamic system. The dynamic systems proposed by Holling are continually affected by stochastic forces in nature such as droughts, heat waves, and floods requiring an understanding of the system's ability to absorb disturbances (Eakin and Luers 2006). Relationships within a complex system are often non-linear and, as such, necessitate an emphasis on thresholds and the uncertainties inherent within the system. Identifying these thresholds can allow for a targeted response which may subsequently assist in absorbing shocks or alterations to the system. Smit and Wandel (2006) argue that in the face of climate change, responses must be multiscale, both spatially and temporally. Because the dynamics of complex systems are shaped at multiple levels, including local, national, and global scales, the increasing importance of these cross-dynamics must be factored into understanding resilience and adaptive capacity. When systems are vulnerable, disturbances such as a warming climate resulting in extreme events such as an urban heat wave can have dramatic consequences, particularly on the most at-risk populations (Bennet *et al.* 2003; Adger 2006; Folke 2006).

In order to begin the necessary dialogue to reduce population vulnerability to extreme heat through addressing issues of scale, Wilhelmi and Hayden (2010) have proposed a research framework to tackle a multiscalar, multidimensional approach to adaptation to extreme heat. They highlight the need to understand local social and cultural factors as well as external macro-scale drivers to address the challenges of connecting people and place in a research framework aimed at reducing adverse outcomes to urban heat stress. The call for both a top-down and bottom-up approach is presented in the framework in Figure 12.1. This figure emphasizes the importance of an integrated systems approach whereby multiple levels of community from government officials to households within neighborhoods are engaged and an iterative process is underscored so that the feedback loops are focal points of the ongoing dialogue among all levels.

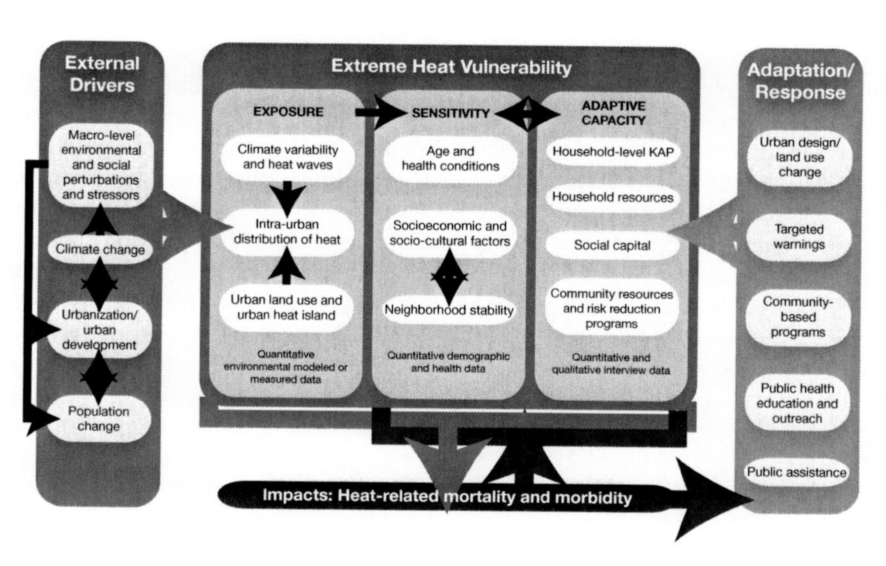

Figure 12.1 Intersection of social and ecological dimensions of urban heat stress in a vulnerability framework.

Source: Wilhelmi and Hayden (2010)

Given previous and recent research on urban heat and human health (Harlan *et al.* 2006; Johnson *et al.* 2009; Reid *et al.* 2009; McGregor *et al.* 2007; O'Neill and Ebi 2009), it is evident that heat stress represents a complex interplay of ecological, social, and medical issues, which are often place and scale-specific. At the local level, characteristics of the built environment, social, and cultural characteristics and a larger context of urban development, availability of community resources and government programs determine differential impacts of heat on subpopulations.

Fewer studies have considered both social capital and the natural environment in assessing urban heat stress. Harlan *et al.* (2008) showed that unequal distributions of human-managed natural resources (vegetation, open space) are key to heat stress vulnerability in Phoenix, Arizona. In the study of urban heat vulnerability in Philadelphia, Pennsylvania and Phoenix, Uejio *et al.* (2011) demonstrated that heat stress risk factors were place-specific in that urban ecology and characteristics of the built environment played a more significant role in Phoenix than in Philadelphia. This finding illustrates the importance of contextualizing vulnerability to heat-related health outcomes in the context of local urban ecology. The extreme heat research framework developed by Wilhelmi and Hayden (2010) emphasizes that contextualizing vulnerability with local-level social, behavioral, and ecological data can influence successful health interventions, heat mitigation, and climate change adaptation.

Conclusion and policy recommendations

As illustrated in this chapter, urban vulnerability to extreme heat and the negative health outcomes result from the interplay of social and ecological dimensions of the urban environment. These complex relationships and the feedbacks between local-level processes and the larger context of global climate, urbanization and changing population structure and health risks need to be considered in planning and implementing adaptation strategies. In addition, the combined top-down and bottom-up approach to heat-health risk reduction would require considerable interactions with various scales of governance. While this recommendation may appear daunting to national, state/provincial, and local officials, a number of measures, both short-term and long-term, can mitigate urban heat and prevent heat-related morbidity and mortality. Because of the interdisciplinary nature of urban heat-health relationships both social and ecological factors need to be addressed.

In terms of physical adaptation responses, cities can increase both vegetation cover, which enhances latent heat flux, and the albedo (reflectivity) of built surfaces, such that a higher proportion of incoming shortwave radiation is reflected rather than absorbed and reradiated (Zhou and Shepherd 2009). Recent research (Oleson *et al.* 2010) demonstrated that white-colored roofs have the potential to significantly cool cities and mitigate some impacts of the UHI effect and global warming, and greenery planted on rooftops acts as an insulator that reduces indoor temperatures and energy consumption (Rosenzweig 2011). Some cities, such as Chicago, are already implementing the "cool roofs" policies

(ICLEI undated). Given budgetary implications, changing the urban landscape to include more green space and cool roofs would certainly require governmental decisions and regulations at different levels and could involve industry as well as the public.

The heat warning systems developed in a number of cities integrate weather forecasting, heat warning, and specific intervention strategies (Sheridan and Kalkstein 2004). In some cases, these warning systems have proven to be successful: it was estimated that 117 people were saved in Philadelphia during heat waves from 1995 to 1998 through a Hot Weather-Health Watch/Warning System (McGregor *et al.* 2007). In addition to linking forecasting and public health interventions, more work needs to be done to ensure that the meteorological thresholds used for heat warnings are based on epidemiological studies of heat-mortality relationships. Also, given the complexity of UHI effects and the social fabric of urban areas, location-specific warnings within a city can more effectively target populations at risk.

In the framework proposed by Wilhelmi and Hayden (2010), the authors argue for a better understanding of differential vulnerability in order to address meaningful interventions. They argue that adaptive capacity cannot be measured solely through the use of demographic data, as underlying vulnerability is often masked at that level of resolution. Localized identification of those at risk from exposure to extreme heat, with a focus on adaptive capacity, is one way of assuring that targeted interventions are meaningful.

Access to cooling is one of the major factors of social vulnerability to extreme heat. Many cities in temperate climates are not equipped with air conditioning systems. In warmer climates, there are segments of the urban population that may not be able to afford to use air conditioning due to the high cost of electricity (Hayden *et al.* 2012). One approach implemented by several cities is the establishment of cooling centers (i.e., air-conditioned public buildings). Given the geographic breadth of studies pointing to social isolation as a risk factor, it is not enough to provide cooling shelters if those at highest risk are unable or unwilling to access the shelter. This possibility then necessitates a concerted community effort to ensure that neighbors, social workers, and home healthcare providers are called upon to ascertain that those who are bedridden or socially isolated are provided for during periods of extreme heat.

As illustrated in many reports and research articles, impacts of heat stress on human health have been observed on all continents and therefore extreme heat remains an important public health concern in cities throughout the world. The negative impacts of extreme heat on human health vary significantly among geographic regions and demographic groups. While extreme heat affects more people in large metropolitan areas with high population density and complex urban morphology, residents in small cities and rural areas (e.g., migrant workers) are affected by extreme heat as well. Projected increases in frequency, duration, and severity of extreme heat events, as well as seasonal warming of summer-month temperatures, can place more vulnerable populations at risk both in large cities and in rural areas worldwide. Further research is needed to better understand how human exposure to temperature extremes discriminates between urban and rural residents as well as among urban residents in different geographic locations. This inadequate appreciation of spatially differentiated factors of extreme heat vulnerability limits the understanding of health risks and reduces the ability to prevent adverse heat health outcomes.

Acknowledgments

The authors would like to acknowledge the comments provided by anonymous reviewers and by Atiqur Rahman of Jamia Millia Islamia University in New Delhi. The work was carried out with support from NASA for the Socioeconomic Data and Applications Center (contract NNG08HZ11C) hosted by CIESIN, Columbia University, and from the National Science Foundation for the National Center for Atmospheric Research.

Notes

1 Studies on the UHI phenomenon using satellite-derived land surface temperature measurements have been conducted using various remote sensing data, such as NOAA Advanced Very High Resolution Radiometer with 1.1 km spatial resolutions (Cao *et al.* 2008), NASA Moderate Resolution Imaging Spectroradiometer (MODIS) (Parida *et al.* 2008), Advanced Spaceborne Thermal Emission and Reflection Radiometer (ASTER) (Tiangco *et al.* 2008), and Landsat Thematic Mapper and Enhanced Thematic Mapper Plus thermal infrared data with 120 m and 60 m spatial resolutions, respectively (Cheung 2002).

2 The full definition is as follows:

> Environmental Justice is the fair treatment and meaningful involvement of all people regardless of race, color, national origin, or income with respect to the development, implementation, and enforcement of environmental laws, regulations, and policies. EPA has this goal for all communities and persons across this Nation. It will be achieved when everyone enjoys the same degree of protection from environmental and health hazards and equal access to the decision-making process to have a healthy environment in which to live, learn, and work.

3 This relationship has been most studied in North America but also holds for many world regions. For example, Weeks *et al.* (2005) found a strong negative relationship between vegetation cover and poverty levels in Accra, Ghana. But in densely settled Asian urban areas the relationship can be the inverse: de Sherbinin (2008) found that for Hanoi, Vietnam, there is a positive relationship between poverty levels and greenness. The more affluent residents tend to live in the urban center, which is less green, and therefore more susceptible to the UHI effect. They presumably also have greater access to space cooling.

References

Adger, W.N. (2006) Vulnerability. *Global Environmental Change* 16(3): 268–281.

Alcoforado, M.J. and Andrade, H. (2008) Global warming and the urban heat island. In *Urban Ecology*, J.M. Marzluff, E. Shulenberger, W. Endlicher, M. Alberti, G. Bradley, C. Ryan, U. Simon, and C. ZumBrunnen (eds). New York, NY: Springer, 249–262.

Battisti, D.S. and Naylor, R.L. (2009) Historical warnings of future food insecurity with unprecedented seasonal heat. *Science* 323: 240–244.

Bennet, E.M., Carpenter, S.R., Peterson, G.D., Cummings, G.S., Zuerk, M., and Pingali, P. (2003) Why global scenarios need ecology. *Frontiers in Ecology and the Environment* 1: 322–329.

Buyantuyev, A. and Wu, J. (2010) Urban heat islands and landscape heterogeneity: linking spatiotemporal variations in surface temperatures to land-cover and socio-economic patterns. *Landscape Ecology* 25: 17–33.

Cao, L., Li, P., Zhang, L., and Chen, T. (2008) Remote sensing image-based analysis of the relationship between urban heat island and vegetation fraction. *International Archives of the Photogrammetry, Remote Sensing and Spatial Information Sciences* XXXVII(B7).

CDC (Centers for Disease Control). (2009) Extreme heat: a prevention guide to promote your personal health and safety. See www.bt.cdc.gov/disasters/extremeheat/heat_guide.asp (accessed 19 June 2012).

Chan, N.Y., Stacey, M.T., Smith, A.E., Ebi, K.L., and Wilson, T.F. (2001) An empirical mechanistic framework for heat related illness. *Climate Research* 16: 133–143.

Cheung, I. (2002) Extreme heat, ground level ozone concentration, and the urban heat island effect in the Washington DC metropolitan area. In *Proceedings for the North America Urban Heat Island Summit sponsored by the Toronto Atmospheric Fund and the U.S. Environmental Protection Agency, Toronto, Canada, May 2002*. See www.cleanairpartnership.org/pdf/finalpaper_cheung.pdf (accessed 19 June 2012).

CIESIN. (2011) Global rural–urban mapping project, version 1 (GRUMPv1): urban extents grid. Palisades, NY: Socioeconomic Data and Applications Center (SEDAC), Columbia University.

Curriero, F.C., Heiner, K.S., Samet, J.M., Zeger, S.L., Strug, L., and Patz, J.A. (2002) Temperature and mortality in 11 cities of the eastern United States. *American Journal of Epidemiology* 155: 80–87.

Davis, R.E., Knappenberger, P.C., Michaels, P.J., and Novicoff, W.M. (2004) Seasonality of climate–human mortality relationships in US cities. *Climate Research* 26: 61–76.

Diffenbaugh, N.S. and Ashfaq, A. (2010) Intensification of hot extremes in the United States. *Geophysical Research Letters* 37: L15701 (doi:10.1029/2010GL043888).

Eakin, H. and Luers, A.L. (2006) Assessing the vulnerability of social–environmental systems. *Annual Review of Environmental Resources* 31: 365–394.

Ebi, K. and Meehl, G.A. (2007) Heat waves and global climate change, the heat is on: climate change and heat waves in the Midwest. In *Regional Impacts of Climate Change: Four Case Studies in the United States*. Arlington, VA: Pew Center on Global Climate Change, 8–21.

Emmanuel, R. and Johansson, E. (2006) Influence of urban morphology and sea breeze on hot humid microclimate: the case of Colombo, Sri Lanka. *Climate Research* 30(3): 189–200.

EPA (United States Environmental Protection Agency). (undated) Environmental justice. See www.epa.gov/compliance/environmentaljustice (accessed 19 June 2012).

EPA (United States Environmental Protection Agency). (2006) *Excessive Heat Events Guidebook*. EPA report. See www.epa.gov/hiri/about/pdf/EHEguide_final.pdf (accessed 19 June 2012).

Filleul, L., Cassadou, S., Medina, S., Fabres, P., Lefranc, A., Eilstein, D., Le Tertre, A., Pascal, L., Chardon, B., Blanchard, M., Declercq, C., Jusot, J.F., Prouvost, H., and Ledrans, M. (2006) The relation between temperature, ozone, and mortality in nine French cities during the heat wave of 2003. *Environmental Health Perspectives* 114(9): 1344–1347.

Folke, C. (2006) Resilience: the emergence of a perspective for social-ecological systems analysis. *Global Environmental Change* 16: 253–267.

Fougères, D. (2007) *Climate Change, Environmental Justice, and Human Rights in California's Central Valley: A Case Study*. See www.ciel.org/Publications/Climate/CaseStudy_CentralValleyCA_Nov07.pdf (accessed 19 June 2012).

Fouillet, A., Rey, G., Laurent, F., Pavillon, G., Bellec, S., Guihenneuc-Joyaux, C., Clavel, J., Jougla, E., and Hemon, D. (2006) Excess mortality related to the August 2003 heat wave in France. *International Archives of Occupational and Environmental Health* 80(1): 16–24.

Frumkin, H., Hess, J., Luber, G., Malilay, J., and McGeehin, M. (2008) Climate change: the public health response. *American Journal of Public Health* 98(3): 435–445.

Ganguly, A.R., Steinhaeusser, K., Erickson, D.J. III, Branstetter, M., Parish, E.S., Singh, N., Drake, J.B., and Buja, L. (2009) Higher trends but larger uncertainty and geographic variability in 21st century temperature and heat waves. *Proceedings of the National Academies of Science* 106(37): 15,555–15,559.

Garssen, J., Harmsen, C., and de Beer, J. (2005) The effect of the summer 2003 heat wave on mortality in the Netherlands. *Eurosurveillance* 10(7): 165–168.

Gosling, S.N., Lowe, J.A., McGregor, G.R., Pelling, M., and Malamud, B.D. (2009) Associations between elevated atmospheric temperature and human mortality: a critical review of the literature. *Climatic Change* 92(3–4): 299–341.

Guest, C.S., Willson, K., Woodward, A.J., Hennessy, K., Kalkstein, L.S., Skinner, C., and McMichael, A.J. (1999) Climate and mortality in Australia: retrospective study, 1979–90, and predicted impacts in five major cities in 2030. *Climate Research* 13: 1–15.

Hajat, S. and Kosatsky, T. (2010) Heat-related mortality: a review and exploration of heterogeneity. *Journal of Epidemiology and Community Health* 64: 753–760.

Hajat, S., Armstrong, B., Baccini, M., Biggeri, A. Bisani, L., Russo, A., Paldy, A., Menne, B., and Kosatsky, T. (2006) Impact of high temperatures on mortality: is there an added heat wave effect? *Epidemiology* 17: 632–638.

Harlan, S.L., Brazel, A.J., Prashad, L., Stefanov, W.L., and Larsen, L. (2006) Neighborhood microclimates and vulnerability to heat stress. *Social Science and Medicine* 63: 2847–2863.

Harlan, S.L., Brazel, A.J., Jenerette, G.D., Jones, N.S., Larsen, L., Prashad, L., and Stefanov, W.L. (2008) In the shade of affluence: the inequitable distribution of the urban heat island. In *Equity and the Environment*, T. I. K. Youn (ed.). Research in Social Problems and Public Policy, volume 15. Bradford, UK: Emerald Group Publishing, 173–202.

Hayden, M.H., Brenkert-Smith, H., and Wilhelmi, O.V. (2011) Differential adaptive capacity to extreme heat: a Phoenix, Arizona case study. *Weather, Climate and Society* 3(4): 269–280.

Hayhoe, K., Cayan, D., Field, C.B., Frumhopff, P.C., Maurer, E.P., Miller, N.L., Moser, S.C., Schneider, S.H., Cahill, K.N., Cleland, E.E., Dale, L., Drapek, R., Hanemann, R.M., Kalkstein, L.S., Lenihan, J., Lunch, C.K., Neilson, R.P., Sheridan, S.C., and Verville, J.H. (2004) Emissions pathways, climate change, and impacts on California. *PNAS* 101: 12,422–12,427.

Helman, R.S. and Habal, R. (2009) Heatstroke. See http://emedicine.medscape.com/article/166320-overview (accessed 19 June 2012).

Holling, C.S. (1973) Resilience and stability of ecological systems. *Annual Review of Ecology and Systematics* 4: 1–23.

Holling, C.S. (1986) The resilience of terrestrial ecosystems: local surprise and global change. In *Sustainable Development of the Biosphere*, W.C. Clark and R.E. Munn (eds). Cambridge, UK: Cambridge University Press, 292–317.

Huang, W., Kan, H. and Kovats, S. (2010) The impact of the 2003 heat wave on mortality in Shanghai, China. *Science of the Total Environment* 408: 2418–2420.

ICLEI (International Council on Local Environmental Initiatives). (undated) *Hot Cities = Dirty Air, Cool Cities = Clean Air: How To Prevent Killer Heat Waves and Save Money at the Same Time.* See www.iclei.org/documents/Global/Progams/CCP/ICLEI_HotCities.pdf (accessed 19 June 2012).

Imhoff, M.L., Zhang, P., Wolfe, R.E., and Bounoua, L. (2010) Remote sensing of the urban heat island effect across biomes in the continental USA. *Remote Sensing of Environment* 114(3): 504–513.

IPCC (2007) *Climate Change 2007: The Physical Science Basis. Contribution of Working Group I to the Fourth Assessment Report of the Intergovernmental Panel on Climate Change.* Solomon, S., D. Qin, M. Manning, Z. Chen, M. Marquis, K.B. Averyt, M. Tignor and H.L. Miller (eds). Cambridge, UK: Cambridge University Press.

Jardine, D.S. (2007) Heat illness and heat stroke. *Pediatrics in Review* 28: 249–258.

Javed, M., Rahman, A., Viet Hoa, P., and Joshi, P.K. (2009) Assessment of night-time urban surface temperature–land use/cover relationship for thermal urban environment studies using optical and thermal satellite data TS 1G—remote sensing for sustainable development. 7th FIG Regional Conference Spatial Data Serving People: Land Governance and the Environment—Building the Capacity Hanoi, Vietnam, 19–22 October.

Johnson, D.P., Wilson, J.S., and Luber, G.C. (2009) Socioeconomic indicators of heat-related health risk supplemented with remotely sensed data. *International Journal of Health Geographics* 8: 57.

Kalkstein, L.S. and Greene, J.S. (2007) An analysis of potential heat-related mortality increases in US cities under a business-as-usual climate change scenario. See www.as.miami.edu/geography/research/climatology/Heat-Mortality_Report_FINAL.PDF (accessed 7 July 2012).

Kalkstein, L.S. and Smoyer, K.E. (1993), The impact of climate change on human health: some international implications. *Experimentia* 49: 44–64.

Karl, T.R. and Knight, R.W. (1997) The 1995 Chicago heat wave: how likely is a recurrence? *Bulletin of American Meteorological Society* 78: 1107—1119.

Kilbourne, E.M. (1997) Heat waves and hot environments. In *The Public Health Consequences of Disasters*, E.J. Noji (sd). Oxford, UK: Oxford University Press, 245–269.

Kinney, P.L., O'Neill, M.S., Bell, M.L., and Schwartz, J. (2008) Approaches for estimating effects of climate change on heat-related deaths: challenges and opportunities. *Environmental Science and Policy*, 11: 87–96.

Klinenberg, E. (2002) *Heat Wave: A Social Autopsy of Disaster in Chicago*. Chicago, IL: University of Chicago Press.

Kovats R. and Hajat, S. (2008) Heat stress and public health: a critical review. *Annual Reviews Public Health* 29: 41–55.

Kuttler, W. (2008) The urban climate: basic and applied aspects. In *Urban Ecology*, J.M. Marzluff, E. Shulenberger, W. Endlicher, M. Alberti, G. Bradley, C. Ryan, U. Simon, and C. ZumBrunnen (eds). New York, NY: Springer, 233–248.

Lee, C. (2002) Environmental justice: building a unified vision of health and the environment. *Environmental Health Perspectives* 110(2): 141–144.

Le Tertre, A., Lefranc, A., Eilstein, D., Declercq, C., Medina, S., Blanchard, M., Chardon, B., Fabre, P., Filleul, L., Jusot, J., Pascal, L., Prouvost, H., Cassadou, S., and Ledrans, M. (2006) Impact of the 2003 heat wave on all-cause mortality in 9 French cities. *Epidemiology* 17: 75–79.

Luber, G. and McGeehin, M. (2008) Climate change and extreme heat events. *American Journal of Preventative Medicine* 35(5): 429–435.

McCarthy, M.P., Best, M.J., and Betts, R.A. (2010) Climate change in cities due to global warming and urban effects. *Geophysical Research Letters* 37: L09705.

McGregor, G.R., Pelling, M., Wolf, T., and Gosling, S. (2007) *Social Impacts of Heat-Waves*. Science Report SC020061/SR6. Rotherham, UK: Environment Agency.

McMichael, A., Campbell-Lendrum, D.H., Corvalan, C.F. Ebi, K.L., Githeko, A., Scherago, J.D., and Woodward, A. (2003) *Climate Change and Human Health*. Geneva, Switzerland: WHO.

Medina-Ramon, M., Zanobetti, A., and Schwartz, J. (2006) Extreme temperature and mortality: assessing effect modification by personal characteristics and specific cause of death in a multicity case-only analysis. *Environmental Health Perspectives* 114: 1331–1336.

Meehl, G.A. and Tebaldi, C. (2004) More intense, more frequent, and longer lasting heat waves in the 21st century. *Science* 305: 994–997.

Michelozzi, P., de Donato, F., Bisanti, L., Russo, A., Cadum, E., DeMaria, M., D'Ovidio, M., Costa, G., and Perucci, C.A. (2005) The impact of the summer 2003 heat waves on mortality in four Italian cities. *European Surveillance* 10(7): pii–556. See www.eurosurveillance.org/ViewArticle.aspx?ArticleId=556 (accessed 19 June 2012).

Nasrallah, H.A., Brazel, A., and Balling Jr., R.C. (1990) Analysis of the Kuwait city urban heat island. *International Journal of Climatology* 10: 401–405.

NWS (NOAA National Weather Service). (2009) National Weather Service glossary. See www.weather.gov/glossary/index.php?letter=h (accessed 19 June 2012).

Oke, T.R. (1987) *Boundary Layer Climates*. New York, NY: Methuen.

Oleson, K.W. Bonan, G.B. and Feddema, J. (2010) The effects of white roofs on the global urban heat island. *Geophysical Research Letters* 37: L03701 (doi:10.1029/2009GL042194).

O'Neill, M.S. and Ebi, K.L. (2009) Temperature extremes and health: impacts of climate variability and change in the United States. *Journal of Occupational and Environmental Medicine* 51(1): 13–25.

O'Neill, M.S., Zanobetti, A., and Schwartz, J. (2003) Modifiers of the temperature and mortality association in seven US cities. *American Journal of Epidemiology* 157: 1074–1082.

O'Neill, M.S., Zanobetti, A., and Schwartz, J. (2005) Disparities by race in heat-related mortality in four US cities: the role of air conditioning prevalence. *Journal of Urban Health* 82: 191–197.

Parida, B.R., Oinam, B., Patel, N.R., Sharma, N., Kandwal, R., and Hazarika, M.K. (2008) Land surface temperature variation in relation to vegetation type using MODIS satellite data in Gujarat state of India. *International Journal of Remote Sensing* 29(14): 4219–4235.

Pascal, M., Laaidi, K., Ledrans, M., Baffert, E., Caserio-Schönemann, C., Le Tertre, A., Manach, J., Medina, S., Rudant, J., and Empereur-Bissonnet, P. (2005) France's heat health watch warning system. *International Journal of Biometeorology* 50(3): 144–153.

Patz, J.A., Campbell-Lendrum, D., Holloway, T., and Foley, J.A. (2005) Impact of regional climate change on human health. *Nature* 438 (17): 310–317.

Reid, C.E., O'Neill, M.S., Gronlund, C., Brines, S., Brown, D., Diez-Roux, A., and Schwartz, J. (2009) Mapping community determinants of heat vulnerability. *Environmental Health Perspectives* 117: 1730–1736.

Robinson, P.J. (2001) On the definition of a heat wave. *Journal of Applied Meteorology* 40:762–775.

Romero-Lankao, P., Wilhelmi, O., Cordova Borbor, M., Parra, D., Behrenz, E., and Dawidowski, L. (forthcoming) Health, development and climate in Latin American cities: old and new challenges. In *The Changing Environment for Human Security: New Agendas for Research, Policy, and Action*, K. O'Brien, L. Sygna, and J. Wolf, (eds). Oslo, Norway: GECHS.

Roth, M. (2007) Review of urban climate research in (sub)tropical regions. *International Journal of Climatology*, 27: 1859–1873.

Ruddell, D.M., Harlan, S.L., Grossman-Clarke, S., and Buyantuyev, A. (2010) Risk and exposure to extreme heat in microclimates of Phoenix, AZ. *Geotechnologies and the Environment* 2(2): 179–202.

de Sherbinin, A. (2008) Integration of poverty and remote sensing data. Paper presented at the Expert Group Meeting on Slum Identification Using Geo-Information Technology, 21–23 May, Enschede, Netherlands.

Schneider, A., Friedl, M.A., and Potere, D. (2009) A new map of global urban extent from MODIS data. *Environmental Research Letters* 4: Article ID044003.

Schuman, S.H. (1972) Patterns of urban heat-wave deaths and implications for prevention: data from New York and St Louis during July 1966. *Environmental Research* 5: 59–75.

Semenza, J.C., Rubin, C.H., Falter, K.H., Selanikio, J.D., Flanders, W.D., Howe, H.L., and Wilhelm, J.L. (1996) Heat-related deaths during the July 1995 heat wave in Chicago. *New England Journal of Medicine* 335: 84–90.

Semenza, J.C., Wilson, D.J., Parra, J., Bontempo, B.D., Hart, M., Sailor, D.J., and George, L.A. (2008) Public perception and behavior change in relationship to hot weather and air pollution. *Environmental Research* 107: 401–411.

Sheridan, S.C. and Kalkstein, L.S. (2004) Progress in heat watch-warning system technology. *Bulletin of the American Meteorological Society* 85: 1931–41.

Smit, B. and Wandell, J. (2006) Adaptation, adaptive capacity, and vulnerability. *Global Environmental Change* 16: 282–292.

Smoyer, K.E. (1998) Putting risk in its place: methodological considerations for investigating extreme event health risk. *Social Science and Medicine* 47: 1809–1824.

Tiangco, M., Lagmay, A.M.F., and Argete, J. (2008) ASTER-based study of the night-time urban heat island effect in Metro Manila. *International Journal of Remote Sensing* 29(10): 2799–2818.

Uejio, C., Wilhelmi, O., Samenow, J., Golden, J., and Mills, J. (2011) Intra-urban spatial patterns of societal vulnerability to extreme heat. *Health and Place* 17: 498–507.

UNDP (United Nations Development Programme). (2008) *Fighting Climate Change: Human Solidarity in a Divided World.* Human Development Report 2007/2008. Oxford, UK: Oxford University Press.

Vandentorren, S., Bretin, P., Zeghnoun, A., Mandereau-Bruno, L., Croisier, A., Cochet, C., Ribéron, J., Siberan, I., Declercq, B., and Ledrans, M. (2006) August 2003 heat wave in France: risk factors for death of elderly people living at home. *European Journal of Public Health* 16(6): 583–591.

Weeks, J., Hill, A.G., Stow, D.A., Getis, A., Agyei-Mensah, S., and Anarfi, J.K. (2005) Intra-urban differentials in poverty and health in Accra, Ghana. Paper presented at the IUSSP International Population Conference, Tours, France, 18–23 July.

White-Newsome, J., O'Neill, M. S., Gronlund, C., Sunbury, T. M., Brines, S.J., Parker, E., Brown, D.G., Rood, R. B., and Rivera, Z. (2009) Climate change, heat waves, and environmental justice: advancing knowledge and action. *Environmental Justice* 2(4): 197–205.

Wienert, U. and Kuttler, W. (2005) The dependence of the urban heat island intensity on latitude: a statistical approach. *Meteorolgische Zeitschrift* 14(5): 677–686.

Wilhelmi, O.V. and Hayden, M.H. (2010) Connecting people and place: a new framework for reducing urban vulnerability to extreme heat. *Environmental Research Letters* 5: 014021 (doi:10.1088/1748–9326/5/1/014021).

Wilhelmi, O.V., Purvis, K.L., and Harriss, R.C. (2004) Designing a geospatial information infrastructure for the mitigation of heat wave hazards in urban areas. *Natural Hazards Review* 5(3): 147–158.

Zhang, K., Wang, R., Shen, C., and Da, L. (2010) Temporal and spatial characteristics of the urban heat island during rapid urbanization in Shanghai, China. *Environmental Monitoring and Assessment* 169: 101–112.

Zhou, Y. and Shepherd, J.M. (2009) Atlanta's urban heat island under extreme heat conditions and potential mitigation strategies. *Natural Hazards* 52(3): 639–668.

13 Power, race, and the neglect of science

The HIV epidemics in sub-Saharan Africa

Eileen Stillwaggon and Larry Sawers

Conventional epidemiology recognizes that epidemics, like individual infections and injuries, arise in a specific ecological context. Epidemics are complex, contingent processes that develop through the interaction of characteristics of the pathogen, the vulnerable population, and environmental factors (Stillwaggon 2009). Whatever the proximate cause of infection or injury, epidemiology examines the multiple factors that influence both individual risk and the vulnerability of populations. In the last half-century, advances in microbiology and medical interventions as well as changes in the political and economic climate have shifted the emphasis of medical research to individual-level rather than population-level factors in theories of disease causation (Rose 1985; Schwartz and Carpenter 1999; Stillwaggon 2006; Susser 1985). Nevertheless, the standard epidemiological approach encompasses the ecological context of both individual and population risk.

Medical researchers who first studied acquired immune deficiency syndrome (AIDS) in sub-Saharan Africa in the 1980s looked at the epidemics through the lens of conventional epidemiology. Although isolated clusters or concentrated epidemics of human immunodeficiency virus (HIV) were found on all continents, primarily among men who have sex with men and needle-sharing drug users, in sub-Saharan Africa the syndrome was seen in men, women, and children. Poor immune status of the population, widespread infectious and parasitic diseases, poorly equipped and poorly staffed medical services, poor nutrition, and chronic exposure to contaminated water, soil, and food made the spread of HIV in Africa if not fully predictable at the start, at least comprehensible in an epidemiological framework. Because of the vulnerable population affected and the diverse presentations of the syndrome, it made sense to search not for one cause, but for the myriad factors that could contribute to the rapid spread of HIV.

By the late 1980s, in a radical departure from standard epidemiology, policymakers and most scholars narrowed their focus to a single proximate cause of HIV infection in sub-Saharan Africa: heterosexual behavior. Since then, the fundamental assumption dominating the AIDS-in-Africa discourse has been that something exceptional about sexual behavior drives the high prevalence of HIV in the region. We call that assumption and the policies that derive from it the behavioral paradigm.

HIV epidemics in sub-Saharan Africa are far larger than anywhere else. By 2009, HIV prevalence in the 9 countries of southern Africa averaged over 17 percent and in sub-Saharan Africa as a whole it was 5 percent. Outside Africa, HIV prevalence averaged only 0.5 percent (UNAIDS 2010). Two-thirds of people living with HIV are in sub-Saharan Africa.

This chapter examines the development of AIDS discourse on sub-Saharan Africa and its implications for policy, the exclusion of much scientific evidence in the prevailing theory of HIV causation, and some of the possible reasons for that departure from conventional epidemiology. We begin by discussing the behavioral paradigm that has dominated HIV policy research on sub-Saharan Africa. Then we examine its latest variant, the concurrency hypothesis. Next is a discussion of ecological explanations for the rapid spread of HIV in sub-Saharan Africa that reflect a standard epidemiological approach. Finally, we explore socio-political factors that have restricted the AIDS-in-Africa discourse to the behavioral paradigm, derailing research and policy from its original epidemiological foundation. Throughout the analysis, we examine the interplay of the ecological determinants of disease and the socio-political obstacles to understanding and controlling its spread.

The behavioral paradigm

In the 1980s in North America and Europe, there were substantial successes in reducing new infections of HIV, especially through promoting sexual behavior change in the gay community, but also by enforcing universal precautions in medical interventions, regulating blood-bank safety, and promoting needle-exchange programs. Transplanting interventions that had succeeded in North America and Europe to sub-Saharan Africa appeared to be a sensible response, but its effect was to displace the broader epidemiological approach of early HIV research in the region.

The ascendancy of the behavioral paradigm that replaced the use of standard epidemiological methods was enabled by long-held Western stereotypes of Africans. By the late 1980s, the presumption that some extraordinary characteristic or characteristics of heterosexual behavior in sub-Saharan Africa explained the high prevalence of HIV in the region was promoted in scholarly and popular literature and became widely accepted among researchers and policy makers (Caldwell and Caldwell 1987; Caldwell *et al.* 1989; Economist 2000; Ford 1994; Rushing 1995; UNFPA 1999). Influential and frequently cited works were characterized by sweeping statements about pan-African sexuality, supported, if at all, by anecdotal evidence dating from the early twentieth century to the 1970s. Through suggestive language and innuendo, they conveyed the impression of Africans bent on self-destruction because of cultural factors that differentiated them from everyone else (inter alia, Caldwell and Caldwell 1987; Caldwell *et al.* 1989). We discuss this issue at greater length below.

Anomalies in the behavioral paradigm

The hypothetical arguments of ethnographers and sociologists about "African sexuality" generated interest in empirical studies to evaluate their validity. Relatively few of those surveys of sexual behavior, however, were conducted in Latin America and Asia, reflecting the presumption that there was something unique about African sexuality. Surveys sponsored by the World Health Organization's Global Program on AIDS, the United States Agency for International Development's (USAID) Demographic and Health Surveys, and numerous others conducted by individual researchers and government statistical offices from 1989 to the present have produced a substantial body of survey research. They contradict the claim that behavior alone can explain the high prevalence of HIV in sub-Saharan Africa. The surveys demonstrate, on the contrary, that within every country studied there is considerable variation in sexual behavior—some people have many partners but most people have very few—and that prevalence of HIV across the globe does not correlate with patterns of risky behaviors, such as early sexual debut, extra- or premarital sex, number of partners in a year, or number of lifetime partners (for example, Cleland *et al.* 1995; Singh *et al.* 2000; Smith 1991; Turner 1993; UNAIDS 1999; Wellings *et al.* 2006).

Although survey research has found that risky sexual behavior is, if anything, less common in sub-Saharan Africa than elsewhere, HIV prevention policy for the region continues to focus almost exclusively on sexual behavior. While individuals who engage in risky sexual behavior are more likely to become infected with HIV within any population, the behavioral approach cannot explain prevalence of HIV in sub-Saharan African countries that is 10 to 250 times that of affluent countries in North America and Europe. The behavioral paradigm also provides no explanation for the greater vulnerability of individual sub-Saharan Africans compared to North Americans and Europeans.

In the 1990s, a second important empirical finding further undermined the validity of the behavioral paradigm. Researchers found that HIV is not a particularly virulent pathogen and that per-act transmission rates are quite low between otherwise healthy adults in heterosexual exposure (Boily *et al.* 2009; Chan 2005; Gray *et al.* 2001; Pilcher *et al.* 2007; Powers *et al.* 2008; Quinn *et al.* 2000; Wawer *et al.* 2005). Moreover, the infectivity of the person with HIV varies as the disease progresses. After an initial, brief spike in infectivity, transmission risk drops precipitously. From then until the onset of AIDS, the risk of transmission is so low that it calls into question the theoretical possibility of an HIV epidemic driven exclusively by heterosexual activity (Pinkerton *et al.* 2000). For an epidemic to be sustained, each infected individual must pass on the virus to at least one other person before death, but for most of the time that people have HIV, the risk of transmission is extremely low. To continue to characterize African sexuality as exceptionally dangerous, defenders of the behavioral paradigm had to respond to survey data that showed unexceptional sexual behavior in sub-Saharan Africa and evidence of the low infectivity of HIV in heterosexual exposures.

The attempt to rescue the behavioral paradigm

By the mid-2000s, the empirical evidence regarding sexual behavior in sub-Saharan Africa and HIV transmission rates, together with challenges to the behavioral paradigm by its few but persistent critics (inter alia, Packard and Epstein 1991; Stillwaggon 2000, 2001, 2002, 2003) led to a new variant of the behavioral paradigm. Conceding that most forms of risky sexual behavior are no more prevalent in sub-Saharan Africa than elsewhere, the promoters of the updated version of the paradigm argued that long-term overlapping partnerships, referred to as concurrency, are much more common in the region (Epstein 2007; Halperin and Epstein 2004, 2007). Furthermore, they argued that those concurrent partnerships spread HIV much more rapidly than sequential multiple partnering (Morris and Kretzschmar 1997, 2000). In a matter of a few years between 2004 and 2006, what had been occasional musings about concurrency in the 1990s by a handful of researchers (Hudson 1993; Watts and May 1992) were transformed into the new conventional wisdom. Since then, proponents of the concurrency hypothesis have continued to promote their cause (Epstein and Morris 2011; Mah and Halperin 2010; Morris 2010; Morris *et al.* 2009, 2010).

The explanation offered for the special efficiency with which concurrency is hypothesized to spread HIV is as follows: if a person infects his or her partner, but neither of them has another partner, then the infection is "trapped" until the partnership dissolves and new partnerships are formed. In contrast, overlapping partnerships can allow the formation of sexual networks through which infection can spread. For there to be an epidemic of HIV, however, those who become infected must have frequent sex with an uninfected partner during the first few months of infection because only during that brief period are per-sex-act transmission rates high enough to create or sustain a heterosexual epidemic (Epstein and Morris 2011; Halperin and Epstein 2004, 2007; Mah and Halperin 2010). Concurrency proponents argue that only long-term overlapping relationships provide the opportunity for sexual exposures frequent enough to generate the rapid spread of HIV.

Modeling concurrency

To show that concurrency spreads HIV more rapidly than other forms of multiple partnering, one must use mathematical models of epidemic dynamics. A pioneering model described in a series of articles published between 1996 and 2000 showed that concurrency could, under very unrealistic assumptions, spread HIV more effectively than serial monogamy (Kretzschmar and Morris 1996; Morris and Kretzschmar 1997, 2000). That model played a pivotal role in turning the concurrency hypothesis into conventional wisdom. The model was mathematically impressive, but also intimidating to those who had neither the temperament nor skills to unravel it. One version of the model found that "when one-half of the partnerships in a population are concurrent, the size of the epidemic after 5 years is 10 times as large as under sequential monogamy" (Morris and Kretzschmar 1997: 641). Astronomical rates of increase in HIV generated by dazzling but impenetrable mathematics convinced many that the concurrency hypothesis was correct and muted the criticism of others.

To generate that tenfold difference, however, the model assumes a per-sex-act risk of transmitting HIV that is approximately 100 times the generally accepted estimate (Boily *et al.* 2009; Chan 2005; Hollingsworth *et al.* 2008), rates of concurrency higher than ever observed for any country, and equal rates of concurrency for men and women (although every survey finds women reporting far less concurrency than men). The model assumed that everyone had sex with every partner every day (up to four times a day, every day). Common sense and a substantial body of research show that the assumption is fantastical. One study, for example, reports on national surveys in five sub-Saharan African countries and shows that between 32 and 59 percent of adults with regular partners report no sex with their regular partner in the previous month (Caraël 1995). Another study of nine countries in the region finds that the frequency of sex for women in their first year of marriage ranged from 2 to 4.4 times per month and was much less frequent in later years of marriage (Brewis and Meyer 2005). Numerous other studies show the same pattern of infrequent sex in long-term partnerships (for citations, see Sawers and Stillwaggon 2010a: 3–4).

With the help of Alan Isaac, we have created a model that is identical to Morris and Kretzschmar's with two exceptions. It replaces their unrealistic rates of transmission with rates provided by widely respected authorities on the subject (Hollingsworth *et al.* 2008). In addition, our model, unlike Morris and Kretzschmar's, incorporates vital dynamics, that is, births and deaths. Simulating our model with the more realistic transmission rate produces HIV prevalence that is the same at every level of concurrency including serial monogamy, not the tenfold difference that Morris and Kretzschmar found. When half of partnerships are concurrent, HIV prevalence rises from 0.05 percent to 45 percent in Morris and Kretzschmar's model. Our model if anything overstates the increase in HIV prevalence, since the inclusion of vital dynamics accelerates the spread of HIV. Nevertheless, with realistic transmission rates, HIV prevalence in our model increases from 0.05 percent to only 0.06 percent.

Morris and Kretzschmar describe their article as a "proof of concept," but the promoters of the concurrency hypothesis used the results of the model to demand an immediate reorientation of HIV prevention policy to address concurrency (Epstein 2007; Morris 2010). The hypothesis is now dictating HIV prevention policy in many countries of sub-Saharan Africa. As early as 2006, the official position adopted by SADC (the consortium of the 15 countries of southern Africa) included concurrency as one of two key drivers (along with lack of circumcision) of the HIV epidemics in the region (SADC 2006; Shelton 2009).

Recent efforts to incorporate realistic assumptions about parameter values into the Morris and Kretzschmar model show that concurrency cannot be an important driver of HIV epidemics in sub-Saharan Africa. Eaton *et al.* adapted the original Morris and Kretzschmar model using transmission rates that reflect the current consensus among researchers and incorporating the variation in infectivity at different stages of infection. That modeling shows that HIV epidemics move to extinction if the prevalence of concurrency (the average of men and women's rates) is 8 percent or less (point prevalence, that is, measured at a point in time; Eaton *et al.* 2010). Their results are a serious blow to the concurrency hypothesis since no survey using currently accepted questionnaire designs (UNAIDS Reference Group on Estimates 2009) has found any country-level point prevalence of concurrency in sub-Saharan Africa or anywhere higher than 8 percent. For example, recent surveys that were the first to use the methodology for measuring concurrency recommended by a group of experts convened by UNAIDS (UNAIDS Reference Group on Estimates 2009) found point prevalence of concurrency to be 5.1 percent in Lesotho (based on our calculations from the survey dataset provided by Measure DHS, www.measuredhs.com) and 3.7 percent in Malawi (National Statistical Office and ICF Macro 2011).

We have made a modification of Eaton *et al.*'s model to correct a problem that is found in most if not all other models of sexual network dynamics and HIV. Previous models simplified their analysis of sexual networks by assuming the same frequency of sex in every partnership. Models have assumed that people with two, three, or four partners have double, triple, or quadruple the frequency of sex as someone who has a single partner. Not only is that intuitively implausible, it contradicts all of the available evidence on coital frequency in multiple partnerships (Sawers *et al.* 2011). We incorporated empirically based assumptions about coital dilution—the lower frequency of sex in secondary partnerships—into Eaton *et al.*'s model. Doing so generates simulated HIV epidemics that move rapidly to extinction at any level of concurrency, including serial monogamy. Thus, properly constructed mathematical models show that the concurrency hypothesis cannot be correct because concurrency does not spread HIV more effectively than other forms of multiple partnering.

Concurrency in sub-Saharan Africa

For the concurrency hypothesis to be valid, concurrency not only has to spread HIV more effectively than other forms of partnering, it must also be substantially more common in sub-Saharan Africa than elsewhere. Our systematic review of studies published between 2004 and 2010 by the most prominent proponents of the concurrency hypothesis, finds that none of the studies they cite provides credible support for the proposition that concurrency is unusually high in sub-Saharan Africa (Sawers and Stillwaggon 2010a).

The proponents of the concurrency hypothesis whose works we examined repeatedly confuse data on concurrent and non-concurrent partnerships. They incorrectly report data from more than a third of the studies they cite. They compare data with different numerators (from different age brackets, for example) or different denominators (all adults, or those who are sexually experienced, or those who are sexually active). Most of the studies they cite cover cities or regions that are not representative of the country of which they are a part or are based on very small and/or non-random samples. Almost all of the studies they cite use definitions of concurrency that UNAIDS and the proponents of the hypothesis themselves argue are poor measures of concurrency and lead to overestimation (UNAIDS Reference Group on Estimates 2009). All of the errors, inaccuracies, imprecise statements, and obfuscation exaggerate the difference between concurrency in Africa and concurrency elsewhere.

In short, after 20 years of trying, the proponents of the concurrency hypothesis have failed to demonstrate that concurrency exposes people to an especially high risk of HIV infection or that concurrency is especially common in sub-Saharan Africa. The notion that exceptional sexual behavior in sub-Saharan Africa explains the HIV epidemics there is without empirical support. Acceptance of the behavioral paradigm, however, has blocked consideration of other factors that contribute to high rates of HIV transmission in sub-Saharan Africa.

Other factors influencing the spread of HIV

The exclusive focus on African sexual behavior abstracts from the many other factors that influence health and vulnerability to any disease for people in the region. The majority of children born to HIV-infected mothers do not acquire the virus at birth. The difference in vertical transmission risks between sub-Saharan Africa and high-income countries was an early clue to the importance of ecological factors in HIV transmission. In the mid-1990s, before anti-retroviral therapy was introduced to prevent mother-to-child transmission of HIV, about 40 percent of infants born to HIV-infected mothers in sub-Saharan Africa were infected at birth, about 25 percent in the US, and about 14 percent in Europe (Fowler and Rogers 1996). That was evidence that differences in immune status of both infected (the mother) and uninfected (the child) individuals had an important effect on HIV transmission risk.

Furthermore, the overwhelming majority of sex acts with an HIV-infected person do not lead to transmission. For most of the time that a person is infected with HIV, his or her risk of infecting a partner is less than 1 in 1000 sex acts if both partners are otherwise healthy. Even in the 1990s, it was known that sexual intercourse was a proximate, but not necessarily sufficient, cause of HIV transmission. Thus, focusing almost all research and policy resources on a sexual explanation for high HIV prevalence in sub-Saharan Africa was a radical departure from conventional evidence-based epidemiology that considers contributing factors. That diversion of research and policy funding continues to the present day.

While the behavioral paradigm has dominated HIV prevention policy and social science HIV research of the past 20 years, some medical researchers continued to employ the conventional epidemiological methodology, exploring the ecological context in which the AIDS epidemic was spreading in sub-Saharan Africa. As we demonstrate, a substantial body of scientific literature indicates that host and ecological factors play an important role in determining an individual's vulnerability to HIV infection and the contagiousness of HIV-infected partners (and mothers). Poor nutrition and parasitic and infectious diseases weaken the immune system and make people more vulnerable to infection with HIV as they would for any disease, however transmitted (Bentwich *et al.* 1995). Moreover, certain diseases, discussed below, are highly prevalent in Africa and provide more efficient transmission routes for HIV during heterosexual and vertical (mother-to-child) exposures.

Nutrition

From 1988 to 1998, when nascent or concentrated HIV epidemics developed into generalized epidemics in sub-Saharan Africa, 30 percent of the population of the region was malnourished (World Bank 1998). Malnutrition increases vulnerability to infectious and parasitic diseases generally, increases HIV viral load and viral shedding, and undermines the integrity of the skin and mucosa, thereby increasing sexual and vertical transmission of HIV (Beisel 1996; Chandra 1997; Fawzi and Hunter 1998; Friis and Michaelsen 1998; John *et al.* 1997; Landers 1996; Nimmagadda *et al.* 1998; Pelletier *et al.* 1995; Semba *et al.* 1994; Stillwaggon 2006). While poor nutrition is a serious problem in parts of Asia and much of Latin America, sub-Saharan Africa was the only world region in which protein and calorie consumption declined from 1970 to 1990 (UNDP 2000).

Malaria

More than 90 percent of all acute malaria infections occur in sub-Saharan Africa. Malaria increases viral load up to ten times for as much as seven weeks after an episode of fever, and that can double heterosexual transmission (Abu-Raddad *et al.* 2006; Hoffman *et al.* 1999; Whitworth *et al.* 2000). The World Health Organization estimates that there are over 200 million cases of malaria in Africa every year and 750,000 deaths (WHO 2008b: viii). An HIV-infected person could have elevated viral load for more than half of every year if repeatedly reinfected.

Schistosomiasis

Virtually all cases of urogenital schistosomiasis occur in Africa, and it afflicts more than 100 million people in the region (WHO 2008a). Schistosome worms (*Schistosoma hematobium*) and their eggs colonize the reproductive tract in men and women, causing inflammation, viral shedding, and genital ulcers that increase transmission of HIV (Attili *et al.* 1983; Feldmeier *et al.* 1995; Leutscher *et al.* 1998; Marble and Key 1995). Women with genital ulcers of schistosomiasis have three times the risk of being infected with HIV as women in the same village without genital ulcers of schistosomiasis (Kjetland *et al.* 2006). According to the World Health Organization (WHO), schistosomiasis is highly prevalent in every country in sub-Saharan Africa, including high HIV prevalence countries such as Botswana, Swaziland, Lesotho, South Africa, and Namibia, and in no other country except Algeria (WHO 2004). A recent double-blind, controlled trial found that treating ascariasis (caused by a soil-transmitted intestinal worm) in HIV-infected persons results in a statistically significant increase in CD4 counts (Walson *et al.* 2008). That suggests that a simple, inexpensive (2 US cents) and effective deworming medication could allow HIV-infected people to be healthier, reducing the risk of infecting their partner, and allowing postponement of antiretroviral therapy.

Non-sexual transmission

In addition to disease co-factors that can increase per-act transmission rates during heterosexual and vertical exposure, we must also recognize that non-sexual modes of transmission could play an especially important role in sub-Saharan Africa and among other poor populations. There are numerous, common medical blood exposures (for example, injections with unsterilized syringes, blood transfusions, catheter and intravenous placements, and internal obstetrical examinations) and non-medical blood exposures (for example, barbering and hairdressing, tattooing, scarification, injections given by non-medical personnel, and intravenous recreational drug use) that can potentially transmit HIV (Brewer *et al.* 2003; Deuchert and Brody 2006; Gisselquist 2008). Some have argued that civil war, high levels of violence, and associated sexual assaults on women accelerate the spread of HIV (Mworozi 1993; Serwadda *et al.* 1985). Nevertheless, they also disrupt the healthcare delivery system and thereby reduce iatrogenic and nosocomial transmission (infection spread by medical treatment or in a health-care facility), so the net effect is unclear (Gisselquist 2004). Even if each one of those possible non-sexual routes of transmission produces only a small share of new infections, together they would play an important role in the epidemics of sub-Saharan Africa and provide an explanation for why at the present time more women than men are infected.

By 2004, most of the information about the connection between HIV and malaria, schistosomiasis, helminthes, and malnutrition was available in the medical literature. There was also ample evidence that HIV transmission through the many kinds of blood exposures were common. Nevertheless, that evidence was ignored by most of the HIV/AIDS community. In light of the conventional epidemiological understanding of disease synergies and the evidence that interactions of specific parasites and infections increase vulnerability to and contagiousness of HIV, the exclusively behavioral focus of AIDS policy reveals a very simplistic notion of disease causation. Moreover, transmission of HIV between men is much more efficient than heterosexual transmission (Chan 2005), but the role of men having sex with men in sub-Saharan HIV epidemics has been almost completely ignored. The single-minded focus not just on sexual behavior but also on heterosexual behavior, and the refusal to consider the ecological environment of HIV in Africa (which increases the transmission efficiency of a virus that is normally not very infectious), have been a costly detour. That has caused a deadly delay in addressing the true drivers of HIV in sub-Saharan Africa and other impoverished populations.

AIDS policy became derailed, but why is it still off track?

We have argued that there is, at best, feeble evidence supporting the behavioral paradigm and that there are viable alternatives for which substantial scientific support is at hand. Why then does the HIV/AIDS research and policy community cling to the behavioral paradigm while ignoring more plausible explanations for sub-Saharan Africa's extraordinarily high HIV prevalence? We argue that, in the decades-long project of addressing AIDS in sub-Saharan Africa, politics won out over science. One of the most important aspects of political dominance is the ability to define problems and to define groups, both the Self and the Other. Westerners (that is, high-income countries) dominate international organizations and bilateral donor organizations in research and policy, and Western views hold sway even among elites in poor countries. Politically dominant organizations, including academia, have defined AIDS discourse.

Racial stereotyping

In the past, racism played an important role in providing the justification for slavery and then colonialism, and old ideas can linger in the public psyche long after they are directly useful. Racial stereotypes continue to pervade Western culture, casting their shadow over scholarship and public policy, even among persons who, on a conscious level, vigorously and sincerely oppose racial discrimination.

Gunnar Myrdal, a Swedish economist writing about race relations in the United States, observed that cultural influences "pose the questions we ask; influence the facts we seek; [and] determine the interpretation we give these facts" (Myrdal 1944: 92). He continued: "Biases in research … are not valuations attached to research but rather they permeate research … [and] insinuate themselves into research in all stages, from its planning to its final presentation" (Myrdal 1944: 1043). The influence of notions of "race" in both the popular mind and in the imagery of science is insidious and difficult to counter because so much of racial stereotyping is in the "unstated assumptions and unthinking responses" (Dubow 1995: 7), rather than in explicit postulates. That is aggravated by the tendency for both academic and journalistic writing about sub-Saharan Africa to consist of a "repertoire of amazing facts" (Coetzee 1988: 13). Writing about sub-Saharan Africans, popular and scholarly, almost always emphasizes how they are different from others, not their commonality with people everywhere. Given the presumption of difference with which authors, readers, reviewers, and editors begin, there is a much lower bar for what constitutes evidence regarding Africans, as long as what is written fits those preconceived notions.

Notions of racial difference pervade the social science literature on AIDS in Africa and were especially explicit during the first 15 years of the epidemic. No one uses the word race, but the notion enters into the discourse as "culture." Two of the most frequently cited anthropological works of the first two decades of the AIDS pandemic employ a foundational metaphor to convey the idea that modern-day sub-Saharan African fertility and sexual behavior choices derive from a religious world view that harks back to the dawn of humankind. The authors propose "a focus on Africa as the domain of *Homo ancestralis* … [to] explain many African anomalies" (Caldwell and Caldwell 1987: 410; see also Caldwell *et al.* 1989). Those articles never say that there are primordial genetic differences that set sub-Saharan Africans apart from everyone else and thus explain "African anomalies," but the use of species terminology and italicized Latin words in the metaphor "*Homo ancestralis*" inescapably makes that point.

The metaphor is especially effective in a discussion of sexual behavior because so much of the racial difference literature of the nineteenth and early twentieth centuries focused on the sexuality of sub-Saharan Africans (Dubow 1995; Gould 1981; Stepan 1982). Racial science in an earlier epoch and popular racial stereotypes that persist to the present day stress sexual differences between the races, and portray sub-Saharan Africans as exotic, strange, and even disturbing (Gilman 1985, 1990, 1992).

In 1989, Caldwell *et al.* used the *Homo ancestralis* metaphor to explain sub-Saharan African AIDS in a social context they had already characterized as primeval. The premise of their interpretation of both fertility preference (Caldwell and Caldwell 1987) and HIV prevalence (Caldwell *et al.* 1989) is that sub-Saharan Africans are so different and their belief system so ancient that they are inscrutable to the Western (read "modern") mind. Their articles were cited in hundreds of scholarly works and policy documents as though they had established their argument empirically. Many other works in the 1980s and 1990s made similar or even less subtle arguments purporting to explain the prevalence of HIV in sub-Saharan Africa by asserting unusual cultural practices or cultural idiosyncrasies for which there was no empirical evidence (Delius and Glaser 2001; Ford 1994; Forster 2001; Rushing 1995). In spite of the lack of evidence, the theme in much AIDS scholarship and policy literature remains that "Africans are not like everyone else."

In the early years of the HIV epidemics, the fear that a heterosexual epidemic could engulf North America and Europe reinforced the desire to posit a sub-Saharan African "Other." High-income countries had very serious epidemics of other sexually transmitted diseases, so the potential for a heterosexual epidemic of HIV seemed quite probable. As the dimensions of the sub-Saharan African epidemics grew increasingly clear, racial stereotypes provided assurance that HIV was still restricted to specific types of people, that is, to the "Other." Westerners could find security in the belief that the African HIV epidemics were in a faraway place and among people with very different sexual behavior. As Roger Chapman describes it, HIV was framed "in terms of the types of people who were getting it, rather than the ways they were becoming infected" (Chapman 2009).

Blatantly racial arguments in the scholarly literature on sub-Saharan African HIV epidemics are rare now in contrast to the 1990s. More common are works that begin from an assumption about sub-Saharan African sexual behavior that is essentially comparative without providing comparative data. The literature abounds with broad, unsupported assertions about presumed differences in behavior that resonate with Western stereotypes of sub-Saharan Africans. We are not suggesting that the proponents of the concurrency hypothesis are attempting to make a racial argument. But the question they pose, as do all proponents of the behavioral explanation of HIV in sub-Saharan Africa, is this: how are Africans different from everyone else?

One example of the characterization of Africans as fundamentally different from everyone else is the repeated emphasis on transactional sex, or the exchange of sex for material gain. Proponents of the concurrency hypothesis frequently use the notion of transactional sex to bolster their arguments about concurrent partnerships in sub-Saharan Africa (Epstein 2007; Halperin and Epstein 2007; Mah and Halperin 2010). Transactional sex, it is even claimed, is the "norm" in sub-Saharan Africa (Shelton 2009). Young women in sub-Saharan Africa are portrayed as poor, materialistic, and thus eager to trade sexual favors for gifts such as clothes, cosmetics, or a cell phone. It is never stated explicitly, but the necessary presumption for the argument to work is that sub-Saharan Africans (a billion people from 2000 ethnic groups) have attitudes condoning sex-for-gifts transactions that are different than elsewhere. The language used to make the argument, on its face, appears to emphasize the distinction between transactional sex and commercial sex work. But the only difference ever mentioned is that transactional sex sometimes involves gifts, not cash. In this way, the "norm" for sex in sub-Saharan Africa is collectively branded as prostitution. None of these discussions of transactional sex in sub-Saharan Africa is placed in a comparative context. Those who talk about transactional sex seem unaware that women receive gifts from lovers in every country on earth, including, for example, the practice of Valentine's Day in Western countries. The transactional sex discourse is one more way to make sub-Saharan Africans "the Other." It is about exceptionalizing their sexual behavior.

Western researchers, editors of academic journals, bilateral donor agencies, and international organizations have the power to define and delimit AIDS discourse in sub-Saharan Africa to the behavioral paradigm. That paradigm has abstracted from every other factor in the ecology of HIV dynamics, and all that remains in the explanation of high rates of HIV in sub-Saharan Africa is sexual behavior. The power to define AIDS discourse in behavioral terms is the most important part of determining research questions and policy responses. Because the question—how are Africans different from everyone else—seems so reasonable to most people, the behavioral paradigm and its latest variant, the concurrency hypothesis, became the conventional wisdom without a demand for credible empirical support. The answer to the question is not nearly as important as the question itself.

Institutional inertia

An ideology of racial superiority provided the justification for carving up Africa, Asia, and Latin America into colonies and spheres of influence. Moreover, in the late nineteenth and early twentieth centuries, the slowing birth rates of Europeans stoked fears in Europe and North America of being overwhelmed by people of color. Western stereotypes of people of color as strange, dangerous, and numerous were recast in the mid-twentieth century as the "population bomb." That emphasis on "too many births" encouraged Western aid to developing countries to focus on sex and reproduction rather than on the need for food, housing, clean water, health care, personal freedom and safety, economic stability, and education, all of which influence the demographic transition, morbidity, mortality, and fertility preference.

The political environment of the 1980s and 1990s further narrowed the development agenda in the United States to little more than population control, so organizations whose mission was to slow population growth played an increasingly prominent role in external assistance to developing countries. By the time that HIV prevention became an objective of US foreign aid policy, the behavioral paradigm was already firmly rooted in AIDS discourse for Europe and North America. Consequently, official and unofficial external assistance to HIV prevention efforts in sub-Saharan Africa focused almost exclusively on changing sexual behavior. Donors in the United States and elsewhere naturally turned to population-control organizations since they had decades of experience in talking about sex and condoms to sub-Saharan Africans. Except for those accustomed to promoting population control, discussing HIV was not within most people's comfort zone. Thus, population control organizations took the lead and ultimately were given wide latitude in shaping HIV prevention in sub-Saharan Africa.

Some sexual behaviors, such as having many partners, clearly raise one's risk of HIV infection. To the extent that the safe-sex message has been effective in encouraging behavior change (the experts do not agree), then the work of population-control organizations has played a role in slowing the spread of HIV in sub-Saharan Africa. But organizations whose institutional mission derives from their origins in population control necessarily view the people and the problem of HIV in sub-Saharan Africa in sexual and reproductive terms, rather than in epidemiological terms, or even more broadly in the context of social and economic determinants of health. That lent institutional momentum to the choice of sexual behavior change as the almost exclusive focus of HIV prevention policy.

An example is instructive: prevalence data suggested that international border crossings in sub-Saharan Africa were hot spots for HIV transmission because truckers were compelled to wait for as much as 10 days to clear customs. They found either that there were no hotels or that it was cheaper to visit prostitutes than pay for a room during their long stays at the border. A number of these border posts are, not surprisingly, in ecological zones with numerous disease vectors, since rivers form the borders in much of the region. Looking at the economic and ecological context, the obvious solution is to reduce costly and wasteful customs regulations that strangle trade and slow economic growth, as well as impeding cross-border trucking in ecological hotspots.

To resolve this source of epidemic spread of HIV at border posts, USAID engaged the resources of three population-control organizations that implement behavior change communication and condom distribution projects. Their solution naturally was behavior change and condom distribution. As stop-gap measures, neither behavior change nor condom distribution was a bad idea. Identifying the problem in such a limited way, however, merely created a permanent need for such programs. They ignored a critical opportunity for addressing a serious economic, political, and ecological problem in a sustainable and creative way. The choice of contract partners (and the power or latitude of USAID to choose those partners) reinforced the dominance of the sexual behavior paradigm that viewed sexual behavior as the principal driver of sub-Saharan Africa's epidemics to the exclusion of broader epidemiological and ecological considerations.

Path dependence and institutional interests

Institutional inertia is not limited to bilateral aid organizations and international population-control organizations that are beneficiaries of billions of dollars spent on HIV prevention programs in sub-Saharan Africa. There are also thousands of sub-Saharan African non-governmental organizations whose *raison d'être*, funding rationale, and daily activities, such as running training programs and focus groups, are based on the behavioral paradigm. In addition, national and subnational governments in sub-Saharan Africa have HIV control agencies, other governmental bureaus involved in HIV prevention, and whole divisions of ministries of health that channel the flow of dollars coming from government budgets and foreign donors. It is difficult for these organizations to consider programs outside the behavioral paradigm because of the institutional momentum that commonly characterizes all private and public bureaucracies. As one researcher who has interviewed most of the donor representatives and agency directors working on HIV prevention in Tanzania put it:

> An organization specialized in abstinence workshops or information, education and communication campaigning, for instance, is unlikely to be a suitable implementing agency for non-behavioural structural prevention measures [such as working on co-factor infections]. A shift in today's prevention strategy would thus entail at least partial reallocation of existing resources. Given the organizations' incomplete convertibility from one activity to another, this reallocation would imply taking funding away from some organizations and giving it to others. As a result, there is a political constituency for keeping preventive priorities unchanged, while, as of today, no clearly circumscribed constituency exists that would push for the adoption of structurally oriented prevention measures.
>
> (Hunsmann 2010)

Those bureaucratic fiefdoms with their thousands of jobs and millions of dollars in salaries then channel funds down to implementing partners. The creation of the Global Fund in 2002 and PEPFAR in 2003 produced a substantial acceleration of this avalanche of money, further cementing the power of vested interests in the behavioral paradigm. To accept that the paradigm is bankrupt would mean the loss of jobs, perks, and salaries of the many people whose careers and skills are linked to that paradigm (Hunsmann 2010).

That path dependence characterizes academic research as well. Researchers in universities in both high- and low-income countries have built their careers around elaborating and refining the behavioral paradigm. A social scientist who spends 20 years studying sexual behavior in sub-Saharan Africa cannot suddenly work in a laboratory studying malaria parasites and HIV replication. The behavioral paradigm is not just an intellectual concept; it is a multibillion dollar project. None of this is intended to say that people working in governmental and non-governmental organizations and academia are not sincerely, perhaps with great passion and energy, trying to turn back the HIV pandemic.

But institutional inertia allows the continued dominance of a faulty paradigm that ultimately undermines hope for success.

Conclusion

In summary, for a quarter of a century, HIV epidemics in sub-Saharan African have been defined almost entirely through the lens of heterosexual behavior despite the lack of evidence that it is a useful approach. There are other, more plausible explanations for the extraordinarily high prevalence of HIV in the region that derive from epidemiological methods and are supported by evidence. Although ecological explanations of HIV epidemics have a more solid basis in scientific evidence, they lack the political momentum of the behavior-in-Africa paradigm and thus they are for the most part either politely ignored or scornfully dismissed. The behavioral paradigm, which is organized around the portrayal of sub-Saharan Africans as different from everyone else, is starkly devoid of nuance. It comprehends neither the politics nor the ecology of HIV spread. It draws attention away from the differences between rich and poor countries in nutrition, clean water, waste disposal, and access to safe healthcare that are important factors in disease transmission but are less entertaining and lack the culturally confirming ring of racial and sexual stereotypes. The HIV/AIDS industry grew up around the behavioral paradigm and cannot easily change course. The dominance of the behavioral paradigm also shields the medical establishment from charges that it has unwittingly spread the infection through poor infection control procedures.

The focus of this chapter is on sub-Saharan Africa, where most people with HIV live. We argue that the region's nutritional and disease burdens are part of the ecological setting for sub-Saharan Africa's extraordinary HIV epidemics. Two of the diseases we point to as critical cofactor infections helping to spread HIV have unusually high burdens in sub-Saharan Africa. Although malaria is found in scores of countries, 85 percent of malarial morbidity and mortality occurs in sub-Saharan Africa where it infects more than a fifth of the population every year (WHO 2008a).

Almost everyone infected with *Schistosoma haematobium* lives in sub-Saharan Africa, and it afflicts more than 10 percent of the population of the region. Furthermore, HIV apparently originated in Africa; the epidemic was building in the region long before HIV was recognized by medical professionals, and long before it arrived in other developing countries. While those factors help to explain why sub-Saharan Africa is more affected than other regions, they do not explain the divergence in incidence between, for example, west Africa and eastern and southern Africa. HIV epidemics are complex, contingent processes with multiple interacting causes (Sawers and Stillwaggon 2010b; Stillwaggon 2009). No one has yet untangled completely the combination of factors that tipped the balance in sub-Saharan Africa, in particular in eastern and southern Africa, but the issues we raise in this chapter at least move the discussion in the appropriate direction.

This chapter has examined the socio-political reasons for the resilience of the behavioral paradigm despite its lack of empirical support, but there is reason for optimism that change is possible. Most people accept the behavioral paradigm because everyone else does. The end of any paradigm is brought about by unexplained anomalies: questions that cannot be answered within the existing intellectual framework (Kuhn 1962). There is a growing realization that HIV prevention policy has largely failed. In some countries in sub-Saharan Africa, HIV prevalence appears to have fallen, but there is no clear agreement among the experts that HIV prevention programs have actually played an important role in that decline. Furthermore, HIV prevalence has increased or remained essentially the same in more than half the countries in the region. The obvious failures and, at best, modest successes of HIV prevention measures are motivating many to search for a new understanding of what drives HIV epidemics in sub-Saharan Africa. Basing inquiry on sound epidemiological principles and asking the right questions about the ecology of risk are the essential first steps for turning back HIV epidemics in sub-Saharan Africa.

References

Abu-Raddad, L., Patnaik, P., and Kublin, J.G. (2006) Dual infection with HIV and malaria fuels the spread of both diseases in sub-Saharan Africa. *Science* 314: 1603–1606.

Attili, V.R., Hira, S., and Dube, M.K. (1983) Schistosomal genital granulomas: a report of 10 cases. *British Journal of Venereal Disease* 59: 269–272.

Beisel, W. (1996) Nutrition and immune function: overview. *Journal of Nutrition* 126: 2611S–2115S.

Bentwich, Z., Kalinkovicj, A., and Weisman, Z. (1995) Immune activation is a dominant factor in the pathogenesis of African AIDS. *Immunology Today* 16: 187–191.

Boily, M.C., Baggaley, R.F., Wang, L., Masse, B., White, R.G., Hayes, R.J., and Alary, M. (2009) Heterosexual risk of HIV–1 infection per sexual act: systematic review and meta-analysis of observational studies. *Lancet Infectious Diseases* 9(2): 118–129.

Brewer, D.D., Brody, S., Drucker, E., Gisselquist, D., Minkin, S.F., Potterat, J., Rothenberg, R.B., and Vachon, F. (2003) Mounting anomalies in the epidemiology of HIV in Africa: cry the beloved paradigm. *International Journal of STDs and AIDS* 14: 144–147.

Brewis, A. and Meyer, M. (2005) Marital coitus across the life course. *Journal of Biosocial Science* 37(4): 499–518.

Caldwell, J. and Caldwell, P. (1987) The cultural context of high fertility in sub-Saharan Africa. *Population and Development Review* 13(3): 409–437.

Caldwell, J., Caldwell, P., and Quiggin, P. (1989) The social context of AIDS in sub-Saharan Africa. *Population and Development Review* 15(2): 185–234.

Caraël, M. (1995) *Sexual Behavior Sexual Behavior and AIDS in the Developing World.* London, UK: Taylor and Francis and WHO: 75–123.

Chan, D.J. (2005) Factors affecting sexual transmission of HIV–1: current evidence and implications for prevention. *Current HIV Research* 3: 223–241.

Chandra, R.K. (1997) Nutrition and the immune system: an introduction. *American Journal of Clinical Nutrition* 66(2): 460S–463S.

Chapman, R. (2009) *Culture Wars: An Encyclopedia of Issues, Voices, and Viewpoints.* Armonk, NY: M. E. Sharpe.

Cleland, J., Ferry, B., and Caraël, M. (1995) Summary and conclusions. In *Sexual Behaviour and AIDS in the Developing World*, J. Cleland and B. Ferry (eds). London, UK: Taylor and Francis for the World Health Organization, 208–228.

Coetzee, J.M. (1988) *White Writing: On the Culture of Letters in South Africa*. New Haven, CT: Yale University Press.

Delius, P., and Glaser, C. (2001) Sexual socialisation in South Africa: a historical perspective. *African Studies* 60(2): 27–54. .

Deuchert, E. and Brody, S. (2006) The role of health care in the spread of HIV/AIDS in Africa: evidence from Kenya. *International Journal of STDs and AIDS* 17(11): 749–752.

Dubow, S. (1995) *Scientific Racism in Modern South Africa*. Cambridge, UK: Cambridge University Press.

Eaton, J., Hallett, T., and Garnett, G. (2010) Concurrent sexual partnerships and primary HIV infection: a critical interaction. *AIDS and Behavior* 15(4): 687–692.

Economist. (2000) South Africa's president and the plague. *The Economist* (25 May). See www.economist.com/node/334597 (accessed 19 June 2012).

Epstein, H. (2007) *The Invisible Cure*. New York, NY: Farrar, Strauss and Giroux.

Epstein, H. and Morris, M. (2011) Concurrent partnerships and HIV: an inconvenient truth. *Journal of the International AIDS Society* 14: 1–11.

Fawzi, W.W. and Hunter, D.J. (1998) Vitamins in HIV disease progression and vertical transmission. *Epidemiology* 9(4): 457–466.

Feldmeier, H., Poggensee, G., Krantz, I., and Helling-Giese, G. (1995) Female genital schistosomiasis: new challenges from a gender perspective. *Tropical and Geographical Medicine* 47(supplement 2): S2–15.

Ford, N. (1994) Cultural and developmental factors underlying the global pattern of the transmission of HIV/AIDS. In *Health and Development*, D. Philips and Y. Verhasselt (eds). London, UK: Routledge, 83–96.

Forster, P. (2001) AIDS in Malawi: Contemporary discourse and cultural continuities. *African Studies* 60(2): 245–261.

Fowler, M. and Rogers, M. (1996) Overview of perinatal HIV infection. *Journal of Nutrition* 126: 2602S–2607S.

Friis, H. and Michaelsen, K.F. (1998) Micronutrients and HIV infections: a review. *European Journal of Clinical Nutrition* 52: 157–163.

Gilman, S. (1985) *Difference and Pathology: Stereotypes of Sexuality, Race, and Madness*. Ithaca, NY: Cornell University Press.

Gilman, S. (1990) "I'm down on whores": race and gender in Victorian London. In *Anatomy of Racism*, D. Goldberg (ed.). Minnesota, MN: University of Minnesota Press, 146–170.

Gilman, S. (1992) Black bodies, white bodies: toward an iconography of female sexuality in the late nineteenth-century arts. In *"Race," Culture and Difference*, J. Donald and A. Rattansi (eds). London, UK: Sage, 171–197.

Gisselquist, D. (2004) Impact of long-term civil disorders and wars on the trajectory of HIV epidemics in sub-Saharan Africa. *Journal of Social Aspects of HIV/AIDS* 1(2): 114–127.

Gisselquist, D. (2008) *Points to Consider: Responses to HIV/AIDS in Africa, Asia, and the Caribbean*. London, UK: Adonis and Abbey.

Gould, S.J. (1981) *The Mismeasure of Man*. New York, NY: W. W. Norton.

Gray, R.H., Wawer, M.J., Brookmeyer, R., Sewankambo, N.K., Serwadda, D., Wabwire-Mangen, F., Lutalo, T., Li, X., vanCott, T., Quinn, T.C., and Quinn, T.C. (2001) Probability of HIV–1 transmission per coital act in monogamous, heterosexual, HIV–1-discordant couples in Rakai, Uganda. *Lancet* 357(9263): 1149–1153.

Halperin, D. and Epstein, H. (2004) Concurrent sexual partnerships help to explain Africa's high HIV prevalence: implications for prevention. *Lancet* 364(9428): 4–6.

Halperin, D. and Epstein, H. (2007) Why is HIV prevalence so severe in southern Africa? *Southern African Journal of HIV Medicine* (March): 19–25.

Hoffman, I.F., Jere, C.S., Taylor, T.E., Munthali, P., Dyer, J. R., Wirima, J., Rogerson, S.J., Kumwenda, N., Eron, J.J., Fiscus, S.A., Chakraborty, H., Taha, T.E., Cohen, M.S., and Molyneux, M. E. (1999) The effect of *Plasmodium falciparum* malaria on HIV–1 RNA blood plasma concentration. *AIDS* 13(4): 487–494.

Hollingsworth, T.D., Anderson, R., and Fraser, C. (2008) HIV–1 transmission, by stage of infection. *Journal of Infectious Diseases* 198: 687–693.

Hudson, C.P. (1993) Concurrent partnerships could cause AIDS epidemics. *International Journal of STD and AIDS* 4(5): 249–253.

Hunsmann, M. (2010) Policy hurdles to addressing structural drivers of HIV/AIDS—a case study of Tanzania. Paper presented at the Annual Conference of the Norwegian Association for Development Research, Oslo, Norway, 25–26 November.

John, G.C., Nduati, R.W., Mbori-Ngacha, D., Overbaugh, J., Welch, M., Richardson, B.A., Ndinya-Achola, J., Bwayo, J., Krieger, J., Onyango, F., and Kreiss, J.K. (1997) Genital shedding of human immunodeficiency virus type 1 DNA during pregnancy: association with immunosuppression, abnormal cervical or vaginal discharge, and severe vitamin A deficiency. *Journal of Infectious Diseases* 175(1): 57–62.

Kjetland, E.F., Ndhlovu, P.D., D., P., Gomo, E., Mduluza, T., Midzi, N., Gwanzura, L., Mason, P.R., Sandvik, L., Friis, H., and Gundersen, S.G. (2006) Association between genital schistosomiasis and HIV in rural Zimbabwean women. *AIDS* 20(4): 593–600.

Kretzschmar, M. and Morris, M. (1996) Measures of concurrency in networks and the spread of infectious disease. *Mathematical Bioscience* 133(2): 165–195.

Kuhn, T. (1962) *The Structure of Scientific Revolutions*. Chicago, IL: University of Chicago Press.

Landers, D. (1996) Nutrition and immune function II: maternal factors influencing transmission. *Journal of Nutrition* 126: 2637S–2640S.

Leutscher, P., Ravaoalimalala, V.E., Raharisolo, C., Ramarokoto, C.E., Rasendramino, M., Raobelison, A., Vennervald, B., Esterre, P., and Feldmeier, H. (1998) Clinical findings in female genital schistosomiasis in Madagascar. *Tropical Medicine and International Health* 3(4): 327–332.

Mah, T.L. and Halperin, D. T. (2010) Concurrent sexual partnerships and the HIV epidemics in Africa: evidence to move forward. *AIDS and Behavior* 14: 11–16.

Marble, M. and Key, K. (1995) Clinical facets of a disease neglected too long. *AIDS Weekly Plus* (7 August): 16–19.

Morris, M. (2010) Barking up the wrong evidence tree. Comment on Lurie & Rosenthal, "Concurrent partnerships as a driver of the HIV epidemic in sub-Saharan Africa? The evidence is limited". *AIDS and Behavior* 14(1): 31–33; discussion 34–37.

Morris, M. and Kretzschmar, M. (1997) Concurrent partnerships and the spread of HIV. *AIDS* 11(5): 641–648.

Morris, M. and Kretzschmar, M. (2000) A microsimulation study of the effect of concurrent partnerships on the spread of HIV in Uganda. *Mathematical Population Studies* 8(2): 109–133.

Morris, M., Kurth, A., Hamilton, D., Moody, J., and Wakefield, S. (2009) Concurrent partnerships and HIV prevalence disparities by race: linking science and public health practice. *American Journal of Public Health* 99(6): 1023–1031.

Morris, M., Epstein, H., and Wawer, M. (2010) Timing is everything: international variations in historical sexual partnership concurrency and HIV prevalence. *PLoS One* 5(11): 1–8.

Mworozi, E.A. (1993) AIDS and civil war: a devil's alliance: dislocation caused by civil strife in Africa provides fertile ground for the spread of HIV. *AIDS Analysis Africa* 3(6): 8–10.

Myrdal, G. (1944) *An American Dilemma: The Negro Problem and Modern Democracy*. New York, NY: Harper and Brothers.

National Statistical Office and ICF Macro. (2011) *Malawi Demographic and Health Survey 2010*. Calverton, MD: Zomba.

Nimmagadda, A., O'Brien, W., and Goetz, M. (1998) The significance of vitamin A and carotenoid status in persons infected by the human immunodeficiency virus. *Clinical Infectious Diseases* 26: 711–718.

Packard, R. and Epstein, P. (1991) Epidemiologists, social scientists, and the structure of medical research on AIDS in Africa. *Social Science and Medicine* 33(7): 771–794.

Pelletier, D.L., Frongillo, E.A., Schroeder, D.G., and Habicht, J.P. (1995) The effects of malnutrition on child mortality in developing countries. *Bulletin of the World Health Organization* 73: 443–448.

Pilcher, C.D., Joaki, G., Hoffman, I.F., Martinson, F.E., Mapanje, C., Stewart, P.W., Powers, K.A., Galvin, S., Chilongozi, D., Gama, S., Price, M.A., Fiscus, S.A., and Cohen, M.S. (2007) Amplified transmission of HIV–1: comparison of HIV–1 concentrations in semen and blood during acute and chronic infection. *AIDS* 21(13): 1723–1730.

Pinkerton, S.D., Abramson, P.R., Kalichman, S.C., Catz, S.L., and Johnson-Masotti, A.P. (2000) Secondary HIV transmission rates in a mixed-gender sample. *International Journal of STDs and AIDS* 11: 38–44.

Powers, K.A., Poole, C., Pettifor, A.E., and Cohen, M.S. (2008) Rethinking the heterosexual infectivity of HIV–1: a systematic review and meta-analysis. *Lancet Infectious Diseases* 8(9): 553–563.

Quinn, T.C., Wawer, M.J., Sewankambo, N., Serwadda, D., Li, C., Wabwire-Mangen, F., Meehan, M.O., Lutalo, T., and Gray, R.H. (2000) Viral load and heterosexual transmission of human immunodeficiency virus type 1. *New England Journal of Medicine* 342(13): 921–929.

Rose, G. (1985) Sick individuals and sick populations. *International Journal of Epidemiology* 14: 32–38.

Rushing, W. (1995) *The AIDS Epidemic: Social Dimensions of an Infectious Disease*. Boulder, CO: Westview.

SADC. (2006) *Expert Think Tank Meeting on HIV Prevention in High-Prevalence Countries in Southern Africa*. Report, 10–12 May. Maseru, Lesotho: Southern African Development Community.

Sawers, L. and Stillwaggon, E. (2010a) Concurrent sexual partnerships do not explain the HIV epidemics in Africa: a systematic review of the evidence. *Journal of the International AIDS Society* 13(34): 1–23.

Sawers, L. and Stillwaggon, E. (2010b) Understanding the southern African "anomaly": poverty, endemic disease and HIV. *Development and Change* 41(2): 195–224.

Sawers, L., Isaac, A.G., and Stillwaggon, E. (2011) HIV and concurrent sexual partnerships: modelling the role of coital dilution. *Journal of the International AIDS Society* 14(44): 1–9.

Schwartz, S. and Carpenter, K. (1999) The right answer for the wrong question: consequences of type III error for public health research. *American Journal of Public Health* 89: 1175–1179.

Semba, R., Miotti, P., Chiphangwi, J.D., Saah, A.J., Canner, J.K., Dallabetta, G.A., and Hoover, D.R. (1994) Maternal vitamin A deficiency and mother-to-child transmission of HIV–1. *Lancet* 343: 1593–1597.

Serwadda, D., Sewankambo, N.K., Carswell, J.W., Bayley, A.C., Tedder, R.S., and Weiss, R.A. (1985) Slim disease: a new disease in Uganda and its association with HTLV-III infection. *Lancet* 2(8460): 849–852.

Shelton, J. D. (2009) Why multiple sexual partners? *Lancet* 374: 367–369.

Singh, S., Wulf, D., Samara, R., and Cuca, Y. (2000) Gender differences in the timing of first intercourse: data from 14 countries. *International Family Planning Perspectives* 26(1): 21–28, 43.

Smith, T.W. (1991) Adult sexual behavior in 1989: number of partners, frequency of intercourse and risk of AIDS. *Family Planning Perspectives* 23(3): 102–107.

Stepan, N.L. (1982) *The Idea of Race in Science: Great Britain 1800–1960*. Hamden, CT: Archon Books.

Stillwaggon, E. (2000) HIV transmission in Latin America: comparisons with Africa and policy implications. *South African Journal of Economics* 68(5): 985–1011.

Stillwaggon, E. (2001) AIDS and poverty in Africa. *The Nation* 272(20): 22–25.

Stillwaggon, E. (2002) HIV/AIDS in Africa: fertile terrain. *Journal of Development Studies* 38(6): 1–22.

Stillwaggon, E. (2003) Racial metaphors: interpreting sex and AIDS in Africa. *Development and Change* 34(5): 809–832.

Stillwaggon, E. (2006) *AIDS and the Ecology of Poverty*. New York, NY: Oxford University Press.

Stillwaggon, E. (2009) Complexity, cofactors, and the failure of AIDS policy in Africa. *Journal of the International AIDS Society* 12(12): 1–9.

Susser, M. (1985) Epidemiology in the United States after World War II: the evolution of technique. *Epidemiological Review* 7: 147–177.

Turner, R. (1993) Landmark French and British studies examine sexual behavior, including multiple partners, homosexuality. *Family Planning Perspectives* 25(2): 91–92.

UNAIDS. (1999) Fact sheet on differences in HIV spread in four African cities. See http://data.unaids.org/Publications/IRC-pub03/lusaka99_en.html (accessed 19 June 2012).

UNAIDS. (2010) *Global Report: UNAIDS Report on the Global AIDS Epidemic, 2010*. Geneva, Switzerland: Joint United Nations Programme on HIV/AIDS.

UNAIDS Reference Group on Estimates. (2009) *Consultation on Concurrent Sexual Partnerships*. Nairobi, Kenya: UNAIDS.

UNDP. (2000) *Human Development Report 2000*. See http://hdr.undp.org/en/reports/global/hdr2000 (accessed 19 June 2012).

UNFPA. (1999) *AIDS Update*. New York, NY: United Nations Population Fund.

Walson, J.L., Otieno, P.A., Mbuchi, M., Richardson, B.A., Lohman-Payne, B., Macharia, S.W., Overbaugh, J., Berkley, J., Sanders, E.J., Chung, M.H., and John-Stewart, G.C. (2008) Albendazole treatment of HIV–1 and helminth co-infection: a randomized, double-blind, placebo-controlled trial. *AIDS* 22(13): 1601–1609.

Watts, C.H. and May, R.M. (1992) The influence of concurrent partnerships on the dynamics of HIV/AIDS. *Mathematical Biosciences* 108(1): 89–104.

Wawer, M.J., Gray, R.H., Sewankambo, N.K., Serwadda, D., Li, X., Laeyendecker, O., Kiwanuka, N., Kigozi, G., Kiddugavu, M., Lutalo, T., Nalugoda, F., Wabwire-Mangen, F., Meehan, M.P., and Quinn, T.C. (2005) Rates of HIV–1 transmission per coital act, by stage of HIV–1 infection, in Rakai, Uganda. *Journal of Infectious Diseases* 191(9): 1403–1409.

Wellings, K., Collumbien, M., Slaymaker, E., Singh, S., Hodges, Z., Patel, D., and Bajos, N. (2006) Sexual behaviour in context: a global perspective. *Lancet* 368(9548): 1706–1728.

Whitworth, J., Morgan, D., Quigley, M., Smith, A., Mayanja, B., Eotu, H., Omoding, N., Okongo, M., Malamba, S., and Ojwiya, A. (2000) Effect of HIV–1 and increasing immunosuppression on malaria parasitaemia and clinical episodes in adults in rural Uganda: a cohort study. *Lancet* 356(9235): 1051–1056.

WHO. (2004) Disease and injury country estimates, burden of disease, table 6: age-standardized DALYs per 100,000 by cause, and member state, 2004. See www.who.int/healthinfo/global_burden_disease/GBD_report_2004update_full.pdf (accessed 1 July 2012).

WHO. (2008a) Current estimated total number of individuals with morbidity and mortality due to *Schistosomiasis haematobium* and *S. mansoni* infection in sub-Saharan Africa 2008. See www.who.int/schistosomiasis/epidemiology/table/en/index.html (accessed 8 November 2008).

WHO. (2008b) *World Malaria Report 2008*. Geneva, Switzerland: World Health Organization.

World Bank. (1998) Nutritional status and poverty in sub-Saharan Africa: findings. Africa Region Number 108 See http://documents.worldbank.org/curated/en/1998/04/12866197/nutritional-status-poverty-sub-saharan-africa (accessed 1 July 2012).

14 Disease as shock, HIV/AIDS as experience

Coupling social and ecological responses in sub-Saharan Africa

Brian King

Introduction

The spread and varied impacts of HIV/AIDS tragically demonstrate the complex and reciprocal relationships between the biophysical and socio-political dimensions of human health. Within sub-Saharan Africa, academic and policy research has worked to detail the consequences of the disease for demographic patterns, national economies, and gender dynamics (Drimie 2003; de Waal and Whiteside 2003; Love 2004; Negin 2005; Barnett and Whiteside 2006; Masanjala 2007; UNAIDS 2008; Bolton and Talman 2010). HIV/AIDS has been called a "major threat to development, economic growth and poverty alleviation in much of Africa" (Whiteside 2002: 313) and "the major development issue facing sub-Saharan Africa" (Drimie 2003: 647). Current studies predict a dire future for agricultural production suggesting that increasing food insecurity will make human populations even more vulnerable to infection and reduce the lifespan of infected individuals (de Waal and Whiteside 2003; Bolton and Talman 2010). Other work has documented the livelihood impacts for individuals and households suffering from the disease while considering its diverse trajectories for economies, extractive industries, and natural resource dependencies (Love 2004; Negin 2005; Barnett and Whiteside 2006; Masanjala 2007). In addition to the substantial social impacts, recent research has shown that the spread of HIV/AIDS is also transforming ecological systems, either in terms of intensifying pressures upon the natural resource base (Hunter *et al.* 2008; Kaschula 2008) or threatening ecosystem functioning (Aldhous 2007; McGarry and Shackleton 2009).

Even with these expansive impacts, HIV/AIDS has proven to be a perversely unequal disease, targeting specific demographic groupings, economic sectors, and biophysical systems (Baylies 2002; Stillwaggon 2006; see also Chapter 13 of the present volume). Understanding the contextually specific, spatially variable, and temporally variable impacts of the disease has been limited by at least four features of existing research.

First, previous scholarship has tended to rely upon national data sets while concentrating upon the impacts of the disease on income and agricultural production (Bachmann and Booysen 2004; Lemke 2005; Barnett and Whiteside 2006; Bolton and Talman 2010). Although essential in understanding broad patterns for social and ecological systems, this focus has reduced attention to the contextual and localized dimensions of the disease. As Hunter *et al.* (2008: 104) explain, fuller understandings of the links between the disease and the natural environment "has been hampered by a shortage of HIV/AIDS data at the individual and household level."

A second feature of existing research has been an emphasis upon the effects following adult mortality that, while critically important, can obscure some of the challenges for those living with HIV, either in terms of securing resources for livelihood production (Barnett and Blaikie 1992), seeking access to antiretroviral medications (Jones 2005), or overcoming social and cultural stigmas attached to the disease (Campbell 2003; Posel *et al.* 2007).

Third, there remains a tendency to focus upon the household as the unit of analysis for much of the existing research, which can reduce attention to the intra-household, gendered, and familial relationships that directly shape the impacts of HIV/AIDS and the specific ways that populations respond to disease transmission and adult mortality (Baylies 2002; Murphy *et al.* 2005). As will be detailed in this chapter, because families and communities in various regions of sub-Saharan Africa remain dependent upon a diverse set of resources to generate income and meet subsistence needs, the impacts of disease must be understood within a gamut of social processes, including the maintenance of land, collection of natural resources, gender and cultural dynamics, and regional political economies. Additionally, the long-term effects of the disease for social and ecological systems means that research and policy must attend to those affected, as well as infected, by HIV/AIDS.

Fourth, some of the existing research lacks longitudinal data that would uncover some of the long-wave impacts of HIV/AIDS for social and ecological systems. In their review of the research on the interactions between HIV/AIDS and the natural environment, Bolton and Talman (2010) make this same point, concluding that localized and detailed assessments of these relationships remain largely anecdotal with few studies addressing the environmental and ecosystem impacts of natural resource use over extended time periods. As Bolton and Talman (2010: 27) suggest, "snapshot information gives a quick glimpse of issues, but without long-term follow-up the view is likely distorted, especially when dealing with ecological, health-related, and socio-economic conditions that are in a state of flux."

The intention of this chapter is to provide a review of the literature on HIV/AIDS in sub-Saharan Africa, concentrating upon major themes employed in understanding the socio-political and ecological dimensions of human health. The first section of the chapter reviews some of the existing work on the social impacts of HIV/AIDS, focusing in particular upon how it has been theorized within livelihood studies, economics, and public health. As I work to show, a feature of much of this research is the reification of the idea of disease as a shock, either to households or livelihoods, or drawing upon the economic tradition of viewing shocks to economic systems (Davies 1996; Ellis 2000; Dercon *et al.* 2005; Kgathi *et al.* 2007). Even though HIV/AIDS has been shown to have unique features (Baylies 2002; Hosegood *et al.* 2007), many of these studies liken the disease to other shocks, such as a famine or flood, in being a large, unpredictable, and irregular disturbance that destabilizes social and ecological systems. The second section provides a review of some of the research on the ecological impacts of HIV/AIDS, which similarly draws upon the idea of a shock in examining how ecosystems and natural resource management are disrupted and negatively impacted by the disease. The goal of this review is to outline key features of existing research because these conceptualizations guide interventions by various actors and institutions. As I work to show in this chapter, the discursive framings of HIV/AIDS have material impacts for both social and ecological systems being reshaped by the disease, as well as the policy interventions that arise in response.

The third section of the chapter argues that theorizing HIV/AIDS a shock might obscure some of the unique dimensions of the disease, and perhaps illness more generally, for social and ecological systems. While there is a logic to theorizing disease as a shock, I argue that this emphasis has produced particular, and in some cases misleading, understandings of human health, especially within the context of the HIV/AIDS epidemic. Unlike other shocks or crises, the effects of HIV/AIDS have been shown to occur over a longer time period, are gradual and incremental, and are generally uneven within communities and regions (Barnett and Blaikie 1992; Love 2004; Hosegood *et al.* 2007; Fassin 2007). Drawing upon scholarship that seeks to address some of the methodological constraints of previous research by attending to the local and contextually specific dimensions of the disease (Campbell 2003), its intra-household and familial features (Madhavan 2007; Schatz and Ogunmefun 2007), and the effects following HIV transmission (Hosegood *et al.* 2007), I argue that the spatial and temporal variabilities of HIV/AIDS warrant renewed consideration. Additionally, given recent improvements in the availability of anti-retroviral therapy (ART)[1] within sub-Saharan Africa, it is possible for more people to live with the disease for extended periods of time (World Health Organization *et al.* 2010). As a result of these circumstances, this chapter argues that HIV/AIDS might be better theorized not as a shock that is large, unpredictable, and irregular, but as a health experience that is spatially and temporally *dynamic*, *disproportionate*, and *dispersed.* I conclude by suggesting how theorizing HIV/AIDS as a coupled socio-ecological experience can contribute to emerging research on human health within the social and natural sciences.

HIV/AIDS as shock to social systems

Research from within geography, agrarian studies, rural sociology, development studies, and related disciplines have drawn upon the concept of a livelihood to understand the complex and reciprocal relationships between social and ecological systems (Long 1984; Chambers 1987, 1997; Scoones 1998; Bebbington 1999, 2000; Ellis 2000; McSweeney 2004; de Haan and Zoomers 2005; King 2011). While differentially understood and operationalized within these fields, livelihoods have been shown to have utility for examining economic neoliberalization, the integration of rural areas into external markets and networks, or the processes shaping social and environmental change across time and space (King 2011). Drawing upon the sustainable livelihoods framework (Scoones 1998; Ashley and Carney 1999), livelihoods are often understood as "the capabilities, assets (including both material and social resources) and activities required for a means of living" (Scoones 1998: 5). Alternatively, Bebbington (1999: 2022) presents a livelihood framework to understand not only survival and adaptation of social actors, but also their power "to act and to reproduce, challenge or change the rules that govern the control, use and transformation of resources." Other work has advanced livelihood styles and pathways (de Haan and Zoomers 2005), livelihood strategies (Long 1984; McSweeney 2004), and livelihood mapping (Carter and May 1999; Kristjanson *et al.* 2005). Regardless of the particular framework employed, livelihoods research has been effective at demonstrating how access to resources and decision-making are simultaneously enabled and constrained by various processes working across spatial and temporal scales.

Much of the research on livelihoods can be classified by its focus upon how social actors are able to access a diverse array of assets, which are theorized as stocks of capital that vary depending upon political economic context and opportunity. The five main categories of capital assets are natural, physical, human, financial, and social capital (see Chapter 3 for an expanded discussion). As Ellis (2000) notes, natural capital is the stocks of the natural resource base (land, water, trees) while physical capital are assets created by economic production activities such as infrastructure, tools, and agricultural technologies. Human capital is the educational level and health status of individuals and populations that facilitate livelihood opportunities and decision-making. Financial capital refers to stocks of cash or credit, and social capital is routinely defined as the social networks and trust operating between individuals and communities. Different livelihood frameworks, whether capitals and capabilities (Bebbington 1999) or sustainable livelihoods (Scoones 1998; Ashley and Carney 1999), share an inclination to examine the availability of varied mixes of capital assets to social actors that can be used to meet basic needs or generate income for investment. Bebbington (1999) asserts the need to address the ways in which individuals combine and transform these capital assets to create livelihoods and the ways people are able to expand their asset bases through their engagement with the state, market and civil society. Finally, he argues that livelihood frameworks must address the ways in which individuals deploy and enhance their livelihood capabilities by changing the dominant rules and relationships governing resource access and distribution. Sustainability is also a feature of many livelihood studies, with research concentrating upon how resource demands or livelihood decision-making can draw down on capital assets without reducing their availability over time. A particular livelihood is seen as sustainable if it is able to "*cope with and recover from stresses and shocks*, maintain or enhance its capabilities and assets, while not undermining the resource base" (Scoones 1998: 5; emphasis added).

Human disease is regularly conceptualized as a shock for livelihood systems that can have significant, and divergent, impacts over time and space (Ellis 2000; Cross 2001; Baylies 2002). Drawing upon his rural livelihoods framework, Ellis (2000) suggests that diseases, along with drought, floods, pests, and civil war, are shocks that can modify access to capital assets through social relationships, institutions, and organizations. This socio-political context, also known as the vulnerability context (cf. Carney 1998), enables and constrains the livelihood strategies in which social actors are able to engage. Unlike stresses, which are smaller, predictable, and more regular, shocks are understood to be large, unpredictable, and irregular disturbances that can reduce or destroy assets. As Ellis (2000: 40) explains, "loss of access rights to land, accident, sudden illness, death, and abandonment are all shocks with immediate effects on the livelihood viability of the individuals and households to whom they occur." Livelihoods research has highlighted the ways that a shock, particularly a famine, triggers coping strategies that result in a specific sequence of activities. As Ellis (2000) outlines, at the onset of a shock, households first pursue new sources of income through diversification. The second coping strategy involves an increased reliance upon social relationships to provide a safety net for livelihood security. Third, temporary migration occurs with the goal of seeking out employment and reducing the pressures upon the household. Fourth, assets, such as livestock are sold to generate income, and fifth, fixed assets such as land or houses are often abandoned. Coping strategies are understood as distinct from adaptive strategies, which are seen as long-term responses to shocks intended to improve livelihood security (Davies 1996; Kgathi *et al.* 2007).

Research on the social impacts of HIV/AIDS has similarly conceptualized the disease as a shock.[2] Drawing upon qualitative data from households in KwaZulu Natal, South Africa, Cross (2001) refers to both poverty and AIDS as an economic shock, suggesting that the shocks resulting from AIDS-related death intensify the challenges of severe poverty. Hunter *et al.* (2008: 104) suggest that "the ill health and mortality resultant of HIV/AIDS can present a difficult-to-manage shock to rural livelihoods particularly in regions characterized by high levels of dependence on collection of proximate natural resources for fuel, sustenance and/or market goods." Based upon research in the Ngamiland district of Botswana, Ngwena and Mosepele (2007) liken AIDS to a shock, detailing its impacts for families dependent upon fishing in the panhandle of the Okavango Delta. In a related study, Kgathi *et al.* (2007) outline the impacts from three shocks for rural livelihoods in the Okavango Delta, specifically the desiccation of river channels, animal diseases, and HIV/AIDS. Defining shocks as large, unpredictable, and irregular disturbances, they conclude that HIV/AIDS differs from other shocks because, among other reasons, it has long-term impacts. Findings from focus groups in the villages of Shorobe, Gudigwa, and Seronga indicate that "the HIV/AIDS epidemic was the most devastating shock they had ever experienced, not comparable with other shocks in Botswana" that resulted in coping strategies such as a reduction in child-bearing, intra-household reallocation of labor, the sale of assets to cover medical and funeral costs, and reliance upon social networks and family members (Kgathi *et al.* 2007: 303).

Representations of disease as a shock is also common within the public health and economics literatures that tend to assert that various diseases, including HIV/AIDS, disrupt demographic and economic systems (Dercon *et al.* 2005; del Ninno and Marini 2005; Russell 2005; Christiaensen *et al.* 2007; Abegunde and

Stanciole 2008). In a longitudinal study from South Africa, Goudge *et al.* (2009) examine coping to illness through household expenditure patterns, concluding that state subsidies can enhance livelihood resiliency to HIV/AIDS. As with studies from other African countries, they find that social networks are important in supporting households dealing with illness. Dercon *et al.* (2005) report from household surveys in Ethiopia that the most commonly reported worst shocks were drought, illness and death, which resulted in reduced consumption. Russell (2005) points to the limited number of longitudinal studies on the economic burden of illness to show how the inclusion of qualitative case study research reveals that the reliance upon household surveys as a primary source of data can underestimate cost burdens. Additionally, drawing from qualitative studies from Sri Lanka he shows that a diversity of assets and access to social networks and formal and informal financial institutions can increase resilience to illness-induced poverty.

Linked to the impacts of shocks such as disease upon livelihood outcomes is the concept of vulnerability, which attends to the degree of exposure or susceptibility to a potentially disruptive event. As Chambers (2006: 33) explains, "Vulnerability has thus two sides: an external side of risks, shocks, and stress to which an individual or household is subject; and an internal side which is defencelessness, meaning a lack of means to cope without damaging loss." As with other elements of livelihoods frameworks, vulnerability has been operationalized by some of the research on the impacts of disease through the classification of households according to vulnerability. For example, Goudge *et al.* (2009) categorized rural households as secure, vulnerable, or highly vulnerable depending upon members' employment status and security, household expenditure and asset portfolios. Based upon these criteria, it is suggested that certain households are better equipped to cope with illness and mortality, while others are more vulnerable due to a lack of capital assets or family support. Some studies suggest that erosion of household assets increases stress (Barany *et al.* 2005) and can trigger a self-reinforcing cycle that pulls the household towards greater vulnerability (Cross 2001). In their extensive review of the links between HIV/AIDS and the natural environment, Bolton and Talman (2010: 33–34) explain:

> Decreased coping ability makes people and communities more vulnerable to HIV/AIDS. HIV/AIDS in turn leads to increased dependence on natural resources, as households lose labor force, land tenure, and traditional knowledge and are less able to maintain their previous livelihoods. This increased reliance on natural resources in turn makes communities even less able to cope, as they become more and more exposed to the vagaries of nature, weather, and availability of resources. Infection with HIV/AIDS also itself decreases coping ability, which may lead to both behavior that increases HIV transmission and also increase natural resource use. The cycle is self-reinforcing and reciprocal.

Research on the social impacts of HIV/AIDS has tended to reify the shock concept from livelihood studies and economics to suggest that the disease contains features that are large, unpredictable, and irregular. These features can be tremendous within various sectors and have cumulative effects that undermine livelihood decision-making and social sustainability. In addition to reshaping social systems, either through livelihoods or through economic and political institutions, HIV/AIDS also has significant impacts for ecological systems throughout sub-Saharan Africa. The next section provides a review of some of the recent research demonstrating some of the biophysical dimensions of HIV/AIDS, which similarly to the literature on the social impacts, draws upon the shock concept. As will be shown, major themes of HIV/AIDS as an ecological shock consider how the disease is disrupting ecosystems, biodiversity, and future trajectories for natural resource management.

HIV/AIDS as shock to ecological systems

In addition to the social impacts arising from HIV/AIDS, the disease has similarly been theorized as a shock to ecological systems. Bolton and Talman (2010) suggest that, among many possible impacts, HIV/AIDS sufferers draw down upon natural resources, exploit natural resources in the short-term through unsustainable practices, shift land use systems including leaving agricultural land fallow, and pressure currently protected natural resource areas due to increased need. The decrease in available food materials due to agricultural decline prompts concern because it has been estimated that those infected with HIV typically require up to 15 percent more energy and 50 percent more protein, as well as more micronutrients (Beisel 2002; Friis 1998 in Kaschula 2008). Increasing pressures upon natural resources collected for consumption or income generation can similarly stress human resistance to HIV/AIDS. Kaschula (2008) reports on the collection of wild foods, specifically vegetables and fruits, from communal areas in South Africa to support household diet. This collection can include South African species that have substantially higher carotenoids and vitamin A, such as *Amaranthus* sp., *Chenopodium album*, and *Bidens pilosa*, which have been shown to play a role in reducing infection risk and slowing the progression of HIV into AIDS (cf. Melikian *et al.* 2001; Beisel 2002). Food security, therefore, extends beyond the viability of agricultural production to include food materials extracted from communal ecosystems. The decreasing availability of food derived from agricultural production or collection can therefore increase vulnerability to infection and accelerate the progression of the disease.

In analyzing the potential impacts of HIV/AIDS on biodiversity, McGarry and Shackleton (2009: 5) identify disease as a shock that "threatens a descent into deeper poverty" that will have concomitant repercussions for biodiversity and ecosystem functioning. Focusing upon the consumption of wild meat in response to AIDS mortality, they find seafood, riverine fish, forest mammals, birds, reptiles, and insects playing a significant role in children's diets. Utilizing the AIDS proxy metrics from the Southern African Development Community Vulnerability Assessment Committee (SADC 2003), they classify households according to their vulnerability to the disease and its effects. Highly vulnerable families were observed to hunt more regularly and consume more wild meat than less vulnerable families, causing McGarry and Shackleton (2009: 6) to conclude that "as the pandemic affects increasing numbers of households and undermines traditional coping strategies, it is conceivable that affected children will range further into forests and rangelands to source sufficient food." The implication is that the spread of HIV/AIDS places intensifying pressure upon the natural environment and will likely lead to degradation and loss of biodiversity.

In a study on the livelihood impacts of HIV/AIDS in the Okavango Delta of Botswana, Kgathi *et al.* (2007) conclude that due to poverty rates in the region, people will not be able to adequately respond to these shocks, suggesting this will invariably lead to the degradation of resources and loss of biodiversity. Similarly, a report in *New Scientist* outlines the biodiversity threats from HIV/AIDS, alarmingly concluding that affected families are forced to "plunder biodiversity to survive" (Aldhous 2007: 7). Other studies indicate another impact for ecosystem management; namely that the impacts for conservation agencies could be severe from the loss of staff and knowledge that accompanies AIDS mortality (Gelman *et al.* 2004; Aldhous 2007). Bolton and Talman (2010) argue that HIV/AIDS can result in increased dependency upon protected natural resources that generates conflict between people, wildlife, and conservation officials. Additionally, they outline a number of ways that the natural environment impacts HIV/AIDS, including a reduction of agricultural viability that reduces macro and micronutrient intake that in turn makes individuals more vulnerable to disease, increased caloric needs due to resource degradation whereby resources are only available at greater distances, and deterioration of potable water.

It should be emphasized that the inevitability of environmental degradation from HIV/AIDS has been questioned by some of the existing research. Frank and Unruh (2008) combine household survey research with land cover change analysis to conclude that AIDS widows in Zambia have strategically invoked the disease to resist land grabbing from recently arriving migrants into the region. In a surprising finding that deviates from much of the literature on gendered land tenure and environmental impacts from the disease, they suggest that stability in land ownership is increasing forest conservation at least in the short to medium term. Beyond the specifics from this one case study, their work challenges assumptions about a unidirectional and necessarily negative ecological impact from HIV/AIDS that warrants further research consideration.

Human vulnerability to a disease such as HIV/AIDS can be compounded by exposure to parasitic and infectious diseases that weaken the immune system. Stillwaggon draws upon existing research to suggest that:

> Recent work in cell biology shows the particular mechanisms by which malnutrition and parasitosis depress both specific and nonspecific immune response by undermining epithelial integrity and the production of NK cells, B cells, and T cells … malnutrition and parasitosis, therefore, make people more susceptible to infectious disease, including HIV.
>
> (Stillwaggon 2006: 41)

Similarly, Stillwaggon and Sawers show in Chapter 13 of the present volume that high incidences of malaria in sub-Saharan Africa might explain the high HIV infection rates compared to other world regions. The conditions that contribute to the spread of malaria, as just one disease that might increase vulnerability to HIV, bear scrutiny from an ecological, as well as a social perspective. Pijanowski *et al.* (2010) argue that the field of landscape ecology can contribute towards understandings of how changing spatial patterns of land cover and land use influence vector-borne and zoonotic diseases. Global climate change is similarly shifting species range distributions so that ecosystems such as the highlands of Kenya are now exposed to diseases including malaria. A number of the chapters in this volume (see for example Chapters 4 and 12) work to demonstrate how climate change is changing biophysical processes in varied ways, including creating new disease vectors. These infectious diseases have the potential to increase human vulnerability to HIV and further intensify the social and ecological impacts of the epidemic.

Returning to the vulnerability concept, there has been an expansion of interest in developing frameworks for addressing vulnerability through coupled social and ecological systems. In a much-cited example, Turner *et al.* (2003) present a vulnerability framework to outline how a hazard, which they define as either a perturbation or stress/stressor, induces harm upon a system or its elements. Perturbations typically originate from outside the system and involve an intense pressure change due to an abnormal variation in the system, such as a hurricane. These differ from stresses that often originate from within the system and are understood as a continuous or slowly increasing pressure generally within the normal range of variability, such as soil degradation. Within this vulnerability framework, a perturbation appears similar to the shock concept within livelihoods research, given that these are understood to be large, unpredictable, and irregular. While the framework is ground-breaking is demonstrating how vulnerability involves "connections operating at different spatiotemporal scales and commonly involving stochastic and nonlinear processes" (Turner *et al.* 2003: 8076), it should be noted that it was not leveraged to specifically address disease.

Clearly, then, HIV/AIDS is having tremendous impacts upon social systems, ecological systems, and the reciprocal relationships between them. As Bolton and Talman (2010) explain, the onset of HIV within a family can produce a reduction in labor that then increases the burden upon other family members and social networks. In some cases, agricultural decline then occurs, triggering a switch to less labor-intensive farming systems and/or reliance upon other types of resources should they be available. In order to cope, household members might reduce food consumption and sell assets to pay for medical bills or other expenses. Nutritional status declines thereby increasing vulnerability to infection or, for already afflicted family members, advancement of opportunistic infection. Following adult mortality, households have been shown to fragment in search of outside employment, generating a reduction in the cultivation of land and increasing reliance upon other natural resources that might be collected in an unsustainable manner. These processes can in turn increase vulnerability to other diseases. As existing research demonstrates, these impacts are variable and highly contextual, yet they also make it clear that HIV/AIDS unfolds in spatially and temporally dynamic ways. The question remains whether classifying HIV/AIDS as a shock or perturbation captures the full gamut of its impacts and the varied processes shaping human vulnerability to disease. While there is some logic in doing so, likening the disease to a shock might fail to capture its divergent and complex impacts for coupled socio-ecological systems. In order to engage with these limitations, the next section argues that rather than conceptualizing HIV/AIDS as a shock, it might be more useful to theorize the disease as a socio-ecological experience.

HIV/AIDS as coupled socio-ecological experience

Even while the concept of a disease as a shock to social and ecological systems remains common in the academic and policy literatures, some studies have suggested HIV/AIDS to be either a unique shock or a different type of event altogether. In their pioneering work on AIDS in Uganda, Barnett and Blaikie (1992: 56) likened AIDS to long-wave disasters such as global warming, explaining that in such cases, "the disaster does not appear to the affected society as a discrete occurrence with recognizable trigger events, which can be used to mobilize action, such as would be the case with an earthquake, a volcanic eruption or a devastating flood." Drawing upon this work in addressing household impacts of AIDS in Zambia, Baylies (2002: 619) provided a compelling explanation for the limitations of conceptualizing HIV/AIDS as a shock. Outlining the varied impacts from the disease, she suggested that "in practice, some of the effects of AIDS mirror those of other shocks. Others, however, are specific to the particular dynamics through which AIDS configures with household characteristics." Additionally, she commented that:

> AIDS is a shock with very specific characteristics. It represents a crisis where all too often the limits of both communal assistance (via mechanisms assumed to apply under the rubric of social capital) and individualistic efforts to secure income generation and household survival are starkly exposed.
>
> (Baylies 2000: 628–629)

Similarly, Hosegood *et al.* (2007: 1256) discuss the shocks from illness and death that were often followed by or occurred at the same time as "other experiences of HIV and AIDS within the household or in inter-connected households."

What then should be made of these differing conceptualizations? Is HIV/AIDS a shock, perturbation, disaster, or event? Does it provide academic and policy clarity to liken the disease to other types of shocks such as droughts, floods, pests, and civil war? Or, drawing upon the vulnerability framework (Turner *et al.* 2003), would not a disease such as HIV/AIDS—which can unfold over a series of years for infected individuals (cf. Morgan *et al.* 2002), and might also be seen as originating from within the system—be more properly conceptualized as a slowly intensifying stress? How should we engage with the socio-political and biophysical dimensions of human diseases, including HIV/AIDS, and what does this contribute to emerging understandings of the ecological and social dimensions of human health? It is important to state that these questions are not simply a matter of academic debate because the discursive framings of HIV/AIDS have substantial impacts for interventions and responses in various settings. As others have noted, an early trend of identifying AIDS as the new variant famine (de Waal and Whiteside 2003) produced policy interventions that under-emphasized other factors shaping food availability within Southern Africa (Murphy *et al.* 2005). Simultaneously, focusing upon the macro-effects of the disease as an economic shock might have the effect of obscuring the gendered dimensions of the disease in terms of which segments of the population are most likely to be infected with HIV (Fassin and Schneider 2003) or bear heavier burdens of caring for sick individuals or orphans (Schatz 2007). Other studies have asserted that the absence of comparative cases of non-affected HIV/AIDS households mean that some impacts ascribed to the disease might be due to other factors altogether (Bolton and Talman 2010). The discursive framings of HIV/AIDS, therefore, have material impacts for both social and ecological systems being reshaped by the disease, as well as the policy interventions that arise in response. Drawing upon Baylies (2002) and some of the other studies that attend to the local and contextually specific dimensions of the disease, its intra-household and familial features, and the effects following HIV transmission, I use the following sections to outline how HIV/AIDS might be better conceptualized not as a shock that is large, unpredictable, and irregular, but as a socio-ecological experience that is spatially and temporally dynamic, disproportionate, and dispersed.

Spatially and temporally dynamic

Unlike other shocks addressed within livelihood studies, HIV/AIDS follows a unique spatio-temporal trajectory. While it is often the case that droughts or floods are shorter-term events, HIV/AIDS can take years to progress within the bodies of infected individuals and through the families and broader social networks of those afflicted. As Barnett and Blaikie (1992) explained, because HIV/AIDS is a long-wave disaster, the different coping stages can be extended and delayed, as individuals and communities take time to identify the presence of the disease. HIV can take months or even years to advance, and so by the time it is symptomatic the eventual scale of the crisis is magnified. In a cohort study from Uganda, Morgan *et al.* (2002) report that the median survival from seroconversion was 9.8 years and that once AIDS developed the median survival was 9.2 months.[3] Barnett and Blaikie (1992: 57) suggested that, depending upon the rate of infection and the mean period between infection and AIDS symptoms, "for every one person with these symptoms there will be many more who are already HIV+. Thus the raw material, the evidence, on which an explanation of the disaster is made, takes at least five years to appear." Similarly, the initial infection and progression of the illness can be quite dynamic depending upon the nutritional status of the individual, as well as access and use of anti-retroviral medications. As previously mentioned, some research suggests that malnutrition and parasitosis make people more susceptible to infectious disease, including

HIV (Stillwaggon 2006). The social and ecological processes shaping these factors are spatially and temporally variable, thereby attesting to the dynamism of human vulnerability to disease. This leads Baylies (2002: 620) to conclude that given the diversity of situations upon which HIV/AIDS intrudes into households and familial networks, its staging is neither automatic nor unilinear.

The continued invocation of HIV/AIDS as a shock to social and ecological systems possibly derives from the lack of longitudinal data in previous studies. As has been reported elsewhere (Bachmann and Booysen 2004; Murphy *et al.* 2005), a reliance upon cross-sectional rather than longitudinal data "blurs the distinction between effects of HIV/AIDS and those of other characteristics of affected individuals and households" (Bachmann and Booysen 2004: 818). Similarly, Bolton and Talman (2010) conclude their review of HIV-environment linkages by calling for future research that utilizes longitudinal data from both afflicted and non-afflicted households. The tendency to rely upon snapshot information might explain why the disease is seen as such an immediate factor shaping rural livelihoods, rather than a cumulative experience (cf. Hosegood *et al.* 2007) that varies spatially and temporally within households, communities, and regions. As previously mentioned, a second reason HIV/AIDS is conceptualized as a shock is that many of the studies tend to be based upon adult mortality (Hunter *et al.* 2008; Kaschula 2008). While clearly important in understanding the impacts of the disease, the emphasis upon adult mortality can miss the complex experiences for families and communities living with HIV/AIDS. Given that the disease can take years to progress, HIV intersects with other social and ecological factors that potentially undermine household security and resiliency. Hosegood *et al.* (2007) conclude that while AIDS causes mortality in young adults and children, illness and death due to other causes compound the consequences of the epidemic for social systems. Future work grounded in longitudinal data attending to the long-term experiences for those living with HIV/AIDS, and for families directly and indirectly impacted, will likely yield a richer understanding of the unique features of the disease.

Disproportionately experienced

Whereas the absence of longitudinal data potentially limits analyses of the spatio-temporal dynamism of HIV/AIDS, the methodological reliance upon the household as a unit of analysis has been critiqued for obscuring the disproportionate effects of the disease (Baylies 2002; Murphy *et al.* 2005). The concentration upon the household as the unit of analysis for much of the work is likely a legacy effect from livelihood studies, in addition to its methodological utility. Within this research tradition, a household is defined as the collection of individuals living under the same roof and sharing meals (Weller and Romney 1988). The benefits of this concept are fairly clear, in that research and policy can work with a seemingly fixed group of people within a particular location. There has been a growth of work within livelihoods research intended to address how livelihood systems are spread across spatial and temporal scales, either intersecting with external trade markets (McSweeney 2004) or development institutions and processes of economic neoliberalism (Bebbington 2000). Similarly, King (2011) argues that livelihood studies can be surprisingly aspatial, and while they often recognize the influence of exogenous processes, they can be imprecise in showing how livelihoods are linked across spatial and temporal scales. Because livelihood studies have helped advance the idea of disease as a shock, the degree to which the household unit shapes understandings of the social and ecological impacts of diseases such as HIV/AIDS bears scrutiny.

As has been explained in the chapter thus far, the diverse effects of HIV/AIDS are often understood as impacting particular individuals within the household, as well as the dominant household livelihood system. It warrants emphasis then that the disease is perversely unequal in terms of who is most likely to be infected and affected. Baylies (2002: 618) explains that AIDS is a differentially experienced type of shock because of its tendency to "cluster within certain households and a pattern of staging which is specific to a given household rather than generally applicable across households within a particular community at a given time." This points to a limitation of the shock concept for the disease because the intra-community, and intra-household, vulnerabilities are differentially experienced across time and space. All of these factors point to the gendered dimensions of disease, which are compounded by other social processes that are contextual and power-laden. Unlike other shocks routinely studied with livelihoods research, a flood for example, the micro-geographies of HIV/AIDS have disproportionate effects within societies. This leads Murphy *et al.* (2005) to conclude that future research must attend to the specific individuals that are infected and/or affected differentially through gender dynamics, age, status, and economic circumstances, as well as the extended family that provides material and non-material support.

South Africa serves as an example of the disproportionate effects of the disease. The country's official HIV infection rate is 18 percent, and with roughly 5.7 million people believed to be infected, this makes the country the site of "the largest HIV epidemic in the world" (UNAIDS 2008: 40). Yet even with the expanse of the epidemic, there are disproportionate effects in terms of who is infected and/or affected. Because women in South Africa are more likely to be infected with HIV than men (Fassin and Schneider 2003), and are differentially impacted by the disease (Drimie 2003; Masanjala 2007), this results in different responses, and variabilities in the social and environmental impacts, than other shocks. Women already responsible for natural resource collection and food production thus bear increasing burdens due to an illness within the household. Older women have been shown to have increasing responsibility for care-giving for AIDS orphans (Schatz and Ogunmefun 2007). Additionally, the death of a male adult can trigger land dispossession for women who have not had historical title to property, thereby dramatically increasing vulnerability to poverty for female-headed families (Bolton and Talman 2010).

Another unique feature of HIV/AIDS is that with greater access to anti-retroviral medications it is now possible to manage the disease for extended periods of time, and in some cases, actually live with HIV. Yet access to ART varies widely within sub-Saharan Africa and is shaped by national governmental policies, corporate resistance to generic medications, cultural and social practices, infrastructure constraints and lack of access, and a host of other challenges. South Africa received international condemnation for resisting access to anti-retroviral medications for years and has only recently instituted a national program to provide universal access (Jones 2005; King 2010). According to the World Health Organization (WHO), within South Africa the proportion of people within South Africa that have access to ART in relation to the estimated number of people in need of treatment is currently 37 percent (World Health Organization *et al.* 2010).[4] This compares to estimates of ART coverage of 39 percent in Uganda, 48 percent in Kenya, 64 percent in Zambia, 76 percent in Namibia, and 83 percent in Botswana. The significant variations in access to ART between these countries, including a number that share a political border, demonstrate how governmental policies and political economies can be determinative in shaping the lived experience of those afflicted with HIV. While HIV/AIDS has expansive impacts for social and ecological systems, these impacts are spatially and temporally disproportionate, suggesting the value of conceptualizing the disease as a socio-ecological experience.

Spatially and temporally dispersed

In addition to being spatially and temporally dynamic and disproportionate, the effects of HIV/AIDS are also notable for being dispersed. Just as the reliance upon the household concept might narrow understandings of the disproportionate impacts of the disease, addressing the effects at the household might restrict an understanding of its dispersed effects. As Murphy *et al.* (2005) show, the household as a unit is fundamentally different from the extended family which includes social networks spread over space. As they explain, "the household is an economic and reproductive unit somewhat fixed in place, which makes it easier to find; but it is distinct from the extended family, which reflects bonds of kin and affection connected over space, sometimes vast distances" (Murphy *et al.* 2005: 269). Given that much of the existing research attends to the importance of social networks in providing gifts and support in response to HIV/AIDS, an overreliance upon the household could miss the broader impacts for communities and regions. Hosegood *et al.* (2007: 1250) suggest that because family networks and social relationships between households extend to different places, the true impacts of illness and death may extend across several residential groups. While HIV/AIDS is clearly disruptive for the particular individuals within a household, its effects can be spread over expansive social networks that are geographically uneven and temporally variable.

Secondly, emerging research is attending to underlying social structures that shape the vulnerability context for human disease. Some of this work is informed by political ecology, which addresses the links between political, economic and social structures and decision-making by local actors. Research on the political ecology of disease (Mayer 1996; Kalipeni and Oppong 1998; Cutchin 2007) and the political ecologies of health (King 2010) have worked to show how inequality within social contexts can produce inequalities in disease vulnerabilities. In returning to Barnett and Blaikie's (1992) early research on AIDS in Uganda, they located individuals and households within broader social networks in order to expose structural conditions that produce disease. As they explained, socio-economic patterns, access to medical care and support networks, gendered power relations, production systems, and the survival strategies of households and communities all contribute in shaping the ways in which the disease affects societies and economies (Barnett and Blaikie 1992). As I have argued in work on HIV/AIDS in South Africa (King 2010), regional variations in HIV are linked to historical processes of place production and racial segregation. The contemporary conditions that influence disease transmission were created by national and provincial agencies to facilitate particular political and economic agendas. This type of analysis helps show how the locations of high incidence of disease are the outcome of social relationships and power dynamics that have been produced over time and through space. These political ecologies of health reveal that HIV/AIDS is uneven, and even though the epidemic is expansive throughout sub-Saharan Africa, its specific effects remain spatially and temporally dispersed.

Conclusion

This chapter provided a review of the literature on the social and ecological impacts of HIV/AIDS in sub-Saharan Africa to demonstrate how the disease is often conceptualized as a transformative shock. Drawing from a number of traditions within the social and natural sciences, HIV/AIDS has been defined as a large, unpredictable, and irregular disturbance that reworks societies and biophysical systems in varied ways. I argued that previous scholarship on HIV/AIDS has tended to concentrate upon the macro-scale, using the household as a unit of analysis for understanding impacts following adult mortality. Additionally, many of these studies are synchronic in their analyses, concentrating upon the more immediate effects for individuals and households, as opposed to the potentially longer-wave impacts. To be clear, previous research has been ground-breaking and essential in generating important findings for how the disease transforms social and ecological systems. My central concern here was to suggest that other studies that engage with the contextually specific dimensions of the disease, its intra-household and familial features, and the effects following HIV transmission, suggest that alternative conceptualizations to a shock warrant attention for research and policy. Considering human disease, and specifically HIV/AIDS, as a shock in which large, unpredictable, and irregular disturbances disrupt livelihood systems and biophysical processes has a certain logic. However, even with the dramatic scope of the HIV epidemic in sub-Saharan Africa, the disease has had diverse effects on social and ecological systems. My intention was to question whether this theorization captures the diversity of its impacts and the varied processes that shape human vulnerability to disease. The effects of HIV/AIDS have been shown to occur over a longer time period than other shocks, and are generally uneven within communities and regions. As a result, nuanced analyses of the complex and reciprocal relationships between the social and ecological impacts suggest that HIV/AIDS might be better theorized as a health experience that is dynamic, disproportionate, and dispersed. Rather than an immediate shock to demographic, economic, and ecological systems, HIV/AIDS is a long-wave experience that impacts societies and environments for years and decades.

Acknowledgments

My sincere thanks to Jamie Shinn who assisted with the literature search on disease as a shock to social and ecological systems, and also to Kayla Yurco, who assisted with the research on access to anti-retroviral (ARV) medications within sub-Saharan Africa. Critical and helpful feedback on earlier versions of this chapter was generously provided by Kelley Crews, in addition to members of the nature society working group at the Pennsylvania State University.

Notes

1 Anti-retroviral therapy involves the treatment of individuals infected with HIV. The standard treatment consists of a combination of at least three drugs that suppress HIV replication and reduce the likelihood of the virus developing resistance (World Health Organization *et al.* 2010).

2 Given the sensitivity of HIV/AIDS in various contexts, it should be noted at the outset that many of these studies utilize proxies for AIDS-related mortality, such as the ones identified by the Southern African Development Community (SADC 2003). These proxies consider for example family mortality within the previous two years, chronic illness prior to death, presence of chronically ill caregivers or recent mortality of care givers, and/or presence of orphans within the home (Kaschula 2008; McGarry and Shackleton 2009).

3 Seroconversion is when antibodies form in response to an infectious organism. Acute HIV infection can appear like viral illnesses such as infectious mononucleosis or flu. HIV can take weeks, or even months, to become symptomatic.

4 The data reported in this section are based upon the 2010 WHO recommendations, which changed the recommendation for requiring ARV therapy from a CD4 cell count of below 200 cells per mm^3 to that of below 350 cells per mm^3. This increased the number of people estimated in need of treatment in low and middle-income countries by 45 percent (World Health Organization *et al.* 2010).

References

Abegunde, D.O. and Stanciole, A.E. (2008) The economic impact of chronic diseases: how do households respond to shocks? Evidence from Russia. *Social Science and Medicine* 66(11): 2296–2307.

Aldhous, P. (2007) The hidden tragedy of Africa's HIV crisis. *New Scientist* (11 July): 6–9.

Ashley, C. and Carney, D. (1999) *Sustainable Livelihoods: Lessons from Early Experience*. London, UK: Department for International Development.

Bachmann, M.O. and Booysen, F.L. (2004) Relationships between HIV/AIDS, income and expenditure over time in deprived South African households. *AIDS Care* 16(7): 817–826.

Barany, M., Holding-Anyonge, C., Kayambazinthu, D., and Sitoe, A. (2005) Fuelwood, food, and medicine: interactions between forests, vulnerability, and rural responses to HIV/AIDS. IFPRI Conference: HIV/AIDS and Food and Nutrition Security. Durban, South Africa, 14–16 April.

Barnett, T. and Blaikie, P. (1992) *AIDS in Africa: Its Present and Future Impact.* New York, NY: Guilford Press.

Barnett, T. and Whiteside A. (2006) *AIDS in the Twenty-First Century: Disease and* Globalization, 2nd edition. New York, NY: Palgrave Macmillan.

Baylies, C. (2002) The impact of AIDS on rural households in Africa: a shock like any other? *Development and Change* 33(4): 611–632.

Bebbington, A. (1999) Capitals and capabilities: a framework for analyzing peasant viability, rural livelihoods and poverty. *World Development* 27(12): 2021–2044.

Bebbington, A. (2000) Reencountering development: livelihood transitions and place transformations in the Andes. *Annals of the Association of American Geographers* 90(3): 495–520.

Beisel, W.R. (2002) Nutritionally acquired immune deficiency syndromes. In *Micronutrients and HIV infection*, H. Friis (ed.). Boca Raton, FL: CRC Press, 23–42.

Bolton, S. and Talman, A. (2010) *Interactions between HIV/AIDS and the Environment: A Review of the Evidence and Recommendations for Next Steps.* Nairobi, Kenya: IUCN ESARO Office.

Campbell, C. (2003) *"Letting Them Die": Why HIV/AIDS Intervention Programmes Fail.* Oxford, UK: James Currey.

Carney, D. (1998) *Sustainable Rural Livelihoods: What Contribution Can We Make?* London, UK: Department for International Development.

Carter, M.R. and May, J. (1999) Poverty, livelihood and class in rural South Africa. *World Development* 27(1): 1–20.

Chambers, R. (1987) *Sustainable Livelihoods, Environment and Development: Putting Poor Rural People First.* IDS Discussion Paper 240. Brighton, UK: IDS.

Chambers, R. (1997) *Whose Reality Counts? Putting the First Last.* London, UK: ITDG Publishing

Chambers, R. (2006) Vulnerability, coping and policy (editorial introduction). *IDS Bulletin* 37(4): 33–40.

Christiaensen, L., Hoffmann, V., and Sarris, A.H. (2007) *Gauging the Welfare Effects of Shocks in Rural Tanzania.* World Bank Policy Research Working Paper 4406. Washington, DC: World Bank.

Cross, C. (2001) Sinking deeper down: HIV/AIDS as an economic shock to rural households. *Society in Transition* 32(1): 133–147.

Cutchin, M.P. (2007) The need for the "new health geography" in epidemiologic studies of environment and health. *Health and Place* 13: 725–742.

Davies, S. (1996) *Adaptable Livelihoods: Coping with Food Insecurity in the Malian Sahel.* London, UK: Macmillan Press.

Dercon, S., Hoddinott, J., and Woldehanna, T. (2005) Shocks and consumption in 15 Ethiopian villages, 1999–2004. *Journal of African Economies* 14(4) 559–585.

Drimie, S. (2003) HIV/Aids and land: case studies from Kenya, Lesotho and South Africa. *Development Southern Africa* 20(5): 647–658.

Ellis, F. (2000) *Rural Livelihoods and Diversity in Developing Countries.* Oxford, UK: Oxford University Press.

Fassin, D. (2007) *When Bodies Remember: Experiences and Politics of AIDS in South Africa.* Berkeley, CA: University of California Press.

Fassin, D. and Schneider, H. (2003) The politics of AIDS in South Africa: beyond the controversies. *British Medical Journal* 326: 495–97.

Frank, E. and Unruh, J. (2008) Demarcating forest, containing disease: land and HIV/AIDS in southern Zambia. *Population and Environment* 29: 108–132.

Friis, H. (1998) The possible role of micronutrients in HIV infection. *SCN News* 17: 11–12.

Gelman, N.B., Oglethorpe, J., and Mauambeta, D. (2004) The impact of HIV/AIDS: how can it be anticipated and managed. *Parks* 15(1): 13–24.

Goudge, J., Russell, S., Gilson, L., Gumede, T., Tollman, S., and Mills, A. (2009) Illness-related impoverishment in rural South Africa: why does social protection work for some households but not others? *Journal of International Development* 21(2): 231–251.

de Haan, L. and Zoomers, A. (2005) Exploring the frontier of livelihoods research. *Development and Change* 36(1): 27–47.

Hosegood, V., Preston-Whyte, E., Busza, J., Moitse, S., and Timaeus, I.M. (2007) Revealing the full extent of households' experiences of HIV and AIDS in rural South Africa. *Social Science and Medicine* 65(6): 1249–1259.

Hunter, L.M., Twine, W., and Patterson, L. (2007) "Locusts are now our beef": adult mortality and household dietary use of local environmental resources in rural South Africa. *Scandinavian Journal of Public Health* 35(3): 165–174.

Hunter, L.M., De Souza, R.M., and Twine, W. (2008) The environmental dimensions of the HIV/AIDS pandemic: a call for scholarship and evidence-based intervention. *Population and Environment* 29(3–5): 103–107.

Jones, P. (2005) "A test of governance": rights-based struggles and the politics of HIV/AIDS policy in South Africa. *Political Geography* 24(4): 419–447.

Kalipeni, E. and Oppong, J. (1998) The refugee crisis in Africa and implications for health and disease: a political ecology approach. *Social Science and Medicine* 46(12): 1637–1653.

Kaschula, S.A. (2008) Wild foods and household food security responses to AIDS: evidence from South Africa. *Population and Environment* 29(3–5): 162–185.

Kgathi, D.L., Ngwenya, B.N., and Wilk, J. (2007) Shocks and rural livelihoods in the Okavango Delta, Botswana. *Development Southern Africa* 24(2): 289–308.

King, B. (2010) Political ecologies of health. *Progress in Human Geography* 34(1): 38–55.

King, B. (2011) Spatialising livelihoods: resource access and livelihood spaces in South Africa. *Transactions of the Institute of British Geographers* 36(2): 297–313.

Kristjanson, P., Radeny, M., Baltenweck, I., Ogutu, J., and Notenbaert, A. (2005) Livelihood mapping and poverty correlates at a meso-level in Kenya. *Food Policy* 30(5–6): 568–583.

Lemke, S. (2005) Nutrition security, livelihoods and HIV/AIDS: implications for research among farm worker households in South Africa. *Public Health Nutrition* 8(7): 844–852.

Long, N. (1984) *Family and Work in Rural Societies: Perspectives on Non-Wage Labour.* London, UK: Tavistock Publications.

Love, R. (2004) HIV/AIDS in Africa: links, livelihoods, and legacies. *Review of African Political Economy* 31(102): 639–648.

McGarry, D.K. and Shackleton, C.M. (2009) Comment: is HIV/AIDS jeopardizing biodiversity? *Environmental Conservation* 36(1): 5–7.

McSweeney, K. (2004) The dugout canoe trade in Central America's Mosquitia: approaching rural livelihoods through systems of exchange. *Annals of the Association of American Geographers* 94(3): 638–661.

Madhavan, S. (2007) Comment: households and HIV/AIDS. *Scandinavian Journal of Public Health* 35(Supplement 69): 155–156.

Masanjala, W. (2007) The poverty–HIV/AIDS nexus in Africa: a livelihood approach. *Social Science and Medicine* 64(5): 1032–1041.

Mayer, J.D. (1996) The political ecology of disease as one new focus for medical geography. *Progress in Human Geography* 20(4): 441–456.

Melikian, G., Mmiro, F., Ndugwa, C., Perry, R., Jackson, J.B., Garrett, E., Tielsch, J., and Semba, R.D. (2001) Relation of vitamin A and carotenoid status to growth failure and mortality among Ugandan infants with human immunodeficiency virus. *Nutrition* 17(7–8): 567–572.

Morgan, D., Mahe, C., Mayanja, B., Okongo, J.M., Lubega, R., and Whitworth, J.A.G. (2002) HIV-1 infection in rural Africa: is there a difference in median time to AIDS and survival compared with that in industrialized countries? *AIDS* 16(4): 597–603.

Murphy, L.L., Harvey, P., and Silvestre, E. (2005) How do we know what we know about the impact of AIDS on food and livelihood insecurity? A review of empirical research from rural sub-Saharan Africa. *Human Organization* 64(3): 265–275.

Negin, J. (2005) Assessing the impact of HIV/AIDS on economic growth and rural agriculture in Africa. *Journal of International Affairs* 58(2): 267–281.

del Ninno, C. and Marini, A. (2005) *Households' Vulnerability to Shocks in Zambia.* Social Protection Discussion Paper 536. Washington, DC: World Bank Special Protection Program.

Ngwenya, B.N. and Mosepele, K. (2007) HIV/AIDS, artisanal fishing and food security in the Okavango Delta, Botswana. *Physics and Chemistry of the Earth* 32: 1339–1349.

Pijanowski, B.C., Iverson, L.R., Drew, C.A., Bulley, H.N.N., Rhemtulla, J.M., Wimberly, M.C., Bartsch, A., and Peng, J. (2010) Addressing the interplay of poverty and the ecology of landscapes: a Grand Challenge Topic for landscape ecologists? *Landscape Ecology* 25(1): 5–16.

Posel, D., Kahn, K., and Walker, L. (2007) Living with death in a time of AIDS: a rural South African case study. *Scandinavian Journal of Public Health* 35(supplement 69): 138–146.

Russell, S. (2005) Illuminating cases: understanding the economic burden of illness through case study household research. *Health Policy and Planning* 20(5): 277–289.

SADC. (2003) *Towards Identifying Impacts of HIV/AIDS on Food Insecurity in Southern Africa and Implications for Response: Findings from Malawi, Zambia and Zimbabwe.* Harare, Zimbabwe: Southern African Development Community FANR Vulnerability Assessment Committee.

Schatz, E.J. (2007) "Taking care of my own blood": older women's relationships to their households in rural South Africa. *Scandinavian Journal of Public Health* 35(supplement 69): 147–154.

Schatz, E. and Ogunmefun, C. (2007) Caring and contributing: the role of older women in rural South African multi-generational households in the HIV/AIDS era. *World Development* 35(8): 1390–1403.

Scoones, I. (1998) *Sustainable Rural Livelihoods: A Framework for Analysis.* Working Paper 72. Brighton, UK: Institute for Development Studies.

Stillwaggon, E. (2006) *AIDS and the Ecology of Poverty.* Oxford, UK: Oxford University Press.

Turner, B.L., Kasperson, R.E., Matson, P.A., McCarthy, J.J., Corell, R.W., Christensen, L., Eckley, N., Kasperson, J.X., Luers, A., Martello, M.L., Polsky, C., Pulsipher, A., and Schiller, A. (2003) A framework for vulnerability analysis in sustainability science. *Proceedings of the National Academy of Science* 100(14): 8074–8079.

UNAIDS. (2008) *Report on the Global AIDS Epidemic.* Geneva, Switzerland: Joint United Nations Programme on HIV/AIDS (UNAIDS).

de Waal, A. and Whiteside, A. (2003) New variant famine: AIDS and food crisis in southern Africa. *Lancet* 362(9391): 1234–1237.

Weller, S.C. and Romney, A.K. (1988) *Structured Interviewing.* Newbury Park, CA: Sage Publications.

Whiteside, A. (2002) Poverty and HIV/AIDS in Africa. *Third World Quarterly* 23(2): 313–332.

World Health Organization with UNAIDS and UNICEF. (2010) *Towards Universal Access: Scaling up Priority HIV/AIDS Interventions in the Health Sector*. Geneva, Switzerland: WHO Press.

15 Challenges and opportunities for future ecologies and politics of health

Brian King and Kelley A. Crews

Human health exists at the interface of environment and society, and so its prospective possibilities will be shaped by the interactions between social and ecological systems. Future research and practice, therefore, need to concentrate upon the ecological dimensions of health and vulnerability, the socio-political dimensions of human health, and the intersections between the ecological and social dimensions of health. One of the hallmarks of this volume is that rather than approaching these factors separately, each of the chapters engages the intersections between the social and ecological dimensions of human health. These contributions help demonstrate that interrogating the nexus of ecologies and politics of health necessitates engagements between the natural and social sciences to develop an integrated realm of theory and practice. In order to identify future trajectories for ecologies and politics of health, it is important to begin by highlighting key features of this volume. First, *Ecologies and Politics of Health* centers upon a broadened conceptualization of human health that extends beyond the absence of disease or the individual actor or location. Rather, we argue that health is produced through the interplay between social and ecological processes operating across multiple spatial and temporal scales. This understanding contributes to addressing human health in ways that attend to social and ecological function, security, well-being, equity, and sustainability. Second, this volume demonstrates that research on human health needs to integrate contributions from both the natural and social sciences to address its social and ecological dimensions. We believe that in order to comprehend the realities of human health, interdisciplinary as opposed to multidisciplinary perspectives are needed.

Given these considerations, this volume serves as a reminder that understanding health systems is hardly straightforward. Studies must necessarily be interpreted within the space, place, time, and context in which they were performed and the data gathered. While generalizability should remain a goal, it is critical to understand how the lessons learned from even the best health research are contingent. Beyond problems presented by data and measurement standards, contradictory findings may in fact point to unresolved theories of disease transmission, or to spurious correlations that, along with unexplained anomalies, offer the opportunity to redesign improperly defined theories, experiments, and studies in a targeted manner suggested by those same contradictions and anomalies. It is these holes in the outdated fabric of mono-disciplinary health research that can provide both an impetus and means for understanding populations and landscapes vulnerable to health threats while leveraging integrated scholarship to promote health and health management, research, and practice. We believe that future research and practice on ecologies and politics of health must attend to four central themes: *scales of interaction*, *vulnerability and difference*, *landscape*, and *representation and discourse*.

Scales of interaction

While the interactions between social and ecological systems are meaningful in shaping health trajectories, interrogating the scales of these interactions remains a challenge for future work. The difficulty in charting a fruitful course through an approach that integrates both ecologies and politics lies in striking the right balance between those two spheres. It is easy, and perhaps too easy, to simply refer to the interaction or integration of any two elements or systems from the drawing board. The construction, observation, inquiry, and reconstruction of human–environment interactions, for example, illustrate centuries of grappling with asymmetries and disproportionate treatment (Glacken 1976; Liverman *et al.* 1998). The challenge put before this volume's authors, and by extension its audience, was to ensure that both spheres of thought were brought to bear upon the topics at hand, rather than focus on equal treatment of the ecological and the socio-political. It was therefore an ontological request with implications perhaps for epistemological, methodological, and material practices and outcomes. It should be noted that a call for balance does not equate to a call for symmetry. There will be times and cases where a true balance can be struck, but more often one might expect to see an 80–20, 60–40, or even 95–5 balance between ecological and socio-political processes. Similar to positions held in the fields of coupled natural–human systems and socio-ecological systems (Liu *et al.* 2007; Ostrom 2008, 2009), we posit it is the coupling itself rather than the ratio that ultimately characterizes the potential power of an integrated approach. Sometimes even acknowledging the influence of the ecological on the socio-political or vice versa is enough to trigger an unexpected and productive line of inquiry, such as with the recognition of the role of helminth loading on "socially transmitted" diseases (Bentwich *et al.* 2008) or institutional construction of disproportionate exposure to environmental hazards and amenities (Lester *et al.* 2001). The line between integrations that are small but significant, and those that are ineffective and superficial, is often prospectively unclear and retrospectively ephemeral.

This fluidity may in part explain why some health scholars have embraced systems theory and derivatives thereof as a vehicle for envisioning functional interactions across ecological and socio-political domains. What such frameworks further offer is a way to forefront multiple levels of organization of health bearers. That is, "health" may be assessed horizontally across levels of organization (e.g., human health as well as ecosystem health) and vertically up through levels of organization (e.g., individual human health to population health or ecosystem health to biome health). Furthermore, these levels may each be observable at multiple spatial, temporal, and spatio-temporal scales, representing the geographic expanse and temporal duration of the health phenomenon of interest (Peterson and Parker 1998; Turner *et al.* 2001). That is, any given level could be assessed at multiple scales, and many phenomena are scale-dependent in their observation. The organizational level of individual human health could be assessed with respect to, for example, cellular immune responses, organ function, endocrine system disruption, and overall individual health. The health of the population of humans inhabiting a particular metro-region could be studied at the individual human level, for only some segments of the population (e.g., age cohorts or those living within a certain proximity of a toxic dumping site), or wall-to-wall across the entire region. The larger point here is that a variety of spatial, temporal and spatio-temporal scales can be employed to understand health phenomena operating at different organizational levels. The corollary to this assertion is that health researchers must be cognizant of these levels and scales both with respect to explicitly reporting their own findings and theories but moreover to understanding the implications for integrating the observable and/or theorized ecologies and politics of health.

Vulnerability and difference

A second theme for future research and practice are the ways in which integrating ecological and social processes assists in revealing the factors shaping vulnerability to disease, ecosystem functioning, and political agency over time and space. As has been demonstrated by decades of scholarship and political activity, vulnerability to infectious disease or exposure to carcinogens that produce ill-health can be highly inequitable. The justice concept has been a topic of interest within various academic and policy realms concerned with the inequities experienced by different social actors. The United States Environmental Protection Agency's Office of Environmental Justice defined environmental justice in 1998 as:

> the fair treatment and meaningful involvement of all people regardless of race, color, national origin, or income with respect to the development, implementation, and enforcement of environmental laws, regulations, and policies … [environmental justice] will be achieved when everyone enjoys the same degree of protection from environmental and health hazards and equal access to the decision-making process to have a healthy environment in which to live, learn, and work.
>
> (EPA 2012)

While much work on environmental justice has effectively focused on the disproportionate exposure of poor and minority populations to environmental hazards, there is a need for future work to expand understandings of the construction of health vulnerabilities over time and space. This progression would emphasize differences between settings and groups to include systematic and comparative research within and between urban and rural populations, as well as within and between industrialized and developing contexts. Health vulnerabilities as a theme for future work would attend to differences in access to and control over resources and economic opportunities, power dynamics shaping decision-making and outcomes, and the social processes that create and perpetuate differential exposure to environmental hazards and amenities. Expansion from primarily focusing on disproportionate exposure to environmental bads ("freedom from"—e.g., pollution) to environmental goods ("freedom to"—e.g., access to green space) would further the move away from the traditional characterization of health as an absence of disease to one that emphasizes a holistic state of well-being.

Beyond blame, other instances of environmental injustice demand equitable and rapid response. The environmental justice movement has often posited that the disparate proximity to environmental hazards and resulting health threats faced by many disadvantaged populations were an inherent part of environmental pollution. That is, decreasing pollution and protecting the environment protects the health of those populations as well. But a daunting challenge lies between the lines of several chapters in this volume: specifically, how to prioritize human and environmental health when they can be competing rather than commensurate goals? Not every health–environment interaction can be constructed as a "win–win" situation, and some vital measures to protect human health in certain times and places might necessarily harm the environment, such as the removal or conversion of disease vector habitats that results in problems such as erosion, poor infiltration, and declining water quantity and quality. Additionally, while socio-economic development is often posited as an integral component for improved human health and well-being, it can also increase exposure to other health hazards. The increasing prevalence of diabetes and obesity in the developing world attests to some of the negative byproducts of development writ large (Hossain *et al.* 2007). Further, improving the health status of the global population might have (unintended) negative consequences for the environment. *Ceteris parabis* (i.e., holding technological developments, fertility and consumption levels, natural resource base, and climatic cycles constant), a healthier population with lower fatalities and a longer lifespan has the potential to increase the demand for natural resources and ecosystem services already experiencing significant pressures.

Landscape

A third theme for future research and practice on ecologies and politics of health is landscape. In considering landscapes of human health, future work should emphasize the interplay of health with changing land use and land cover (LULC), climate, and migration patterns. The very mobility of humans and many other disease vectors necessarily makes health a transboundary issue that cannot be managed in geographic isolation. Porous borders facilitate the movement of people and microbes and, as climate change and disease outbreaks have each shown, political boundaries are not ecological boundaries. Biogeochemical, atmospheric, and hydrologic cycling, as well as sediment and biotic transport, have rarely conformed to political boundaries (save for when landscape features such as rivers or rifts are used to demarcate political territories), and environmental policy has long recognized the importance of transboundary issues with respect to negative externalities (Vig and Kraft 2000). But the changing face and speed of both technology and transportation are exacerbating the nature and spread of these externalities (Sheppard 2000). That is, the opening of such boundaries has only increased, though differentially, the transboundary nature of many diseases. However, the roles that states play in shaping the discourses and management of health, of the environment, and of populations must remain at the forefront of health research but only through their integration with ecological understandings of health. Another consideration is the incorporation of LULC in integrated eco-social studies on human health. Health concerns arise at the interfaces of biophysical environment, LULC systems, and their respective drivers of change and feedbacks. Additionally, improvements in geospatial technologies have diverse and problematic implications for defining, studying, and managing human health. Enhanced monitoring of landscapes of disease takes on different forms of surveillance and policy intervention, with concomitant impacts for human populations. The varied types of political responses to human disease and health planning more generally, therefore, can have unintended consequences for ecologies and politics of health.

A continued challenge for future health research and policy is the tendency to examine only one disease or health threat in isolation from others. Health research is necessarily broad and the intent to focus on, for example, one infectious disease to better assess its function is understandable. Yet no disease functions in complete isolation from other diseases, and the health landscape is an arena for the interactions of many other factors impacting the human ability to thrive in the face of health challenges. Scholarship and policy interventions that ignore these interacting factors, such as nutrition status and immune response, can lead to inaccurate and disabling results. At times these inaccuracies are fed by, entangled with, and even promote cultural, ethnic, and racial stereotypes. These stereotypes may mislead health investigations: at times, buried in these dialectics are accusations of blame or intent, positing that the health threat is on some level deserved. While prevention of injury (intentional or not) and disease (infectious or not) should remain a part of any health remediation strategy, placing blame hardly furthers healthy outcomes and may in fact lead to disengagement by affected, managerial, or even donor communities.

Representation and discourse

Finally, future research on ecologies and politics of health needs to concentrate upon how the discursive framing of a disease, and hence human health more generally, has material import for how it is understood, defined, and managed. As has been noted elsewhere, these subaltern health narratives potentially challenge conventional disease orthodoxies produced by the biomedical model, or representations of disease that are created by national and international agencies (King 2010). Several contributions in this volume demonstrate that the discursive framings of HIV/AIDS are powerful in shaping understandings of human health, the impacts of the disease for social and ecological systems, and the policy interventions that arise in response. Other contributions in this volume have shown that vector management strategies employed by the state are often informed by divergent politico-technological and cultural-racial discourses that inform perceptions of vector ecology. The particular features of these case studies are critical and informative; however, we believe they also signal a general lesson for future research on ecologies and politics of health.

Bound in this tension between the ecological and the socio-political is an undercurrent of epistemological differences in how "truth" is acknowledged as opposed to being constructed or discovered (Latour 1993), critical when considering these social constructions of disease ecologies. Clearly, contributions in this volume indicate where power and access, for example, differentially shape the resulting disease landscape(s). But the counterpoint raised, and touched on briefly in the first chapter of this volume, is that there are also some biophysical "truths": excessive exposure to radiation, regardless of political power structures that may have exacerbated or mitigated that exposure level, triggers a biochemical response in living tissue regardless of whether that damage is sensed by that individual, recorded by observers, or even acknowledged by the cognitive power structures or stakeholders. Thus in addition to the previously discussed ontological challenge of a two-pronged approach to health, there also exists an epistemological challenge in how and what truths can be known and potentially agreed upon among scholarly subfields that address health. Axiomatically, this challenge thus also indicates the likelihood of moving to a protocol of recognizing multiple not only possible but potentially co-existing health systems. Put another way, at any given time in any given space there could be multiple disease landscapes operating with varying levels of interaction. Rather than the "crisp" (health) landscape definitions associated with one output map, for example, disease incidence instead requires recognition of both (i) socio-politically varying landscapes (where, for example, certain populations would be more or less vulnerable to environmental hazards and amenities regardless of the actual location, perhaps due to nutritional deficiencies associated with malnutrition), and (ii) ecologically "fuzzy" landscapes (where stochastic biophysical processes may only accurately yield probability maps of disease vulnerability rather than deterministic ones). Granted, the latter especially presents an acknowledged increase in computational and conceptual complexity, but is more realistic and further perhaps begins to bridge the epistemological gaps inherent in ecological and social approaches to health studies.

This last remaining challenge might also be well characterized as methodological or practical in nature: how can a health system, already approached in a variety of disciplinary approaches, also bridge the divide between health researchers and health practitioners, between health oversight agencies and stakeholders, between the local and the global? There are, notably in tropical diseases, some practicing health workers who also publish their findings in either academic or gray/policy literatures. And there are some stakeholders who have become leaders in outreach and education with regard to their particular health concern. But systematically the communication lines among these groups and their perspectives evidence less interaction than even the disparate disciplines from which they have emerged. How best to meet that call is unclear, though case studies illustrate examples of what, in some times and places, has failed to resonate. Top-down, state-sponsored management systems in which residents and citizens are not allowed or cannot reach access to state knowledge has been shown in several chapters to be ineffective. In contrast, what has proven at least partially successful in some settings is the cross-fertilization among actors, such as with focus groups or neighborhood gatherings where specialists and stakeholders interact. But the reality is that improving communication in health matters can be as spatially and temporally contextualized as an understanding of human health itself.

Taken together, *scales of interaction*, *vulnerability and difference*, *landscape*, and *representation and discourse* are essential components for future research and practice on the social and ecological dimensions of human health. We believe these themes are best advanced through the effective integration of the strongest contributions from the natural and social sciences, drawing upon interdisciplinary or even transdisciplinary perspectives on human health. A central contribution for future research and practice is the necessity of integration between social and ecological systems, historical and contemporary conditions, and multiple scales of interaction. Human health exists at the interface of environment and society, and only through interdisciplinary and integrative research and practice will it be possible to achieve equitable, sustainable, and healthy futures.

References

Bentwich, Z., Teicher, C.L., and Borkow, G. (2008) The Helminth HIV connection: time to act. *AIDS* 22(13): 1611–1614.

EPA (Environmental Protection Agency). (2012) Environmental justice. See www.epa.gov/environmentaljustice (accessed 9 April 2012).

Glacken, C.J. (1976) *Traces on the Rhodian Shore: Nature and Culture in Western Thought from Ancient Times to the End of the Eighteenth Century*. Berkeley, CA: University of California Press.

Hossain, P., Kawar, B., and El Nahas, M. (2007) Obesity and diabetes in the developing world: a growing challenge. *The New England Journal of Medicine* 356(3): 213–215.

King, B. (2010) Political ecologies of health. *Progress in Human Geography* 34(1): 38–55.

Latour, B. (1993) *We Have Never Been Modern*. Translated by C. Porter. Cambridge, MA: Harvester Wheatsheaf and the President and Fellows of Harvard College.

Lester, J.P., Allen, D.W., and Hill, K.M. (2001) *Environmental Injustice in the United States: Myths and Realities*. Boulder, CO: Westview Press.

Liu, J., Dietz, T., Carpenter, S.R., Albierti, M., Folke, C., Moran, E., Pell, A.N., Deadman, P., Kratz, T., Lubchenco, J., Ostrom, E., Ouyang, Z., Provencher, W., Redman, C.L., Schenider, S.H., and Taylor, W.W. (2007) Complexity of coupled human and natural systems. *Science* 317: 1513 (doi:10.1126/science.1144004).

Liverman, D., Moran, E.F., Rindfuss, R.R., and Stern, P.C. (eds). (1998) *People and Pixels*. Washington, DC: National Academy Press.

Ostrom, E. (2008) Frameworks and theories of environmental change. *Global Environmental Change* 18: 249–252.

Ostrom, E. (2009) A general framework for analyzing sustainability of social-ecological systems. *Science* 325(5939): 419–422.

Peterson, D.L. and Parker, V.T. (eds). (1998) *Ecological Scale: Theory and Applications.* Columbia University Press: New York.

Sheppard, E. (2000) The spaces and times of globalization: place, scale, networks, and positionality. *Economic Geography* 78(3): 307–330.

Turner, M.G., Gardner, R.H., and O'Neill, R.V. (2001) *Landscape Ecology in Theory and Practice: Pattern and Process.* New York, NY: Springer Press.

Vig, N.J. and Kraft, M.E. (2000) *Environmental Policy: New Directions for the Twenty-First Century.* Washington, DC: CQ Press.

Index